Evapotranspiration:
Modeling and Simulation

Evapotranspiration: Modeling and Simulation

Edited by **Elizabeth Lamb**

New York

Published by Callisto Reference,
106 Park Avenue, Suite 200,
New York, NY 10016, USA
www.callistoreference.com

Evapotranspiration: Modeling and Simulation
Edited by Elizabeth Lamb

International Standard Book Number: 978-1-63239-332-6 (Hardback)

Printed in the United States of America.

Contents

Permissions

List of Contributors

Preface

The world is advancing at a fast pace like never before. Therefore, the need is to keep up with the latest developments. This book was an idea that came to fruition when the specialists in the area realized the need to coordinate together and document essential themes in the subject. That's when I was requested to be the editor. Editing this book has been an honour as it brings together diverse authors researching on different streams of the field. The book collates essential materials contributed by veterans in the area which can be utilized by students and researchers alike.

This book contains several well researched topics associated with various aspects and properties of evapotranspiration. These fields are the leading technologies that measure the extremely spatial ET from the planet's surface. The analyses explain technicalities of ET simulation from incompletely vegetated surfaces and stomatal conductance behavior of natural and agricultural ecosystems. Evaluation strategies that use weather based ways such as soil water balance, temperature-radiation based methods and many others have been explained. Suggested guiding principles for applying operational satellite-based energy balance models and overcoming common challenges have also been included.

Each chapter is a sole-standing publication that reflects each author's interpretation. Thus, the book displays a multi-facetted picture of our current understanding of application, resources and aspects of the field. I would like to thank the contributors of this book and my family for their endless support.

Editor

Guidelines for Remote Sensing of Evapotranspiration

Christiaan van der Tol and Gabriel Norberto Parodi
University of Twente, Faculty of ITC
The Netherlands

1. Introduction

This chapter describes the possibilities, the limitations and the future of remote sensing of evapotranspiration (ET). The principles behind the techniques of remote sensing of ET are presented systematically. The mathematical formulations of the key equations are used to highlight the critical parts and the variables that remote ET is most sensitive to. The focus will be on the input data. Which input data do we definitively need, and with what accuracy? How can we select the best methodology to estimate ET spatially? A number of new developments will be introduced, and priorities for the near future formulated.

There is no global, validated ET product available today. We can find products of other components of the terrestrial water cycle, like rainfall and soil moisture, but not of ET. This means that remote estimation of ET is custom made, and that it requires specific skills. At first glance, this is surprising, because the idea of remote sensing of evapotranspiration is more than three decades old (Jackson et al., 1977; Jackson et al., 1987; Seguin, 1988). In this chapter we hope to clarify the reasons why the operational dissemination of remote sensing evapotranspiration products lags behind.

The one fundamental problem with estimating ET is that it cannot be measured directly. This is well illustrated by borrowing an allegory from the evangelist Billy Graham: "I've seen the effects of the wind but I've never seen the wind". This quote is certainly true, in a literal sense, for evapotranspiration. Evapotranspiration affects the water and energy balances, and it is these effects of evapotranspiration which are observed. For example, evapotranspiration reduces soil moisture content and its cools the land surface. By studying changes in soil moisture with lysimeters, or by studying patterns in land surface temperature with remote sensing, ET is estimated. The problem is that ET is not the only factor affecting the water and energy budgets of the surface. Other processes and physical properties play important roles as well. For this reason, quite a large number of input variables are needed to achieve reasonably accurate estimates of ET; estimates that are at least better than rule-of-thumb a priori estimates. This is the case for field techniques, and even more so for remote techniques. In remote sensing an additional issue is that the input data are not from just one source. Often data of different satellite platforms are needed in conjunction with meteorological data from ground stations.

In the process of collecting the data, the user faces practical and scientific problems. The practical problems of the user include the data collection, merging the data in a GIS database, and programming the algorithm for the calculation of evapotranspiration. Each of these steps requires expertise and access to dedicated GIS software packages, because most algorithms are not available 'on the shelf'. The scientific problem is that the procedure requires merging of data measured at different time and spatial scales. The merging of data for the calculation of evapotranspiration is a typical example of a data assimilation problem. Ideally, both the accuracy and the representativeness of each of the input data are taken into account in the calculation of the final product: the spatial map of evapotranspiration. Although algorithms for the calculation of evapotranspiration are available in the literature and on the internet, there is no consistent data assimilation procedure attached that calculates the accuracy and reliability of the final product.

In recent years there have been a number of initiatives to build global products of evapotranspiration (Vinukullu et al., 2011; www.wacmos.org), addressing the above mentioned issues. The priorities for the near future are to establish a consistent way of merging the input data, improve the data assimilation techniques and to validate algorithms. In addition, new sources of data may be introduced. A promising tool is the use of satellite based laser altimetry for surface roughness estimation (Rosette et al., 2008).

It is the aim of this chapter to focus on principles that the algorithms have in common, and on the input data. Reviews of the history of remote sensing of ET can be found in the scientific literature (Courault et al., 2005; Glenn et al., 2007; Gowda et al., 2007; Kalma et al., 2008). In addition to these reviews, we would like to provide some anchor points and guidelines for the selection of a methodology for estimating ET in basin hydrology. We will quantify and evaluate the error of each of the input data, and show how this error propagates into the final result. For this analysis we will use theoretical considerations, a remote sensing model, and a selection of field data.

2. Principles of remote sensing algorithms for evapotranspiration

Although remote sensing of evapotranspiration has evolved since the first initiatives in the 1970's, the fundamental principle has remained the same. All remote sensing based evapotranspiration estimates make use of the thermal and visible bands and the formulation of the energy balance of the surface. The instantaneous latent heat flux of evaporation is calculated as a residual of the energy balance, and this latent heat flux is in turn converted into an evapotranspiration rate after time integration. An inherent problem of this approach is that the errors in the various terms of the energy balance are affecting the latent heat flux in a manner that is difficult to predict. For this reason it is necessary to evaluate the different terms of the energy balance individually.

In the evaluation of the remote sensing algorithms presented in this section, we will discuss the terms, and indicate at what spatial and temporal resolution the data can be collected. It will become clear that the land surface temperature is the most important state variable. It plays a crucial role in sensible heat flux, ground heat flux and the balance of long wave radiation. Apart from the collection of accurate land surface temperature data, important selection criteria for a methodology are the heterogeneity of the land cover, the topography and the spatial resolution (sampling) of the remote data.

Neglecting the energy used in the process of photosynthesis, the instantaneous energy balance equation (EBE) over crops reads:

$$R_n = G + H + \lambda E \tag{1}$$

R_n is the net radiation remaining in the system, G the ground heat flux, H the sensible heat flux and λE is the latent heat flux that is the energy consumed in evapotranspiration (all in

W m^{-2}). Radiation fluxes are positive when directed towards the land surface, the other fluxes are positive when pointed away from the surface. The partition of energy between the terms is largely controlled by the availability of water or moisture in the system. When moisture is not restricted, λE reaches a maximum and H is small.

In order to estimate ET, Eq. 1 is solved for λE. When applied to remote sensor retrievals, R_n is solved entirely from a combination of radiation counting at sensor level and few ground information. Ground or soil heat flux is a minor component in densely vegetated areas, but a large term in non-vegetated or sparsely vegetated areas (Heusinkveld et al., 2004). The importance of a better evaluation of the soil heat flux is gaining attention, mainly to ensure the EBE closure in such areas. The evaluation of H is the major difficulty. There are several models and approaches to solve for H (SEB models) and a number of parameters and assumptions are still under debate. The remote sensing models for ET mainly differ in the way H is treated.

In the following sections, the individual terms of the EBE (Eq. 1) will be discussed in further detail. A theoretical description is presented for each term in the EBE, in combination with a discussion on the feasibility of data acquisition from remote and ground sources.

2.1 Net radiation

Net radiation R_n is the dominant term in the EBE, since it represents the source of energy that must be balanced by the thermodynamic equilibrium of the other terms. The net radiation can also be expressed as an electromagnetic balance of all incoming and outgoing radiation reaching and leaving a flat horizontal and homogeneous surface as:

$$R_n = S\downarrow -S\uparrow +L\downarrow -L\uparrow \qquad (2)$$

Where S is the shortwave radiation, nominally between 0.25 to 3 μm and L is the long wave radiation, nominally between 3 to 100 μm. The arrows show the direction of the flux entering '$\downarrow$' or leaving '$\uparrow$' the system.

Equation 2 is very convenient from the data acquisition point of view since each term can either be obtained from available models, or directly from instruments at ground stations or remote platforms. As remote sensors are positioned looking to Earth, they measure outgoing radiation only. The incoming fluxes must be either modelled or derived through alternative methodologies.

The instantaneous incoming shortwave radiation (also called global radiation), $S\downarrow$, is commonly measured at ground stations by means of pyranometers or solarimeters. These instruments usually work in the shortwave broadband range (usually 0.305 - 2.4 μm). This range comprises almost 96% of the spectral interval of the solar irradiance. Recently there are remote sensing products and clearinghouses that account for the incoming and outgoing shortwave and long wave radiation. The use of them may reduce the need of permanently operational ground radiometers.

The outgoing shortwave radiation is the portion of the shortwave reflected back to the atmosphere. It is characterized by the albedo. The reflectance is the ratio between the reflected and the incoming radiation in a certain wavelength over an arbitrary horizontal plane. The integrated value over all visible bands defines the albedo, r_0. Since albedo is a reflective property of the material, it can be evaluated from remote sensors multi-spectral bands, and the integration to full shortwave range is approached by a linear model that might include the atmospheric correction. The shortwave radiation balance reads:

$$\Delta S = S \downarrow - S \uparrow = (1 - r_0) \cdot S \downarrow \tag{3}$$

For all bodies, the total incident radiation is either reflected by the body, absorbed by it or transmitted through. This is expressed by the Kirchoff's law:

$$1 = \rho_\lambda + \tau_\lambda + \alpha_\lambda \tag{4}$$

where ρ_λ is the reflectivity, τ_λ the transmissivity and α_λ the absorptivity. A blackbody is defined as a body that absorbs all the radiation that receives. A blackbody is a physical abstraction that does not exist in nature. To keep a body temperature constant, it should emit the same radiation that absorbs. As a consequence a property of blackbodies, the absorptivity, is equal to the emissivity, and both are equal to 1, while reflectivity and transmissivity are equal to zero.

Terrestrial materials behave more as grey bodies, meaning that part of the received radiation is reflected back to the atmosphere, or in other words, not all the energy that receives is absorbed. In order to keep the temperature constant, the absorbed radiation should equal the emission, so again emissivity is equal to absorptivity. Because the reflectivity is not zero, emissivity of real bodies is smaller than 1.

The longwave radiation terms are calculated with Planck's equation extended to real bodies. A blackbody having a kinetic temperature T_0 [K] emits in a single wavelength a radiation that corresponds to:

$$L_\lambda^{bb} = \frac{3.74 \cdot 10^8}{\lambda^5} \cdot \frac{1}{\left[\exp\left(\dfrac{1.44 \cdot 10^4}{\lambda \cdot T_0} \right) - 1 \right]} \tag{5}$$

Where L_λ^{bb} is the blackbody energy emission [W m^{-2} µm^{-1}] and λ is the wavelength [µm]. The kinetic temperature is the temperature as it would be measured by a standard thermometer in contact with the surface of the body. Emissivity ε_λ at a chosen wavelength is the ratio of the radiation emitted by a real body at temperature T_0 to the radiation emitted by a blackbody at the same temperature. By definition, a blackbody has a constant emissivity equal to one for all wavelengths, whereas the real emissivity varies with wavelength. For natural bodies, the thermal emission can then be written as:

$$L_\lambda(T_0) = \varepsilon_\lambda \cdot L_\lambda^{bb}(T_0) \tag{6}$$

Integration of L over all wavelengths leads to:

$$L(T_0) \uparrow = \int_0^\infty \varepsilon_\lambda \cdot L_\lambda^{bb}(T_0) \cdot d\lambda = \sigma \cdot \varepsilon_0 \cdot T_0^4 \tag{7}$$

where $\sigma = 5.67 \times 10^{-8}$ W m^{-2} K^{-4} is the Stefan-Boltzmann constant and ε_0 is a broadband surface emissivity. A remote sensor working within a spectral range of the thermal channels measures only a portion of $L(T_0)\uparrow$. The outgoing longwave radiation at any sensor channel is calculated by integration over the spectral range of the sensor:

$$L_i^{sat}(T_0) \uparrow = \int_i \varepsilon_\lambda \cdot L_\lambda^{bb}(T_0) \cdot d\lambda \tag{8}$$

The surface temperature T_0 is retrieved from Eq (8), once the surface emissivity ε_λ in the considered thermal channel is estimated. Once 'T_0' is obtained and ε_0 estimated, $L\uparrow$ is retrieved from Eq 7. Before the application of Eq (8), an atmospheric correction process is needed to derive $L_i^{sur}(T_0)\uparrow$ at the surface, because $L_i^{sat}(T_0)\uparrow$ as measured at the satellite sensor is affected by atmospheric interference. Atmospheric correction in the thermal range and in the shortwave is out of the scope of this chapter. We only mention that $L_i^{sur}(T_0)\uparrow$ can be obtained from $L_i^{sat}(T_0)\uparrow$ and using atmospheric correction model, in which water vapour and aerosol concentrations are the main input variables.

The incoming long wave radiation cannot be derived directly from remote sensors. It can either be determined from ground data or derived after atmospheric modelling. It varies with cloudiness (water vapour), air temperature and atmospheric constituents. For clear skies, the notion of effective thermal infrared emissivity of the atmosphere or apparent emissivity of the atmosphere (ε_a') introduces an overall emission value for all constituents. If the air temperature T_a at screen level is available, $L\downarrow$ is estimated as:

$$L\downarrow = \sigma \cdot \varepsilon_a' \cdot T_a^4 \tag{9}$$

There are several models simple to evaluate ε_a'. The apparent emissivity of the atmosphere is usually estimated with equations based on vapour pressure and temperature at standard meteorological stations. For clear skies a common formulation, among others, is (Brutsaert, 1975):

$$\varepsilon_a' = 1.24 \left(\frac{e_a}{T_a} \right)^{\frac{1}{7}} \tag{10}$$

Where T_a is the air temperature [K] and e_a is the vapour pressure [mbar], everything measured at screen level. A portion $L\downarrow$ reaching the Earth surface is reflected back to the atmosphere. Since the surface is opaque the transmissivity is zero, the reflection of $L\downarrow$ can be evaluated with Kirchoff law. As ε_0 describes the emissivity of a body in the thermal range, (1- ε_0) accounts for the reflection. The final expression for R_n becomes:

$$R_n = (1 - r_o) \cdot S\downarrow + \varepsilon_a' \cdot \sigma \cdot T_a^4 - (1 - \varepsilon_0) \cdot \varepsilon_a' \cdot \sigma \cdot T_a^4 - \varepsilon_0 \cdot \sigma \cdot T_0^4 \tag{11}$$

Eq. 11 is valid for instantaneous observations. The conversion to a daily value is briefly discussed in Sect 2.4.

It is not always necessary to carry out the calculations of Eqs 3-11 manually. Some organizations provide atmospherically corrected components of the radiation platforms directly. For example LandSaf (landsaf.meteo.pt) provides MeteoSat Second Generation (MSG) products of atmospherically corrected $S\downarrow$ and $L\downarrow$ with a 15 minute resolution and daily albedo for South America, Africa and Europe. An emissivity product will be released soon. Validation over ground based measurements for a site in Spain over sparse vegetation shows that these products are rather reliable (Fig 1.).

As an alternative to the use of satellite data, a computation of the radiation terms from synoptic weather stations is also possible. The recommendations by the FAO (Doorenbos and Pruitt, 1977; Allen et al., 1998) could be followed. The daily short wave radiation $S\downarrow_{day}$ [MJm^{-2}day^{-1}], is measured at agrometeorological stations with pyranometers and integrated to daytime hours. In most areas in the world, only sunshine hours are measured with periheliometers. In that case, the daily incoming shortwave radiation $S{\downarrow}_{day}$ can be obtained from the following empirical relationship:

$$S_{day} \downarrow = \left(a_s + b_s \cdot \frac{n}{N}\right) \cdot S_{0,day} \downarrow \qquad (12)$$

where a_s is the fraction of the extraterrestrial radiation reaching the ground in a complete overcast day (when $n=0$), $a_s + b_s$ the fraction of the extraterrestrial radiation reaching the ground in a complete clear day ($n=N$), n the duration of bright sunshine per day [hours], N the total daytime length [hours], $S_{0,\ day}\downarrow$ is the terrestrial radiation [MJm^{-2} day^{-1}]. Local instrumentation can be used to calibrate a_s and b_s for local conditions.

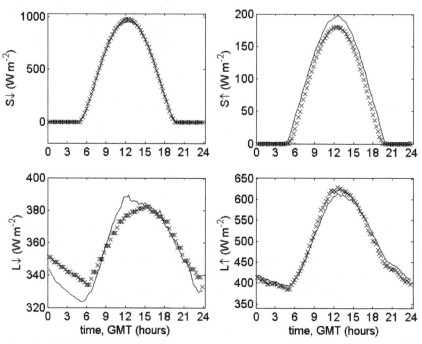

Fig. 1. Comparison between MeteoSat Second Generation radiation products (symbols) with 5-minute interval ground based measurements (lines) for a pixel with sparse vegetation in central Spain, for 5 July 2010.

The net daily shortwave radiation ΔS_{day} is estimated as in Eq. (3), assuming an average daily (sun hours only) albedo r_{0day}. The daily longwave radiation exchange between the surface and the atmosphere is very significant. Since on average the surface is warmer than the atmosphere and $\varepsilon_0 > \varepsilon_a'$, there is usually a net loss of energy as thermal radiation from the ground. The daily net shortwave radiation ΔL_{day} [W m^{-2}] between vegetation and soil on the one hand, and atmosphere and clouds on the other, can be represented by the following radiation law:

$$\Delta L_{day} = -f \cdot \varepsilon'_{a,day} \cdot \sigma \cdot (T_{a,mean} + 273.15)^4 \qquad (13)$$

Where $\varepsilon'_{a,day}$ [-] is the daily net emissivity between the atmosphere and the ground, f a cloudiness factor and $T_{a,mean}$ is the mean daily air temperature at screen level [°C]. Parameter $\varepsilon'_{a,day}$ can be estimated from data from meteorological stations as:

$$\varepsilon'_{a,day} = a_e + b_e \cdot \sqrt{e_{d,mean}/10} \tag{14}$$

Where a_e is a correlation coefficient (ranging from 0.34 to 0.44, with a default of 0.34), b_e a correlation coefficient (ranging from -0.14 to –0.25 with a default of -0.14), $e_{d,mean}$ the average vapour pressure at temperature [mbar]. If true $e_{d,mean}$ is not available, then it can be calculated from daily average relative humidity RH_{mean} and mean air temperature $T_{a,mean}$ [°C]:

$$e_{d,mean} = \frac{RH_{mean}}{100} \cdot e_{s,mean} \quad \text{and} \quad e_{s,mean} = 6.108 \cdot \exp\left[\frac{17.27 \cdot T_{a,mean}}{T_{a,mean} + 237.15}\right] \tag{15}$$

The cloudiness factor f is equal to 1 in case of a perfect clear day and 0 in a complete overcast day. In case the station has solar radiation data from pyranometers f can be calculated as:

$$f = a_c \cdot \frac{S\downarrow_{day}}{S\downarrow_{clear,day}} + b_c \quad \text{or} \quad f = a_c \cdot \frac{S\downarrow_{day}}{(a_s + b_s \cdot) \cdot S_0\downarrow_{day}} + b_c \tag{16}$$

Where a_s, b_s, a_c and i_c are calibration values to be estimated through specialized local studies which involve measuring longwave radiation values. Average values for a_c and b_c in arid and humid environments can be found in Table 1:

Climate	a_c	b_c	a_s	b_s
Arid	1.35	-0.35	0.25	0.50
Humid	1.00	0.00	0.25	0.50

Table 1. Typical values the coefficients a_c, b_c, a_s and b_s for arid and humid climates (Maidment, 1992).

If only data on sunshine hours data are available, then:

$$f = \left(a_c \cdot \frac{b_s}{a_s + b_s}\right) \cdot \frac{n}{N} + \left(b_c + \frac{a_s}{a_s + b_s} \cdot a_c\right) \tag{17}$$

2.2 Sensible heat flux

The sensible heat flux (H) is the exchange of heat through air as a result of a temperature gradient between the surface the atmosphere. Since the surface temperature during the day is usually higher than the air temperature, the sensible heat flux is normally directed upwards. During the night the situation may be reversed. Close to the surface, the sensible heat transport takes place mostly by diffusive processes, whereas at some distance away from the surface turbulent transport becomes more important.

The mathematical formulation of the sensible heat flux is based on the theory of mass transport of heat and momentum between the surface and the near-surface atmospheric environment (surface boundary layer). All existing remote sensing algorithms for turbulent sensible heat flux use the analogy of Ohm law of resistance driven by a gradient of temperature:

$$H = \rho_a \cdot c_p \cdot \frac{T_s - T_a}{r_{ah}} \qquad (18)$$

where ρ_a is the density of moist air [kg m^{-3}], c_p is the air specific heat at constant pressure [J kg^{-1} K^{-1}], r_{ah} is the aerodynamic resistance to heat transport between the surface and the reference level [s m^{-1}] and $T_s - T_a$ is the driving temperature gradient between the surface (with temperature T_s) and the reference height (with temperature T_a).

Equation 18 shows that the estimation of sensible heat flux has two main elements: a temperature difference between two heights and the corresponding resistance. As a first approximation we can conclude that the error in the sensible heat is linearly proportional to the error in the temperature gradient, and linearly proportional to the error in the inverse of the resistance. Equation 1 shows that this error (W m^{-2}) is directly transferred to the latent heat flux. We will now show that this is only approximately true, because the equation is not linear and the aerodynamic resistance itself depends on the temperature gradient. We will show that because of this, r_{ah} can only be solved iteratively.

Understanding the physical concepts involved in the calculation of sensible heat flux, and in particular the aerodynamic resistance, is essential for an evaluation of remote sensing techniques. The evaluation of r_{ah} is the most complicated issue of all in the whole EBE procedure for AET estimates. It is our experience that lack of or incomplete knowledge of the entire formulation, image pre-processing and atmospheric correction processes leads to severe flaws in the intermediate and final outputs. Many researchers are still seeking for alternatives, procedures and methods to improve the accuracy of ET estimates form the EBE – RS approach. The actual parameterization is not optimal in the sense that some sensitive information can only be strictly evaluated under controlled experimental research, and not in a routine fashion.

Near the ground two phenomena take place simultaneously in the transfer of heat between the surface and the atmosphere: free convection produced by temperature gradient T_s-T_a and forced convection by the dragging forces of the wind. Then, the estimation of the turbulent heat fluxes requires a description of the turbulent wind profile near the surface. The starting point of the analysis is the wind profile in a neutral atmosphere (no convection, and T_s=T_a). In this situation and for an open site, the horizontal wind speed u [m s^{-1}] varies logarithmically with height above the ground z [m]:

$$u(z) = A \cdot \ln(z) + B \qquad (19)$$

B is usually replaced by A .ln (z_{om}) where z_{om} is the aerodynamic roughness length of the surface for momentum transport and represents the value of z for which Eq 19 predicts $u(z)$ = 0 (see also Fig 2):

$$u(z) = A \cdot \ln\left(\frac{z}{z_{om}}\right) \qquad (20)$$

In Eq 20, A must have the dimension of velocity and it should be independent of z since the profile description is given by the logarithmic term. Over plant communities of uniform height h, the turbulent boundary layer behaves as if the vertically distributed elements of the community were located at a certain distance d from the ground. Parameter d is called the zero plane displacement or displacement height level of the flow. It acts as a correction to the level where $z=0$, and thus z in Eqs 19 and 20 should be directly replaced by $z-d$ in vegetated areas:

$$z_{\text{shift}} = z - d \tag{21}$$

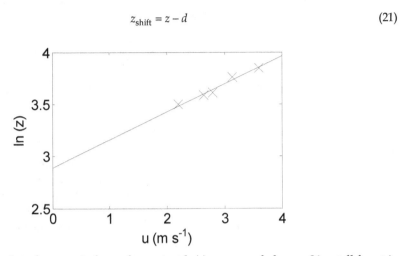

Fig. 2. Example plot of mean wind speed u against $\ln(z)$, measured above a 31 m tall forest in The Netherlands in August 2009. The intercept with the vertical axis leads to $z_{0m}+d = 17.9$ m.

The displacement height d is usually rather large: it ranges between 60 to 80 percent of the plant community height. Common values for well developed wheat are found in Verma and Bartfield (1979). A relation $d= 0.67h$ is usually adopted for other vegetation types (Allen et al., 1998). An exact estimation can be carried out when a wind speed measurements at three or more heights are available. A plot of $\ln(z-d)$ versus wind speed should then give a straight line for the correctly calibrated value of d. Strictly speaking, d also depends on plant density; for sparse vegetation, d is often neglected. In many occasions, insufficient data are available to accurately predict its values for discrete crop canopies (Verma and Bartfield, 1979). If no valid field data are present, then it is suggested to leave d out of the equations altogether.

The logarithmic wind profile in neutral conditions forms the basis for calculating the aerodynamic resistance for heat transport (and transport of water vapour) by mechanical turbulence. To include thermal turbulence (or convection or buoyancy), the Monin-Obukhov theory is needed in addition (Obukhov, 1946). We will now summarize the equations leading to the aerodynamic resistance of mechanical transport, followed by the modification for non-neutral conditions, when convection plays a role.

The transfer of momentum in the direction of the flux takes place through molecular and turbulent eddy activity. Random vertical movements of the air cause air with different horizontal wind speeds to mix. This causes a momentum sink at the surface in a form of shear stress, τ. For convenience the shear stress is expressed as a function of a scalar u_*, usually called eddy velocity or friction velocity [m s^{-1}]:

$$\tau = \rho_a u_*^2 \tag{22}$$

Since wind is produced by turbulent eddy motion, it is postulated that A in Eq 20 is proportional to the speed of the internal eddies. Then it can be demonstrated that:

$$A = \frac{u_*}{k} \tag{23}$$

Where k is Von Kármán's constant, experimentally found to be 0.41. The final expression of the wind profile under neutral atmosphere is:

$$u(z) = \frac{u_*}{k} \cdot \ln\left(\frac{z}{z_{0m}}\right) \tag{24}$$

Then the gradient of wind speed with height can be expressed as:

$$\frac{du}{dz} = \frac{u_*}{k \cdot z} \tag{25}$$

In the theory of estimating the aerodynamic resistance from the wind profile, only vertical transport is considered. In the atmosphere the steepest gradients of heat, wind speed and humidity are found in the vertical direction. Horizontal variation is present in the order of tens of kilometres (Brutsaert, 2005), but these are considered negligible. This implies that horizontal advection effects (for example between pixels) are not considered in the remote sensing approach, a serious restrictions in patchy (wet and dry) environments.

In analogy to horizontal wind speed, heat and water vapour also have vertical profiles near the surface. Vertical mixing then causes a transport of heat and vapour too, resulting in vertical fluxes of sensible and latent heat. The three fluxes, of momentum (F_u), heat (F_h) and vapour (F_v), can be expressed as the covariance of vertical wind speed (w') and concentration of the admixture (u', T' and q'):

$$\begin{aligned} F_u &\equiv -\tau = \rho \cdot \overline{w' \cdot u'} \\ F_h &\equiv -H = \rho \cdot c_p \cdot \overline{w' \cdot T'} \\ F_v &\equiv -\lambda E = \rho \cdot \overline{w' \cdot q'} \end{aligned} \tag{26}$$

These equations can be linked with the approach of electrical analogy (Eq 18) by approximating the covariance to simply the product of the vertical gradient of the quantities at two different heights (Brutsaert, 2005). A dimensionless parameter C is needed to fit the equality. This can be done for all three quantities. For example, for heat:

$$\overline{w' \cdot T'} = C_h \cdot \left(\overline{u_2} - \overline{u_1}\right) \cdot \left(\overline{T_4} - \overline{T_3}\right) \tag{27}$$

In this case, C_h will depend on the heights 1, 2, 3 and 4. It is convenient to choose the heights 4 and 3 equal to 2 and 1:

$$\overline{w' \cdot T'} = C_h \cdot \left(\overline{u_2} - \overline{u_1}\right) \cdot \left(\overline{T_2} - \overline{T_1}\right) \tag{28}$$

Similarly, for momentum transport (shear stress):

$$\overline{w' u'} = C_d \cdot \left(\overline{u_2} - \overline{u_1}\right)^2 \tag{29}$$

The coefficient C_d can be calculated by combining Eqs 22, 24, 26 and 29. Using $z_1 = z_{0m}$, and considering that at $z = z_{0m}$, the wind speed is zero:

$$C_d = \left[k \Big/ \ln\left(\frac{z_2}{z_{0m}}\right) \right]^2 \tag{30}$$

In neutral conditions it can be assumed that $C_h = C_v = C_d$, and thus:

$$H = -\rho \cdot c_p C_d \cdot u_2 \cdot \left(\overline{T_2} - \overline{T(z_{0m})}\right) \tag{31}$$

The appearance of the average temperature at height z_{0m} in Eq 31 is inconvenient. It can be eliminated by assuming a logarithmic wind profile for temperature too, by defining a scalar roughness height for heat transfer z_{0h} at which the extrapolated temperature profile fitted through $\overline{T_2}$ and $\overline{T(z_{0m})}$ becomes T_0, i.e. the kinematic surface temperature. Using this we finally express the aerodynamic resistance r_{ah} in neutral conditions as:

$$r_{ah} = \frac{\ln\left(\frac{z_2}{z_{0m}}\right)\ln\left(\frac{z_2}{z_{0h}}\right)}{k^2 \cdot u} = \frac{\ln\left(\frac{z_2}{z_{0h}}\right)}{k u_*} \tag{32}$$

The roughness height, z_{0h}, changes with surface characteristics, atmospheric flow and thermal dynamic state of the surface (Blümel, 1999; Massman, 1999). It can be shown that:

$$z_{0h} = z_{0m} / \exp\left(kB^{-1}\right) \tag{33}$$

where B^{-1} is the inverse Stanton number, a dimensionless heat transfer coefficient.
Free convection might alter the forced convective eddies generated by wind turbulence. During daytime or when temperature decreases with height, convection amplifies the vertical eddy motions (unstable condition). During the night or when inversion conditions occur, and temperature increases with height, the horizontal eddy motions are enhanced (stable conditions).
Mechanical turbulence and buoyancy coexists in a form of a hybrid regime known as mix-convection. Monin and Obukhov showed that these conditions eventually lead to an alteration of the wind and temperature profiles (Brutsaert, 1982). The Monin-Obukhov similarity theory uses dimensional analysis to correct the wind profile produced by buoyancy effects in such conditions. A non-dimensional correction factor for momentum transfer $\varphi_m(\xi)$ is introduced to correct the wind profile gradient for conditions different from neutral, in which ξ is the ratio of thermal to mechanical turbulence:

$$\frac{du}{dz} = \frac{u_*}{kz} \cdot \varphi_m(\xi) \tag{34}$$

They introduced semi-empirical functions to correct the wind profile depending on the stability, based on dimensional analysis, of the form:

$$u_* = \frac{\overline{u(z)}}{k} \cdot \left[\ln\left(\frac{z - d_0}{z_{0m}}\right) - \Psi_m\left(\frac{z - d_0}{L}\right) + \Psi_m\left(\frac{z_{0m}}{L}\right) \right] \tag{35}$$

$$H = \frac{T_s - \overline{T(z)}}{k \cdot u_* \cdot \rho \cdot c_p} \cdot \left[\ln\left(\frac{z - d_0}{z_{0h}}\right) - \Psi_h\left(\frac{z - d_0}{L}\right) + \Psi_h\left(\frac{z_{0h}}{L}\right) \right] \tag{36}$$

Where L is defined as the Monin-Obukhov length ($L = z \cdot \xi$) [m], calculated as:

$$L = -\frac{\rho_a \cdot C_p \cdot u_*^3 \cdot \overline{T(z)}}{k \cdot g \cdot H} \tag{37}$$

Where g is the gravity constant (9.81 m s^{-2}). Semi-empirical expressions for the stability corrections Ψ_h and Ψ_m can be found in the literature, for example Paulson (1970) and Brutsaert (1982). It is important to realize that L depends on air temperature and sensible heat flux, while sensible heat flux and air temperature in turn depend on L. For this reason, an iterative procedure is needed to calculate L, u_* and H using Eqs 35-37.

2.3 Ground heat flux

The ground heat flux has received relatively little attention compared to the other terms. This is often justified, because ground heat flux is usually the smallest of all terms. Moreover, the 24-hour sum of ground heat flux is close to zero, because the heat absorbed during the day is released during the night.

At the moment of a satellite overpass, ground heat flux is not necessarily negligible. At midday it usually varies from 10% of net radiation for dense vegetation to 45% of net radiation for bare soil (Clothier et al., 1986). Often a vegetation cover dependent ratio between G and R_n is assumed at satellite overpass (Kustas et al., 1990).

If more accurate estimates of ground heat flux are required, for example in areas with sparse vegetation, then remote estimates of ground heat flux are possible with the method of Van Wijk and De Vries (1963). For this method, diurnal cycles of land surface temperature and net radiation are needed (Verhoef, 2004; Murray and Verhoef, 2007); this means that time series of data of a geostationary satellite are required.

An equation for ground heat flux can be derived the thermal diffusion equation, assuming a periodic land surface temperature:

$$G(t) = \Gamma \sum_{k=0}^{n} \sqrt{k\omega/2} \cdot \left(A_k \cdot \sin(\omega k t) + B_k \cdot \cos(\omega k t) \right) \tag{38}$$

where Γ is the thermal inertia of the soil (J m^{-2} K^{-1} s$^{-1/2}$), which depends on texture and soil moisture, t is time (s), $\omega = (2\pi/N)$ is the radial frequency (s^{-1}), N the length of the time series [s], A and B integration coefficients [°C], and n the number of harmonics. The coefficients A and B are fitted against the observed land surface temperature time series, for a chosen number of harmonics. The thermal inertia Γ can be estimated from soil texture and soil moisture, or calibrated against night time radiation, by assuming that night time radiation equals the night-time ground heat flux.

2.4 Latent heat flux

Latent heat flux is finally calculated as a residual of the energy balance (Eq. 1). Because H, G and R_n are instantaneous measurements, it is necessary to find a procedure to integrate to daily totals. A common way to carry out this integration, is by making use of the evaporative fraction, Λ. The evaporative fraction (Brutsaert and Sugita, 1992) is the energy used for the evaporation process divided by the total amount of energy available for the evaporation process:

$$\Lambda = \frac{\lambda E}{\lambda E + H} = \frac{\lambda E}{R_n - G} \tag{39}$$

It is assumed that the evaporative fraction remains constant throughout the day.

$$\Lambda_{inst} = \Lambda_{24hrs} \tag{40}$$

Assuming that the ground heat flux integrated over 24-hours is negligible, the evapotranspiration rate over 24 hours can be calculated as:

$$E_{24hrs} = \frac{8.64 \cdot 10^7 \, \Lambda_{inst}}{\lambda \rho_w} R_{n,24hrs} \tag{41}$$

Where $\lambda = 2.0501-0.00236 \, T_{water}$ MJ kg^{-1} (T in °C), $\rho_w = 1000$ kg m^{-3} and R_{n24} is the average net radiation over 24 hours [W m^{-2}].
The assumption of a constant evaporative fraction may lead to underestimates of daily evaporation, because the evaporative fraction in reality has a diurnal cycle with a concave shape (Gentine, et al., 2007). The concave shape is caused by changes in weather conditions (wind, advection, humidity), a phase difference between ground heat flux and net radiation, and stomatal regulation. There is an alternative to the assumption of constant evaporative fraction if hourly weather data are available. It may then be assumed that the ratio of actual to reference evaporation is constant over the day; hourly values of reference evaporation can be calculated (Allen et al., 2007). The ratio of actual to reference evaporation is more stable, because it eliminates the effects of diurnal variations in weather conditions.

3. Data requirements and sensitivity

Every remote sensing based SEB model requires a sequence of dedicated ground and remote sensing data to properly operate. Efforts increasingly focus on the remote estimation of the necessary variables, but ground data are still needed in addition.
All models require net radiation and land surface temperature retrieved from remote sensing. The additional required information varies among algorithms. As an example we list the input needed for the remote sensing model SEBS (Su, 2002). This model explicitly solves Eqs 35-37. It also includes an algorithm to estimate kB^{-1} from vegetation cover fraction.
SEBS requires the following data, most of which cannot be retrieved from remote sensing, but is obtained from ground-based meteorological data instead:
1. Reference height z_{ref} [m]: height from the ground where measurements of temperature, wind, pressure and specific humidity are made [m].
2. Air Temperature at reference height (T_a) [°C].
3. Specific humidity [kg.kg^{-1}] or relative humidy [%], for calculation of emissivity of the sky.
4. Wind speed at the reference height (u_{ref}) [m.s^{-1}].
5. Air Pressure at reference height [Pa].
6. Air Pressure at land surface and reference height [Pa].
7. The planetary boundary (PBL) height h_i [m], required for the calculation of stability. It can be estimated by radiosounding or using atmospheric model outputs. By default h_i=1000 m.
8. A map of vegetation heights, or alternatively, classes associated with vegetation height values, a map of Leaf Area Index (LAI) from which vegetation height is estimated.

All meteorological input must be instantaneous information collected at the time of satellite overpass, interpolated and re-sampled to the pixel size. Other models require similar input. Two source models do not use the concept of kB^{-1}, but require separate resistances for soil and vegetation. Although the exact input data varies per algorithm, the most important are those related to the calculation of sensible heat flux, in particular the surface-air gradient and the corresponding aerodynamic resistance r_{ah}. The success or failure of a SEB relies on the skills of the research team to extract realistic values for these two variables. For this reason, we will discuss these in more detail in the following sections.

3.1 The temperature gradient

For the temperature gradient we need to estimate both the air and the land surface temperature. The issue is that sensible heat flux is proportional to a difference between two temperatures which are obtained from two different sources in the same vertical. For this reason great care should be taken to retrieve both temperatures accurately.

For the air temperature at reference height, interpolated data of meteorological stations are commonly used. We need the air temperature well above the canopy, in the atmospheric surface layer, for which the aerodynamic resistance is defined. The standard measurement height in meteorological stations of 2 m cannot be used for vegetation taller than this height. Thus a conversion of temperatures from the meteorological stations to a higher reference height is needed. Another option is to use temperature profiles disseminated by organizations like EUMetsaf (www.eumetsat.int).

For the surface temperature it is necessary to take a closer look at the concepts first. As discussed before, the radiometric temperature is the temperature as it is retrieved from a remote radiometer by inverting Stefan-Boltzmann's law, assuming a bulk emissivity for the thermal spectrum range of the radiometer. The kinematic temperature is the real, contact temperature. A third definition is needed here: the aerodynamic temperature, which is hypothetic temperature obtained when extrapolating the vertical profile of air temperature to the depth z_{0h}. The aerodynamic temperature is a conceptual model parameter that is close to the kinetic temperature, but they are not equal. The reason is that kinematic temperature varies between the elements of the surface within a remote sensing pixel. For example, sunlit and shaded parts of the soil and canopy may have rather different temperatures. This is particularly the case in a heterogeneous landscape, where bare soil, vegetated and paved areas are mixed. It is even the case in a homogeneous land cover, where leaf temperatures may differ depending on their vertical position in the crown. This is illustrated in Fig 3, showing the diurnal variations of contact temperatures of a needle forest in the Netherlands, the Speulderbos site, measured during a field campaign in The Netherlands on 16 June 2006 (Su et al., 2009). The lines are ensembles of 8 soil surface and 9 needle temperature sensors, mounted at different heights of the canopy of just a few trees.

The heterogeneity of the soil and canopy temperatures will affect the radiometric surface temperature. The radiometric temperature is predominantly affected by the upper, visible, part of the canopy. Lower canopy layers also contribute to the outgoing upward radiation, but their contribution will be relatively low due to re-absorption of radiation. The radiometric temperature also depends on the solar angle and the observation angle of the satellite.

A new model to analyse these effects is the model SCOPE (Soil Canopy Obvservation of Photosynthesis and the Energy balance). This model is a radiative transfer model combined with an energy balance model for homogeneous vegetation (Van der Tol et al., 2009). With SCOPE one can analyse the relation between the sensible heat flux, the kinematic

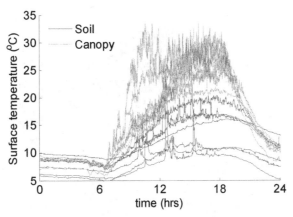

Fig. 3. Diurnal cycle of 8 soil and 9 needle contact temperature measurements (with NTC sensors) at the Speulderbos needle forest site in the Netherlands, on 16 June 2006.

temperatures of different elements in the canopy and the radiometric temperature. An example of an analysis carried out with SCOPE is shown in Fig 4. This figure shows simulated radiometric temperature of sparse but homogeneous crop as a function of the satellite observation azimuth (the counter-clockwise rotation angle from the top in the graph) and zenith angle (distance from the centre of the graph). We can see pronounced differences in the observed radiometric temperatures. Cleary visible is the hotspot, the situation where the solar zenith and azimuth angles equal those of the sensor. In the hotspot the radiometric temperature is higher than outside the hotspot. Radiometric temperatures are also higher at lower zenith angles compared to NADIR observations (vertically downward, in the centre of the graph). The differences in temperature are up to 2 °C, indicating that care should be taken of the observation angle relative to the solar angle. It is also possible to exploit the differences in radiometric surface temperature observed at different angles in order to separate soil and canopy kinetic temperatures (Timmermans et al, 2009).

How severe is an error of 2 °C for the estimation of sensible heat flux? Equation 18 shows that the error in sensible heat flux is proportional to the ratio of the temperature gradient to the resistance. This means that for the same error in the temperature gradient, the error in the sensible heat flux will be larger if the aerodynamic resistance is low than if the aerodynamic resistance is high.

In order to consider the sensitivity more precisely, we take the example of a situation where the aerodynamic resistance is low: the Speulderbos forest site in The Netherlands. This site is equipped with a 46-m tall eddy covariance measurement tower. Because of the low aerodynamic resistance, the sensitivity to temperature is expected to be relatively high. For this site, we calculated the friction velocity and the sensible heat flux with Eqs 35-37, using a canopy height of 30 m, the measured wind speed and temperature at 45 m height, radiometric temperature measured with a long-wave radiometer, and assuming that z_{0m} = 0.12 h and d = 0.67 h.

Figure 5 shows the results for 15-18 July 2009. The day-time friction velocity matches well with the measurements, showing that the calculation of aerodynamic resistance was

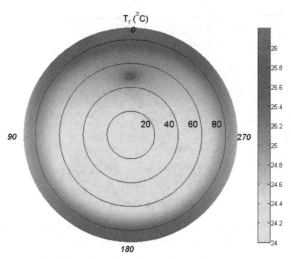

Fig. 4. Hemispherical graph of simulated radiometric surface temperature of a thinned maize crop with a LAI of 0.25, as a function of viewing zenith angle and viewing azimuth angle (relative to the solar azimuth). Zenith angle varies with the radius, the azimuth angle (in italic) increases while rotating anticlockwise from north. The solar zenith angle was 48° (after Van der Tol et al., 2009).

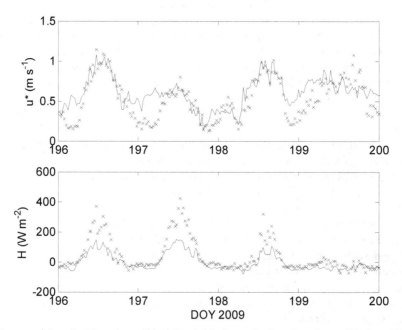

Fig. 5. Measured (symbols) and modeled (line) friction velocity u^* and sensible heat H flux versus Julian day number (14-19 July 2009) for an eddy covariance tower in the Speulderbos forest site, The Netherlands.

accurate. During the night (stable conditions), the performance is worse, but this is not a large problem because the night-time sensible heat flux proved very small. However, there is a 50% error in afternoon sensible heat flux. Can this be related to an error in the surface temperature? Figure 6 shows the result of a sensitivity analysis to surface temperature. A consistent bias was added to the measured time series (x-axis), and the resulting root mean square error (RMSE) of friction velocity and sensible heat flux was calculated (y-axis). The RMSE reaches a minimum when surface temperature is 0.5 °C above the measured value, but it rises to unacceptably high values of the absolute temperature bias is greater than 2 °C. In this example, field data of radiometric surface temperature were used. What if remote sensing data are available? Satellite products are available at either high temporal (geostationary satellites) or at high spatial resolution (polar orbiting satellites). Data are available at a spatial resolution of 3-5 km and a temporal resolution of 15 minutes (MeteoSat or GOES) to 1 km resolution at a daily time scale (AVHRR, MODIS or MERIS), or 60 m with a repetition time of weeks to months (LANDSAT, ASTER). The low temporal resolutions are not really useful, because of the dynamic nature of the turbulent heat fluxes. The daily revisits are useful provided that reasonable assumptions are made about the diurnal cycles of the fluxes (see Sect 2.4). The orbits are designed to overpass at the same solar time every day. The 15-minute intervals are ideal, but the spatial resolution makes the estimation of an effective aerodynamic resistance difficult, as we will see later.

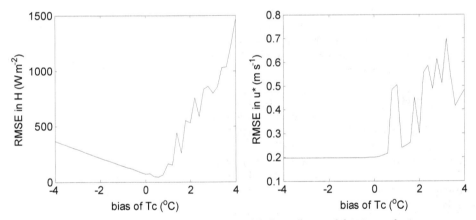

Fig. 6. Root mean square error of modelled sensible heat flux and friction velocity versus a forced bias in the observed radiometric surface temperatures for the Speulderbos forest site, for 14-19 July 2009.

A final issue that needs to be considered is the topography. In areas with large elevation differences, the interpolation technique for air temperature data is crucial. Errors of several degrees in the air temperature are easily introduced if an incorrect adiabatic lapse rate is used.

It is possible to circumvent the problem of estimating the temperature gradient by using an image based calibration (Bastiaanssen et al., 1998), in which assumptions are made for the energy balance state at the hottest and the coolest pixel in the image. In the first versions of this approach the calculated fluxes depended on the size of the image that was selected and the on assumption that the hottest pixel is dry, but more recent developments do not suffer from this drawback. The Mapping Evapotranspiration with Internalized Calibration (METRIC) model (Allen et al., 2007) uses reference evapotranspiration of alfalfa to calibrate the relation between the temperature gradient and the measured surface temperature. In the METRIC model it is assumed that the evaporation in the wettest pixel is 5% above the reference evapotranspiration, and the evaporation of the driest pixel is estimated with a soil-vegetation-atmosphere model. This has the additional advantage that the evaporation values are bound to a realistic minimum and a realistic maximum rate. The METRIC model also accounts for topography by correcting radiation for slope and aspect and temperature for elevation using a local lapse rate.

3.2 Sensitivity to the aerodynamic resistance

The roughness length z_{0m} (and often displacement height is linked to it) is recognized as the main source of error in the remote estimate of ET. Currently, there are several methods that can be used to approach a good z_{0m} (see Table 2).

When near surface wind speed and vegetation parameters (height and leaf area index) are available, the within-canopy turbulence model proposed by Massman (1999) can be used to estimate aerodynamic parameters, d, the displacement height, and, z_{0m}, the roughness height for momentum. This model has been shown by Su et al. (2001) to produce reliable estimates of the aerodynamic parameters. If only the height of the vegetation is available, the relationships proposed by Brutsaert (1982) can be used. If a detailed land use classification is available, for example based on LandSat images, the tabulated values of Wieringa (1993) can be used. By using literature values, errors in the canopy height of the order of decimetres to several metres are likely to occur, and errors in the roughness length in the order of decimetres.

Method	Input needed	Remark
z_{0m} = 0.136 h	Vegetation height map (h)	
from Lookup table (LUT)	Vegetation map & z_{0m} LUT	
From vegetation index	Vegetation index maps	
z_{0m} from modelling	Landuse & veg. structure	
LIDAR	Experimental. Costly.	Costly method
Retrievals from wind profiles	Wind speed profiles	Point values only

Table 2. Methods for the estimation of z_{0m} (After: Su, 2002).

When all of the above information is not available, then the aerodynamic parameters can be related to vegetation indices derived from satellite data. However in this case, care must be taken, because the vegetation indices saturate at higher vegetation densities and the relationships are vegetation type dependent. For example, characteristic of the land surface are sometimes calculated from indices like the Normalized Difference Vegetation Index (NDVI), but there is no reason why NDVI should have a universal relation with surface roughness. A grass field may have a similar NDVI to that of a forest, but a roughness length that is an order of magnitude smaller. For this reason, literature data or ground-truth data are indispensible for an accurate estimate of the surface roughness. A relation between NDVI and surface roughness can only be made for low vegetation, normally irrigated. In that case a non-linear relation with vegetation structure is first established by assigning a maximum value and a minimum value of height corresponding to values of NDVI.

We illustrate the sensitivity of the aerodynamic resistance model with the data set of the Dutch forest site introduced in the previous section. Now, the height of the forest was varied between 5 and 45 m, and the RMSE of sensible heat flux and friction velocity evaluated (Fig 7). Note that the vertical scale in Fig 7 is much smaller than in Fig 6, which indicates that for forest, the model is less sensitive to errors in the canopy height than errors in surface temperature.

Sparse canopies require special attention. In sparse canopies the temperature differences between canopy and soil may be over 20 °C. In addition, no canopy height can be defined, which makes it difficult to estimate the roughness length z_{0m}, and normally d is neglected. A solution to these problems is to program a two-source model (e.g. Norman et al., 1995). An alternative solution is to modify the parameter kB^{-1} to incorporate the differences in surface temperature implicitly in the value of z_{0h} (Verhoef et al., 1997). We will illustrate the latter solution with a simple example.

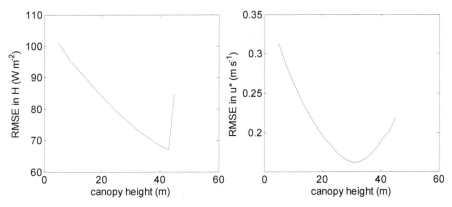

Fig. 7. Root mean square error of modelled sensible heat flux H and friction velocity u^* velocity versus assumed canopy height of the Speulderbos forest site, for 14-19 July 2009.

The sparse canopy of our example is a study site in the province of León, Spain. The vegetation cover fraction is 11%, consisting of patches of 6-m tall *Quercus ilex* and *Quercus pyrenaica*. Data of an eddy covariance flux tower are used for validation of the satellite product. For this site, a roughness length of z_{0m}=0.2 m was assumed, and a displacement height of d=0. The friction velocity and sensible heat flux were again calculated from Eqs 35-37. For z_{0h}, a value of 0.02 m was initially assumed (kB^{-1} = 2.3), and for wind speed, the field measurements at the flux tower were used. For net radiation and surface temperature, 15-minute interval MeteoSat Second Generation (MSG) satellite data were used. The top panels in Fig 8 show the results of the satellite based algorithm. The friction velocity observations are accurately reproduced, but the modelled sensible heat flux is extremely high, even double the net radiation. The overestimate is solved when we reduce z_{0h} by four orders of magnitude (kB^{-1} = 11.5). The reduction in z_{0h} needed to match the model with the observations is large. This problem was discussed earlier after the Hapex-Sahel measurement campaign (Verhoef et al., 1997). It was then concluded that the whole concept of kB^{-1} is questionable. It is indeed recommended to avoid the use of kB^{-1}, and this can be done in two ways: (1) by using more complicated two-source models for sparse vegetation, or (2) to use image-based calibration to relate surface temperature to a temperature gradient between two heights well above z_{0h}. The second approach is used in models such as SEBAL (Bastiaanssen et al., 1998) or METRIC (Allen et al., 2007).

A model exists to estimate the kB^{-1} from vegetation density (Su et al., 2001). This model is used in the remote sensing algorithm SEBS (Su, 2002). However, care should be taken with any kB^{-1} model for areas where no detailed information on cover or other field data are available for calibration.

In the future, global maps of surface roughness may become timely available. Through synthesis of LiDAR with high resolution optical remote sensing, the roughness parameters have been successfully estimated spatially (Tian et al., 2011). Surface maps produced with laser satellites (NASA's ICESat and the future ICESat2) are also promising tools for estimating roughness (Roxette et al., 2008).

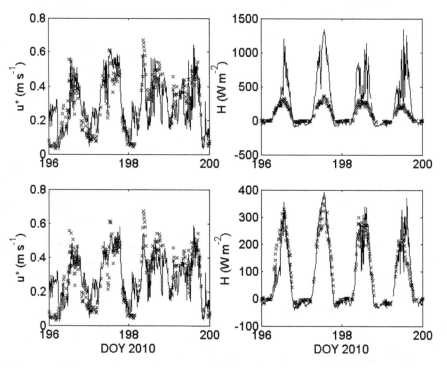

Fig. 8. Measured (symbols) and modeled (line) friction velocity u^* and sensible heat H flux versus Julian day number (14-19 July 2010) for an eddy covariance tower in the sparsely vegetated area of Sardon, Spain. Top graphs: using z_{0m} = 0.2 and kB^{-1} = 2.3. Bottom graphs: using z_{0m} = 0.2 and kB^{-1} = 11.5.

4. Conclusions

All remote sensing algorithms for ET make use of the energy balance equation (EBE). In this equation, latent heat flux is calculated as a residual of the energy balance. Net radiation can be estimated from remote sensing products relatively easily. Ground heat flux can only be retrieved with geostationary satellites for sparsely vegetated areas or bare land. It is usually a minor term in vegetated areas that causes relatively small errors in the final ET product.

The most critical component of the energy balance is the sensible heat flux. In the calculation of the sensible heat flux, both the temperature difference (land surface temperature minus the air temperature) and the aerodynamic resistance need careful attention.

In areas with high elevation differences, the errors in temperature are usually so high, and temperature correction using local lapse rates is necessary. In flat areas, a local sensitivity analysis is recommended. For forest, the accuracy of the temperature gradient should be better than 2 °C in order to achieve reasonable results. In sparse vegetation two source models are preferred over single-source models, because in the latter, parameterization of z_{0h} on operational basis is no better than a wild guess. If a two-source model is not an option, then image based calibration using reference evaporation is a good alternative in these areas. Accurate roughness information (z_{0m}) is required; the information is preferably verified and monitored on the ground. Satellite laser altimetry provides a promising tool for better roughness estimates in the near future.

5. References

Allen, R.G., Pereira, L.S., Raes, D. & Smith M. (1998). Crop Evapotranspiration. Guidelines for computing crop water requirements. *FAO Publication* 56, Rome, Italy, pp. 300

Allen, R.G., Tasumi, M. & Trezza, R. (2007). Satellite-based energy balance for mapping evapotranspiration with internalized calibration (METRIC) – model. *J. Irr. Drainage Engineering ASCE* 133 (4), pp. 380-394

Bastiaanssen, W.G.M., Menenti, M., Feddes, R.A. & Holtslag, A.A.M. (1998). A remote sensing surface energy balance algorithm for land (SEBAL). 1. Formulation. *J. Hydrol.* 212-213, pp. 198–212

Blümel, K. (1999). A simple formula for estimation of the roughness length for heat transfer over partly vegetated surfaces, *J. Appl. Meteorol.*, 38, pp. 814–829

Brutsaert, W. (1975). On a derivable formula for long-wave radiation from clear skies. Water Resour Research 11 (5), pp. 742-744, doi:10.1029/WR011i005p00742

Brutsaert, W. (1982). Evaporation into the atmosphere. *D. Reidel Publishing Co.*, Dordrecht, The Netherlands, pp. 300

Burtsaert, W., (2005). *Hydrology - an introduction.* Cambridge University Press: Cambridge, UK, pp. 605

Brutsaert, W. & Sugita, M. (1992). Application of self-preservation in diurnal evolution of the surface energy budget to determine daily evaporation. *J. Geophysical Res* 97, pp. 18377-18382

Courault, D., Seguin, B. & Olioso, A. (2005). Review on estimation of evapotranspiration from remote sensing data: From empirical to numerical modeling approaches. *Irrigation Drain Syst* 19, pp. 223–249

Clothier, B.E., Clawson, K.L., Pinter, P.J. Jr., Moran, M.S., Reginato, R.J. & Jackson, R.D. (1986). Estimation of Soil Heat Flux from Net Radiation During Growth of Alfalfa. Agric. *For. Meteorol.* 37, pp. 319-329

Doorenbos, J. & Pruitt, W.O. (1977). Guidelines for predicting crop water requirements, *Irrigation and Drainage Paper* 24, FAO, Rome, Italy, pp. 179.

Gentine P., Entekhabi D., Chehbouni A., Boulet G. & Duchemin B. (2007). Analysis of evaporative fraction diurnal behaviour. *Agric. Forest Meteorol.*, 143 (1-2), pp. 13-29

Gieske, A.S.M. (2003). Operational solutions of actual evapotranspiration. In: Understanding water in a dry environment. Hydrological processes in arid and semi-arid zones. I. Simmers (Ed.), *International Association of Hydogeologists*, Balkema Publishers, The Netherlands

Glenn, E.P., Huete, A.R., Nagler, P.L., Hirschboeck, K.K. & Brown, P. (2007). Integrating remote sensing and ground methods to estimate evapotranspiration. *Crit Rev Plant Sci* 26(3), pp. 139–168, doi:10.1080/07352680 701402503

Gowda, P.H., Chavez, J.L., Colaizzi, P.D., Evett, S.R., Howell, T.A. & Tolk, J.A. (2007). Remote sensing based energy balance algorithms for mapping ET: current status and future challenges. *Trans Am Soc Agric Biol Engineers* 50(5), pp. 1639–1644

Heusinkveld, B.G. Jacobs, A.F.G., Holtslag, A.A.M. & Berkowicz, S. M. (2004). Surface energy balance closure in an arid region: role of soil heat flux. *Agric. Forest Meteorol.* 122(1-2), 21-37, doi:10.1016/j.agrformet.2003.09.005

Heusinkveld, B.G., Jacobs, A.F.G., Holtslag, A.A.M. & Berkovicz, S.M. (2004). Surface energy balance closure in an arid region: role of soil heat flux. *Agric. Forest Meteorol.* 122 (1-2), pp. 21-37

Jackson, R.D., Reginato, R.J. & Idso, S.B. (1977). Wheat canopy temperatures: A practical tool for evaluating water requirements. *Water Res. Res* 13, pp. 651–656

Jackson R.D., Moran, M.S., Gay, L.W. & Raymond, L.H. (1987). Evaluating Evaporation from Field Crops Using Airborne Radiometry and Ground-based Meteorological Data. *Irrig. Sci.* 8, pp. 81-90.

Kalma, J.D., McVicar, T.R. & McCabe, M.F. (2008). Estimating land surface evaporation: A review of methods using remote sensed surface temperature data. *Surveys in Geophysics*, 29, pp. 421-469

Kustas, W.P. & Daughtry, C.S.T. (1990). Estimation of the Soil Heat Flux/Net Radiation Ratio from Spectral Data. *Agric. For. Meteorol.* 49, pp. 205-223

Maidment, D. (1992). *Handbook of Hydrology.* Ed: D.R. Maidment, McGraw-Hill, Inc.

Massman, W.J. & Weil, J.C. (1999). An analytical one-dimensional second-order closure model of turbulence statistics and the Lagrangian time scale within and above plant canopies of arbitrary structure. *Boundary-Layer Meteorol.* 91, pp. 81-107

Murray, T. & Verhoef, A. (2007). Moving towards a more mechanistic approach in the determination of soil heat flux from remote measurements I. A universal approach to calculate thermal inertia. *Agric Forest Meteorol.* 147, pp. 80–87

Norman, J.M., Kustas, W.P. & Humes, K.S. (1995). Source approach for estimating soil and vegetation energy fluxes in observations of directional radiometric surface temperature. *Agric. Forest Meteorol.* 77 (3-4), pp. 263-293

Obukov, A. (1946). Turbulence in an atmosphere with a non-uniform temperature. *Bound. Layer Meteorol.* 2, pp. 7-29

Paulson, C.A. (1970). The mathematical representation of wind speed and temperature profiles in the unstable atmospheric surface layer. *J. Applied Meteorol.* 9 (6), pp. 857-861

Reginato R., Jackson, R. & Pinter, J.Jr. (1985). Evapotranspiration Calculated from Remote Multispectral and Ground Station Meteorological Data. *Remote Sens. Environ.* 18, pp. 75-89

Rosette, J.A.B., North, P.R.J. & Suárez, J.C. (2008). Vegetation height estimates for a mixed temperate forest using satellite laser altimetry. *Int. J. of Rem. Sens.*, 29(5), pp. 1475-1493.

Seguin, B. (1988). Use of Surface Temperature in Agrometeorology. In: *Applications of Remote Sensing to Agrometeorology.* Ed F. Toselli, pp. 221-240

Su, Z. (2002). The Surface Energy Balance System (SEBS) for estimation of turbulent heat fluxes. *Hydrol. Earth Syst. Sci.* 6, pp. 85-99

Su, Z., Schmugge, T., Kustas, W.P. & Massman, W.J., (2001). An evaluation of two models for estimation of the roughness height for heat transfer between the land surface and the atmosphere. *J. Appl. Meteorol.* 40, pp. 1933–1951

Su, Z., Timmermans, W. J., van der Tol, C., Dost, R., Bianchi, R.,Gómez, J. A., House, A., Hajnsek, I., Menenti, M., Magliulo, V., Esposito, M., Haarbrink, R., Bosveld, F., Rothe, R., Baltink, H. K., Vekerdy, Z., Sobrino, J. A., Timmermans, J., van Laake, P., Salama, S., van der Kwast, H., Claassen, E., Stolk, A., Jia, L., Moors, E., Hartogensis, O. & Gillespie, A. (2009). EAGLE 2006 – Multi-purpose, multi-angle and multi-sensor in-situ and airborne campaigns over grassland and forest, *Hydrol. Earth Syst. Sci.*, 13, pp. 833–845

Tian, X. Li, Z.Y., Van der Tol, C., Su, Z., Li, X., He, Q.S., Bao, Y.F., Chen, E.X. & Li, L.H. (2011). Estimating Zero-Plane Displacement Height and Aerodynamic Roughness Length using Synthesis of LiDAR and SPOT-5 data, *Remote Sens. Environ.* 115, 2330-2341.

Timmermans, J., Verhoef, W., Su, Z. & Van der Tol, C. (2009). Retrieval of canopy component temperatures through Bayesian inversion of directional thermal measurements. *Hydrol. Earth Syst. Sci.*, 13, pp. 1249–1260

Van der Tol, C., Verhoef, W., Timmermans, J., Verhoef, A. & Su, Z. (2009). An integrated model of soil-canopy spectral radiances, photosynthesis, fluorescence, temperature and energy balance, *Biogeosciences* 6, pp. 3109–3129

Van Wijk, W.R., De Vries, D.A. (1963). Periodic temperature variations in a homogeneous soil. In: W.R. Van Wijk (Ed.), Physics of Plant Environment, North-Holland, Amsterdam, pp. 102–143

Verhoef, A., 2004. Remote estimation of thermal inertia and soil heat flux for bare soil. Agric. Forest Meteorol. 123, pp. 221–236

Verhoef, A. (2004). Remote estimation of thermal inertia and soil heat flux for bare soil. *Agric. Forest Meteorol.* 123, pp. 221–236

Verhoef, A., De Bruin, H.A.R. & Van Den Hurk, B.J.J.M. (1997). Some Practical Notes on the Parameter kB−1 for Sparse Vegetation. *J. Appl. Meteor.* 36, pp. 560–572, doi: 10.1175/1520-0450(1997)036<0560:SPNOTP>2.0.CO;2

Verma, S.B. & Bartfield, B.J. (1979). Aerial and Crop Resistances Affecting Energy Transport. In: *Modification of Aerial Environment of Crops.* Ed. B.J. Bartfield & J.F. Gerber, SAE, pp. 230-248

Vinukollo, R.K., Wood, E.F., Ferguson, C.R., & Fisher, J.B. (2011). Global estimates of evapotranspiration for climate studies using multi-sensor remote sensing data:

Evaluation of three process-based approaches. *Remote Sens. Environ.* 115, pp. 801-823

Wieringa, J. (1993). Representative Roughness Parameters for Homogeneous Terrain, *Boundary-Layer Meteorol.* 63, pp. 323–363

2

Estimation of the Annual and Interannual Variation of Potential Evapotranspiration

Georgeta Bandoc

University of Bucharest, Department of Meteorology and Hydrology
Center for Coastal Research and Environmental Protection
Romania

1. Introduction

Knowledge of ecological factors for all natural systems, including human-modified natural systems, is essential for determining the nature of changes in these systems and to establish interventions that must be achieved to ensure optimal functioning of these systems.

The purpose of this chapter is to identify annual and interannual variations of potential evapotranspiration, in conjunction with climate changes in recent years, on the coastal region of Sfântu Gheorghe – Danube Delta. Under natural conditions, evapotranspiration flows continuously throughout the year, representing a main link in the water cycle and an important heat exchange factor affecting ecosystems. Potential evapotranspiration is the maximum amount of water likely to be produced by a soil evaporation and perspiration of plants in a climate.

Real balance between the amount of precipitation fallen named P and the amount of water taken from the atmosphere as vapour, called potential evapotranspiration PET is of particular importance in characterizing climate, representing an expression of power absorption by the atmosphere and expressing quantity water on soil and vegetation that request (Henning & Henning, 1981).

The difference between precipitation (P) and potential evapotranspiration (PET), i.e. $P - PET$ known as ΔP is denoted by excess precipitation to PET (E) or deficit of precipitation to PET (D) if the difference is positive or, respectively, negative. The intensity of water loss through evaporation from the soil or by transpiration from the leaf surface is largely determined by vapour pressure gradient, i.e. the vapour pressure difference between leaf and soil surface and atmospheric vapour pressure (Berbecel et al, 1970).

The vapour pressure gradient is determined, in turn, by the characteristics of air and soil factors, such as: radiant energy, air temperature, vertical and horizontal movements of the air saturation deficit, the degree of surface water supply evaporation, plant biology and soil characteristics.

Heat factor also has a significant influence on evapotranspiration as temperature, on one hand, intensify of water vapour increases and, on the other hand, increases air capacity to maintain water vapour saturation state, reducing atmosphere's evaporated power (Eagleman, 1967).

2. General issues related to estimate

Potential evapotranspiration evidence and interannual variations of *PET* potential evapotranspiration and water balance, climate charts are used based on measurements from weather stations hydrothermal (Walter & Lieth, 1960; Walter, 1955, 1999; Köppen, 1900, 1936 etc.).

In 2005, Oudin et al. compile lists 25 methods for estimating potential evapotranspiration based on a series of meteorological parameters (Douglas et al, 2009).

Estimation of potential evapotranspiration can be done using the indirect method based on air temperature readings and diagrams and on Thornthwaite's tables (Thorntwaite, 1948; Donciu, 1958; Walter & Lieth, 1960).

Recent studies use Penman's equation for this purpose, Penman (Penman, 1946), Penman - Monteith (Thomas, 2000a, 200b; Choudhury, 1997; Allen et al., 1998; Chen et al, 2005). Also, in determining the potential evapotranspiration, other formulas have been used with results almost similar with the ones of direct measurements, such as formulas of Bouchet (Bouchet, 1964), Turc (Turc, 1954), Hargreaves (Hargreaves & Samani, 1982), Papadakis (1966), Hamon (1963), Priestley – Taylor (1972), Makkink (1957) (Lu et al, 2005) and Blaney-Criddle (1950) (Ponce, 1989).

Potential evapotranspiration *PET* is of great temporal variability and thus an estimation can be done based on heat and water vapour from the atmosphere (Dugas et al, 1991; Celliar and Brunet, 1992; Rana & Katerji, 1996; Droogers at al, 1996; Frangi et al, 1996; Linda et al, 2002 as cited in Chuanyan et al, 2004).

Another model of estimation for *PET* is based on soil moisture and rainfall (model Century) (Metherell et al 1992; Zhou et al, 2008 as cited in Liang et al, 2010). On the interaction of global precipitation and air temperature estimates can be done for potential evapotranspiration (Raich & Schlesinger, 1992; Buchmann, 2000; Andréassian et al, 2004; Li et al, 2008a, 2008b; Casals et al, 2009).

Estimation of potential evapotranspiration can be achieved also based on satellite measurements related to air humidity and wind characteristics, but only in case of high-resolution satellite images (Irmak, 2009).

In the estimation of *PET* remote sensing methods are applied (Chaudhury, 1997; Granger, 1997; Stefano & Ferro, 1997; Caselles et al, 1998; Stewarta et al, 1999). These methods are based using geographic information system using GIS spatial modeling (Baxter et al, 1996; Srinivasan et al 1996; Moore, 1996; Cleugh et al, 2007, Tang et al, 2010).

Other studies use numerical modeling to simulate various weather variables in a particular location, variables used to calculate potential evapotranspiration (Kumar et al, 2002; Smith et al, 2006; Torres et al 2011).

3. Research on characteristics of coastal area of potential evapotranspiration

The location where this study has been made is the south-east of Salt and marine field is bordered by the Black Sea coast in the east, marine low deltaic plain in the west and north-west and the arm of Sfântu Gheorghe –Danube Delta (fig. 1). To the south of arm of Sfântu Gheorghe is the marine plain Dranov , Sfântu Gheorghe secondary delta and Sacalin Island. In the context of global climate change, interannual evolution analysis, annual and multiannual magnitudes that characterize the climate of a region are of particular interest (Palutikov et al, 1994; Chattopadhyay et al, 1997; Kouzmov, 2002; Oguz et al, 2006). This interest increases when it is a coastal region where sea atmosphere - interactions induce very specific issues.

Fig. 1. Study area location

The Danube Delta combines the temperate semi-arid climate space typical for the Pontic steppes. The aquatic very wide plane spaces, differently covered by vegetation and intrerrupted by the sandy islands of the marine fields, make up an active area specific to the delta and to the adjacent lagoons but totally different from that belonging to the Pontic stepps. This active area reacts upon the total radiation intercepted by the general circulation of atmosphere, resulting in a mosaic of microclimates (Vespremeanu, 2000, 2004).

For determining how climate changes affect the interannual potential evapotranspiration in the Sfântu Gheorghe costal area it was started, primarily from the fact that PET potential evapotranspiration has strong fluctuations in time and space as a direct consequence of the variation factors leads. Thus, in order to achieve the intended purpose of this chapter, interannual and annual potential evapotranspiration values were determined according to Thornthwaite's method, both for the period 1961 - 1990, taken as a reference period and analyzed for the studied period 2000 - 2009. Interannual differences $P - PET$ as well as annual amounts of the differences of the same sign, $\sum(P - PET)^{+}$ and $\sum(P - PET)^{-}$ as well as the annual review, are important climatic indicators. The determination of the efficiency of precipitation was done by calculating the difference $P - PET$ taken as reference period 1961 - 1990 and for the period under study from 2000 to 2009. Positive differences indicate excess water from rainfall, water shortages $\sum \Delta P^{+}$ and the negative ones indicate deficit of precipitation, water requirements from the atmosphere $\sum \Delta P^{-}$.

It was also determined the precipitation deficit offset by previously accumulated surpluses and deficits of precipitation uncompensated by previous surpluses.

To identify climate changes in coastal Sfântu Gheorghe area and deviations from the average annual values of air temperature and precipitation, diagrams were drawn, type Walter and Leith, to identify dry periods and also different indices and specific factors were calculated such as: Martonne arid index (I_{ar}), retention index offset (I_{hc}), the amount of rainfall in the period with t ≥ 10 ° C temp ($P_{t \geq 10^0 C}$) rainfall amount of soil loading in the months from November to March (P_{XI-III}), the amount of summer rainfall in July and August ($P_{VII-VIII}$), Lang precipitation index for the period with t ≥ 10 ° C ($L_{t \geq 10^0 C}$), precipitation index for summer Lang ($L_{VI-VIII}$) and Lang precipitation index for spring season (L_{III-V}) and annual and interannual precipitation deficits (D) and excess (E) respectively , comparing to potential evapotranspiration of 10 mm, 20 mm, 30 mm etc. These indices and ratios were calculated based on meteorological measurements for the period 1961 - 1990, taken as a reference period for the 2000 - 2009 period under study. In this chapter, climate charts are playing an important role in the knowledge of the climate changes in the studied area and also helps in determining the precipitation – evapotranspiration, and hence the temperature deficit or surplus in the form of precipitation from evapotranspiration. Dryness site layout is determined in this study. Curve surplus or deficit of precipitation from evapotranspiration is crucial in environmental hydrothermal annual and interannual knowledge of an area.

Climate chart includes curved surfaces and values of temperature, precipitation at the scale 1/5 and 1/3 and potential evapotranspiration after Thornthwaite, interannual, annual and for certain periods (the amount of rainfall during the period from November - March, yet soil load, and the summer period July - August). The diagram also contains interannual surpluses and deficits and total rainfall to *PET*, the deficits in compensated and uncompensated previous surpluses, Walter - Lieth dry period, the annual aridity index, retention index offset, Lang rainfall index, calculated for the period temperature $t \geq 10^0 C$, for summer and spring time.

At the bottom of the chart months of the year and intra-annual values ΔP are indicated to express the character of moisture or dryness of the climate in different months, the monthly differences in classification categories E and D for each 10 mm, 20 mm, 30 mm etc. On the diagram, for 1 degree of temperature correspond 5 mm, 3 mm respectively of precipitation. Scale 1/5 was chosen in order to maximally achieve principle of the rainfall curve to be above the temperature when precipitation *PET* outperforms, and below it, when *PET* exceeds precipitation. Scale 1/3 was chosen to determine the dry period after the Walter - Lieth, which lasts as long as the rainfall curve is well below that of temperature.

It is important to know to what extent and interannual deficit of precipitation to *PET* during the growing season is offset by the surplus of precipitation to *PET* during the loading of the soil with water from precipitation (late autumn - winter). In this way deficit or surplus annual and interannual of effective precipitation is obtained comparing to *PET*.

In case of no loss of water through surface runoff and water infiltration or gains, the excess water is retained during loading or accumulated in the soil, and it is called full hydrologic soil (Chiriță et al, 1977).

The entire accumulated surplus of precipitation is the main reserve of soil water in the vegetation is gradually consumed and evapotranspiration together with new fallen rains (Donciu, 1983).

For the studied area, where the climate is characterized by periods of dryness, the water reserve accumulated in the soil is gradually depleted by evapotranspiration and biomass formation. This last amount of water should be considered as an element of water balance, important in the quantitative ratio (Chiriță et al, 1977).

Until finishing the accumulated precipitation of the soil in each month, water loss through evapotranspiration and precipitation is compensated by the previous reserve accumulation. Once this reserve is ended, precipitation deficit starts for the area studied. Evapotranspiration consumes current rainfall, leaving an additional demand of the atmosphere, dissatisfied with the precipitation.

The deficit of precipitation in this period presents the quantitative nature of PET's dry climate and soil and thus the existence of a period of severe water available to vegetation.

4. Results and discussion

The analysis of average monthly PET value as obtained for Sfântu Gheorghe, was a functional correlation of these values with the mean monthly air temperature T (Bandoc & Golumbeanu, 2010). For both analyzed periods, the correlations are straightforward.

From the calculation of correlation coefficient r and determination coefficient r^2 between potential evapotranspiration air temperature values of this coefficient $r = 0,98$, $r = 0,97$ and $r^2 = 96,04\%$, $r^2 = 94,04\%$ resulted, for the reference period 1961 - 1990 and for 2000 - 2009 period under study (fig. 2).

From the climate charts made for coastal Sfântu Gheorghe area (fig. 2, fig. 3, fig. 4, fig. 5, fig. 6, fig. 7, fig. 8 and fig. 9) for the analyzed periods, the result is a series of changes comparing to the duration of dryness reference interval 1961 - 1990.

Analizing the data the drought period for 2000-2009 was found to be 7 months which compared to the reference 1961-1990 (average drought) period of 6 months shows a increase of one month of drought per year.

Arid annual index Martonne calculation to determine the ratio between the amount of rainfall and temperatures $\left(I_{ar} = \dfrac{P}{T+10} \right)$ showed that for the period 2000 - 2009, there was a decrease in the value of the index with 17,71 % which leads to increased awareness of dryness for the studied area (fig. 10).

Rain index called Lang index or Lang factor of the period with temperatures $\geq 10\ ^0C$ ($L_{t \geq 10^0 C}$), spring (L_{III-V}) and summer ($L_{VI-VIII}$) determined as a ratio of the average monthly precipitation values and P values of monthly average air temperature $T \left(L = \dfrac{P}{T} \right)$.

The results obtained for these intervals revealed that the index $L_{t \geq 10^0 C}$ values decreased by 20,22 % for the period with t ≥ 10 ^{0}C, for spring period L_{III-V} rainfall index fell 26,05 %, while during the summer $L_{VI-VIII}$ value of this index was 37,20 % compared to the reference period 1961 - 1990 (fig. 10).

Offset fluid index $\left(I_{hc} = \dfrac{\sum \Delta P^+}{\sum \Delta P^-} \right)$ expresses the extent of precipitation deficits are

compensated by the surpluses. Values lower than the 1 $\left(I_{hc} \leq 1 \right)$ expressed precipitation deficits unabated. Following determination of the index for the two periods analyzed, that index values are 0,24 for the reference period 1961 - 1990 and 0,15 for the period 2000 - 2009. From the two values determined using the formula (0,24 and 0,15), for the past 10 years interval, results that the fluid compensation index decreased by 37,5 % compared to the reference period 1961 - 1990.

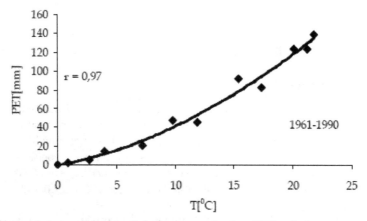

Fig. 2. Correlation between the potential evapotranspiration *PET* and air temperature *T* in the coastal region Sfântu Gheorghe for reference period 1961 - 1990

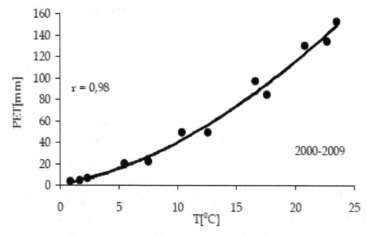

Fig. 3. Correlation between the potential evapotranspiration *PET* and air temperature *T* in the coastal region Sfântu Gheorghe for the period 2000 – 2009

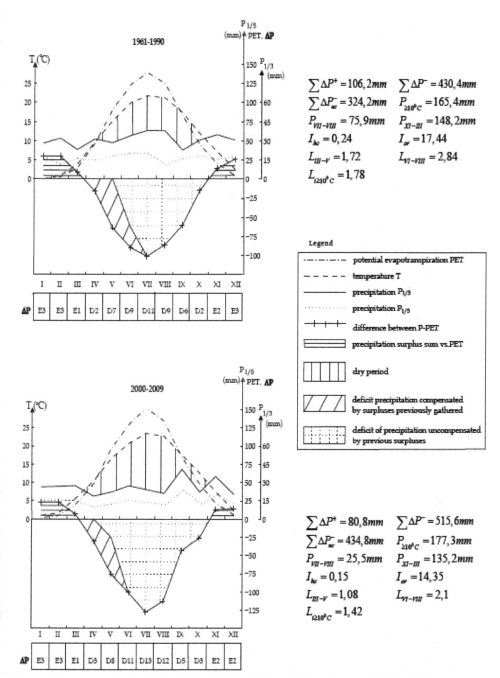

$$\sum \Delta P^+ = 106,2mm \qquad \sum \Delta P^- = 430,4mm$$
$$\sum \Delta P^-_{ac} = 324,2mm \qquad P_{210^0C} = 165,4mm$$
$$P_{VII-VIII} = 75,9mm \qquad P_{XI-III} = 148,2mm$$
$$I_{he} = 0,24 \qquad I_{ar} = 17,44$$
$$L_{III-V} = 1,72 \qquad L_{VI-VIII} = 2,84$$
$$L_{t210^0C} = 1,78$$

$$\sum \Delta P^+ = 80,8mm \qquad \sum \Delta P^- = 515,6mm$$
$$\sum \Delta P^-_{ac} = 434,8mm \qquad P_{210^0C} = 177,3mm$$
$$P_{VII-VIII} = 25,5mm \qquad P_{XI-III} = 135,2mm$$
$$I_{he} = 0,15 \qquad I_{ar} = 14,35$$
$$L_{III-V} = 1,08 \qquad L_{VI-VIII} = 2,1$$
$$L_{t210^0C} = 1,42$$

Fig. 4. Climate diagrams for reference interval 1960 – 1990 and interval 2000-2009 with characteristics sizes determined for reviewed site

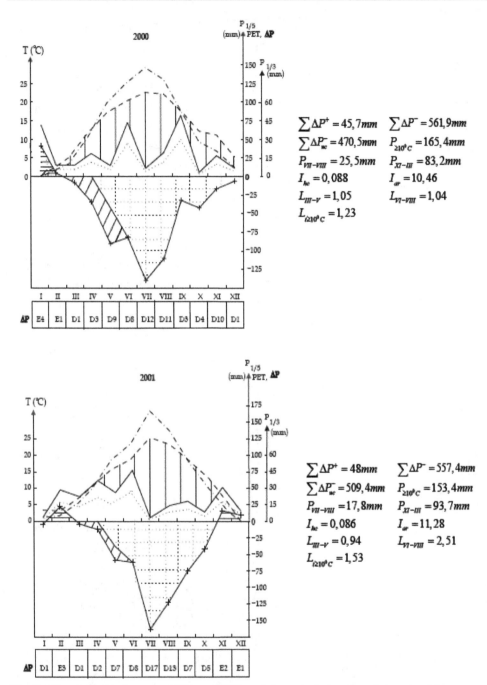

$\sum \Delta P^+ = 45,7mm$ $\sum \Delta P^- = 561,9mm$
$\sum \Delta P_{ac}^- = 470,5mm$ $P_{210^oC} = 165,4mm$
$P_{VII-VIII} = 25,5mm$ $P_{XI-III} = 83,2mm$
$I_{hc} = 0,088$ $I_{ar} = 10,46$
$L_{III-V} = 1,05$ $L_{VI-VIII} = 1,04$
$L_{t210^oC} = 1,23$

$\sum \Delta P^+ = 48mm$ $\sum \Delta P^- = 557,4mm$
$\sum \Delta P_{ac}^- = 509,4mm$ $P_{210^oC} = 153,4mm$
$P_{VII-VIII} = 17,8mm$ $P_{XI-III} = 93,7mm$
$I_{hc} = 0,086$ $I_{ar} = 11,28$
$L_{III-V} = 0,94$ $L_{VI-VIII} = 2,51$
$L_{t210^oC} = 1,53$

Fig. 5. Climate charts for years 2000 and 2001 and characteristics sizes determined for reviewed site

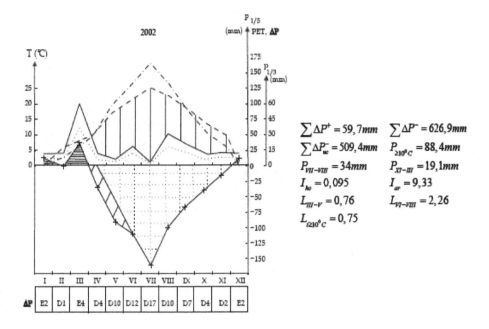

$$\sum \Delta P^+ = 59,7mm \quad \sum \Delta P^- = 626,9mm$$
$$\sum \Delta P^-_{we} = 509,4mm \quad P_{\geq 10^0 C} = 88,4mm$$
$$P_{VII-VIII} = 34mm \quad P_{XI-III} = 19,1mm$$
$$I_{he} = 0,095 \quad I_{ar} = 9,33$$
$$L_{III-V} = 0,76 \quad L_{VI-VIII} = 2,26$$
$$L_{\geq 10^0 C} = 0,75$$

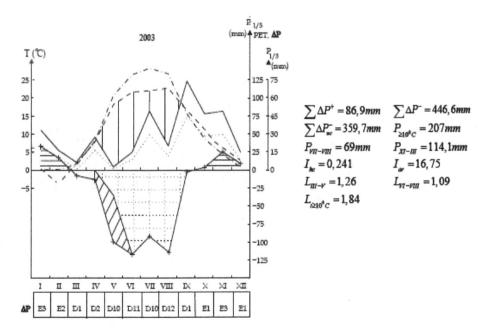

$$\sum \Delta P^+ = 86,9mm \quad \sum \Delta P^- = 446,6mm$$
$$\sum \Delta P^-_{we} = 359,7mm \quad P_{\geq 10^0 C} = 207mm$$
$$P_{VII-VIII} = 69mm \quad P_{XI-III} = 114,1mm$$
$$I_{he} = 0,241 \quad I_{ar} = 16,75$$
$$L_{III-V} = 1,26 \quad L_{VI-VIII} = 1,09$$
$$L_{\geq 10^0 C} = 1,84$$

Fig. 6. Climate charts for years 2002 and 2003 and characteristics sizes determined for reviewed site

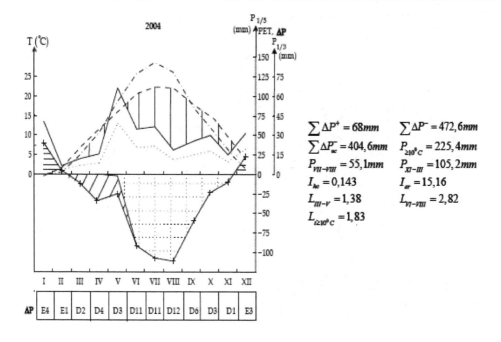

$$\sum \Delta P^+ = 68mm \qquad \sum \Delta P^- = 472,6mm$$
$$\sum \Delta P^-_{ac} = 404,6mm \qquad P_{\geq10^{\circ}C} = 225,4mm$$
$$P_{VII-VIII} = 55,1mm \qquad P_{XI-III} = 105,2mm$$
$$I_{hc} = 0,143 \qquad I_{ar} = 15,16$$
$$L_{III-V} = 1,38 \qquad L_{VI-VIII} = 2,82$$
$$L_{t\geq10^{\circ}C} = 1,83$$

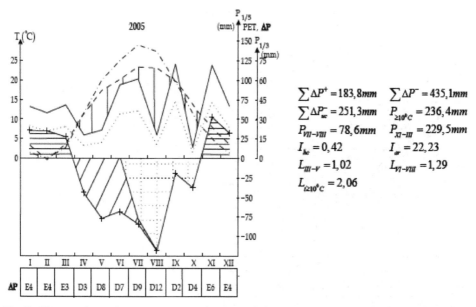

$$\sum \Delta P^+ = 183,8mm \qquad \sum \Delta P^- = 435,1mm$$
$$\sum \Delta P^-_{ac} = 251,3mm \qquad P_{\geq10^{\circ}C} = 236,4mm$$
$$P_{VII-VIII} = 78,6mm \qquad P_{XI-III} = 229,5mm$$
$$I_{hc} = 0,42 \qquad I_{ar} = 22,23$$
$$L_{III-V} = 1,02 \qquad L_{VI-VIII} = 1,29$$
$$L_{t\geq10^{\circ}C} = 2,06$$

Fig. 7. Climate charts for years 2004 and 2005 and characteristic sizes determined for reviewed site

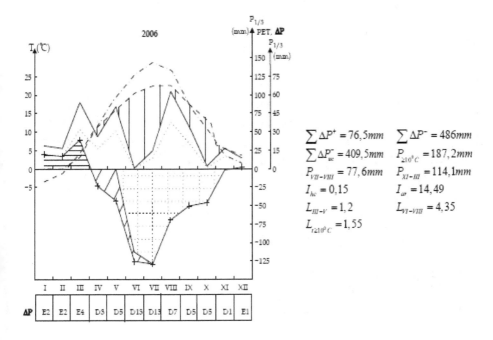

$$\sum \Delta P^+ = 76,5mm \qquad \sum \Delta P^- = 486mm$$
$$\sum \Delta P_{ac}^- = 409,5mm \qquad P_{\geq 10^0 C} = 187,2mm$$
$$P_{VII-VIII} = 77,6mm \qquad P_{XI-III} = 114,1mm$$
$$I_{hc} = 0,15 \qquad I_{ar} = 14,49$$
$$L_{III-V} = 1,2 \qquad L_{VI-VIII} = 4,35$$
$$L_{t\geq 10^0 C} = 1,55$$

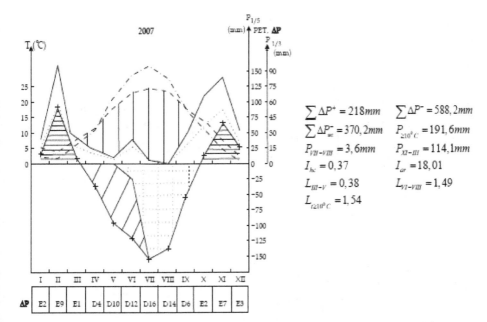

$$\sum \Delta P^+ = 218mm \qquad \sum \Delta P^- = 588,2mm$$
$$\sum \Delta P_{ac}^- = 370,2mm \qquad P_{\geq 10^0 C} = 191,6mm$$
$$P_{VII-VIII} = 3,6mm \qquad P_{XI-III} = 114,1mm$$
$$I_{hc} = 0,37 \qquad I_{ar} = 18,01$$
$$L_{III-V} = 0,38 \qquad L_{VI-VIII} = 1,49$$
$$L_{t\geq 10^0 C} = 1,54$$

Fig. 8. Climate charts for years 2006 and 2007 and characteristic sizes determined for reviewed site

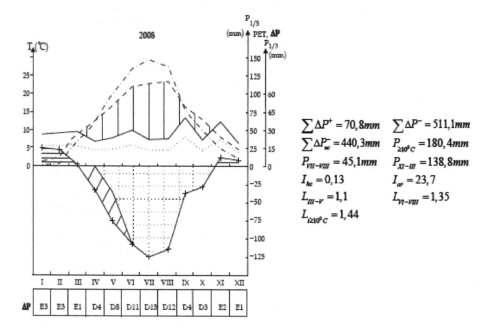

$$\sum \Delta P^+ = 70,8mm \qquad \sum \Delta P^- = 511,1mm$$
$$\sum \Delta P^-_{uc} = 440,3mm \qquad P_{\geq 10^0 C} = 180,4mm$$
$$P_{VII-VIII} = 45,1mm \qquad P_{XI-III} = 138,8mm$$
$$I_{he} = 0,13 \qquad I_{ar} = 23,7$$
$$L_{III-V} = 1,1 \qquad L_{VI-VIII} = 1,35$$
$$L_{t \geq 10^0 C} = 1,44$$

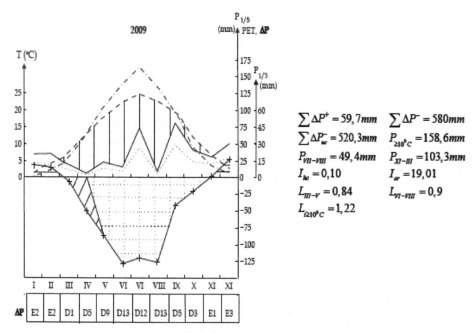

$$\sum \Delta P^+ = 59,7mm \qquad \sum \Delta P^- = 580mm$$
$$\sum \Delta P^-_{uc} = 520,3mm \qquad P_{\geq 10^0 C} = 158,6mm$$
$$P_{VII-VIII} = 49,4mm \qquad P_{XI-III} = 103,3mm$$
$$I_{he} = 0,10 \qquad I_{ar} = 19,01$$
$$L_{III-V} = 0,84 \qquad L_{VI-VIII} = 0,9$$
$$L_{t \geq 10^0 C} = 1,22$$

Fig. 9. Climate charts for years 2008 and 2009 and characteristic sizes determined for reviewed site

All the obtained values places the deltaic coast Sfântu Gheorghe in area with a dry climate (Bandoc, 2009).

Regarding the average annual values of the variation of potential evapotranspiration, we can say that, for the period 2000 - 2009 is an increase PET value to the annual average of the reference period 1961 - 1990 at a rate of 7 %. Highest increases were registered in 2002, 2007 and 2009, years in which temperatures were recorded over annual average values of the reference period.

The observed values of PET in these years are on average 11 % higher than the reference period 1961 - 1990, while during other years the annual increases are in the range 0,07 ... 1 6 % for the period 2000 - 2009 (fig. 11).

Concluding, it can be stated that for Sfântu Gheorghe coastal region there is a significant increase in the potential evapotranspiration PET for the last 10 years compared to the reference 1961-1990.

The method used to calculate potential evapotranspiration is Thorntwaite's method, using average monthly air temperature values. Based on the values obtained for PET using the method of Thornthwaite (Thornthwaite diagram), one can say that there are significant variations in PET for the period under study from 2000 to 2009 compared with the reference period 1961 - 1990, both as annual values and mean interannual values (fig. 12).

The interannual distribution of PET in the period 2000 - 2009 shows that these values were, in most months in each year of the analyzed interval over the average interannual values of the reference period 1961 - 1990. It appears that for the months of July and August all PET values are over the annual average calculated for the same month of the reference period 1961 - 1990. For instance, for the months of July in 2000-2009 period compared to the the the reference values in 1961-1990, PET values are above the multiannual July average (fig.12). Notable years for July values are 2001, 2007 and 2009 where the increase above the multiannual monthly average were 20.14%, 13.66% and 17.98% respectively.

In the same time the following indices were calculated: monthly differences $P - PET$, annual amounts of differences with the same sign $\Sigma(P - PET)^{+}$ and $\Sigma(P - PET)^{-}$, as well as the yearly balance $\Sigma(P - PET)_{A}$, all these being important climatic indices. Calculations for the two analyzed periods led to the following results regarding water deficit and excess from precipitation presented below:

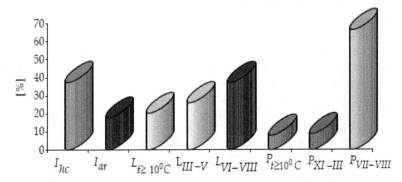

Fig. 10. Increases of the average annual percentage values of main indices for the period 2000 - 2009 for the studied site comparing to the specific values of the reference period 1961 – 1990.

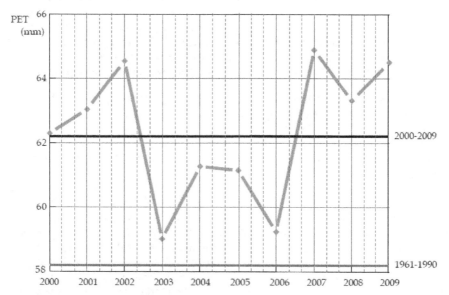

Fig. 11. Changes in annual and multiannual average values of *PET* for the period 2000 -
2009. Comparison with the 1961 - 1990 annual average for the chosen location.

$$\Sigma(P-PET)^-_{1961-1990} = 430,4mm \; ; \; \Sigma(P-PET)^-_{2000-2009} = 515,2mm \; ;$$

$$\Sigma(P-PET)^+_{1961-1990} = 106,2mm \; ; \; \Sigma(P-PET)^+_{2000-2009} = 80,8mm$$

The annual balance sheet $\Sigma(P-PET)_{A:2000-2009}$ shows a significant increase, with 31,6 %
of the water deficit comparing to the period 1961 - 1990 for which the balance reference
value is $\Sigma(P-PET)_{A:1961-1990} = -330,2mm$.
The obtained values show that there is an increase in the deficit for the last 10 years by
19,7 % compared to the reference period and a decrease of 23,9 % in terms of excess rainfall
for the period 2000 - 2009 (fig. 13).
For emphasizing very clear each month's character, at the bottom of the chart climate values
ΔP were given indicating each month's category in terms of surplus E or deficit D of
precipitation versus potential evapotranspiration. Thus, there are determined the
interannual values for the period 2000 - 2009 as well as average multiannual values for the
two periods under study.
Based on measurements one could build a mosaic of surpluses E and deficits D of
precipitation variation comparing to potential evapotranspirationfor in the period 2000-
2009, comparison with average multianual of E and D of the periods 2000-2009 and 1961-
1990 intervals (fig. 14).
Values for excess precipitation comparing to potential evapotranspiration reached a
maximum of $E9$ (>80 mm) and $E7$ (>60 mm) in February and November 2007 respectively,
values much higher than multiannual average of the reference period when the values were
$E3$ and $E2$ (see fig. 14).

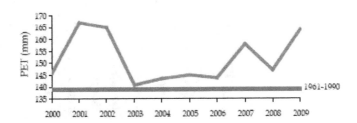

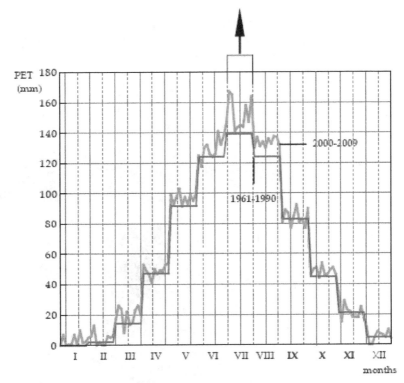

Fig. 12. Interannual distribution of *PET* in the period 2000 - 2009 comparing to the annual average of the reference period 1961 - 1990 for the studied area.

In addition, a reduction of the months with surplus between 2000 - 2009 for the years 2000, 2001, 2003 and 2004 can be seen. Also, there is a reduction in the number of months with a precipitation surplus for 2000, 2001, 2003 and 2004. In these years the precipitation excedent over *PET* period narrowed to 2 months in 2000 and 3 months in 2001, 2002, 2003 compared to 5 months in the reference 1961-1990 period (fig. 14).

As for the precipitation - potential evapotranspiration deficit it can be stated that the deficits suffered a significant increase compared to the reference period. Thus, there can be noticed maximum values of deficits *D17* (>160 mm) to be recorded in 2001 and 2002.

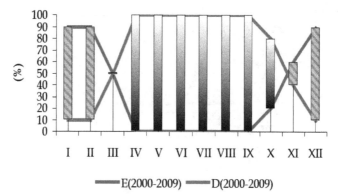

Fig. 13. Percent interannual variations of deficits D and surpluses E of precipitation to potential evapotranspiration for the period 2000 - 2009.

It appears that while the deficit intervals of the average multiannual values is seven months, the interannual period with deficit intervals is a few months longer between 2000 - 2009. Thus, in 2000, 2001 and 2004 this period has increased by three months and two months respectively compared to that of reference period (fig. 14).

D1		D2		D3	D4	D5	D6		D7		D8		D9
D10		D11		D12	D13	D14	D15		D16		D17		
E1		E2		E3	E4	E5	E6		E7		E8		E9

	I	II	III	IV	V	VI	VII	VIII	IX	X	XI	XII
2000	E5	E1	D1	D4	D9	D9	D14	D12	D4	D5	D2	D1
2001	D1	E3	D1	D2	D7	D8	D17	D13	D7	D5	E2	E1
2002	E2	D2	E4	D4	D10	D12	D17	D10	D7	D4	D2	E2
2003	E3	E2	D1	D2	D10	D11	D10	D12	D1	E1	E3	E1
2004	E4	E1	D2	D3	D3	D11	D11	D12	D6	D3	D1	E3
2005	E4	E4	E3	D3	D8	D7	D9	D12	D2	D4	E6	E4
2006	E2	E2	E4	D3	D5	D13	D13	D7	D5	D5	D1	E1
2007	E2	E9	E1	D4	D10	D12	D16	D14	D6	E2	E7	E3
2008	E3	E3	E1	D4	D8	D11	D13	D12	D4	D3	E2	E1
2009	E2	E2	D1	D5	D9	D13	D12	D13	D5	D3	E1	E3
1961-1990	E3	E3	E1	D5	D7	D9	D11	D9	D6	D2	E2	E3
2000-2009	E3	E3	E1	D3	D8	D11	D13	D12	D5	D3	E2	E2

Fig. 14. Distribution of surpluses E and deficits D of precipitation comparing to potential evapotranspiration in the period 2000 - 2009; comparison with average multiannual of E and D of the periods 2000 - 2009 and 1961 – 1990.

Analysis of reference period in terms of deficit and surplus, highlights that the studied area is characterized by a lack of $D3$ compared to the same period last years when the average value increased to a deficit of $D4$, which means a 17,06 % increase in the deficit.

5. Conclusions

The research results concerning yearly and monthly potential evapotranspiration in the Sfantu Gheorghe coastal area, synthetized in this chapter revealed for years 2001 to 2009 changes in the humidity periods, an increase in air temperature (Busuioc et al, 2010), a diminished atmospheric precipitation amount and also an increase of precipitation to potential evapotranspiration deficit compared to 1961-1990 reference period.

All these changes lead to high vulnerability and low adaptive capacity to adverse impacts from climate change of this area (Liubimtseva & Henebry, 2009).

Thus, by drawing Walter and Leith diagrams, significant increase of dryness periods and decrease of moisture periods were observed with implications upon potential evapotranspiration and upon the shore phytocoenoses.

There are also changes in the length of the periods with precipitation surplus and deficit compared to potential evapotranspiration that means increasing periods of deficit and decreasing periods of surplus.

The following calculated characteristic measurements include the delta coast in Sfântu Gheorghe in arid climate and climatic changes show that the period 2000 - 2009 led to a trend towards increasing aridity: *Martonne arid index (I_{ar}), retention index offset (I_{hc}), the amount of rainfall in the period with temperature $T \geq 10$ ° C ($P_{t \geq 10^0 C}$), the amount of rainfall the soil load in the months from November to March (P_{XI-III}), the amount of summer rainfall July and August ($P_{VII-VIII}$), Lang precipitation index for the period with $t \geq 10$ °C ($L_{t \geq 10^0 C}$), Lang precipitation index for the summer season ($L_{VI-VIII}$) and Lang precipitation index for the spring season (L_{III-V}).*

From the differences in monthly $P - PET$ calculation of amounts $\sum (P - PET)^+$, $\sum (P - PET)^-$ of the precipitation deficit offset by previously accumulated $\sum \Delta P^+$, surpluses and deficits of precipitation uncompensated by previous surpluses $\sum \Delta P_{uc}^-$ and the annual balance $\sum (P - PET)_A$ for the period under study year 2000 - 2009 and for the reference period 1961 - 1990, there was a deficit increase and a decrease of excess water from precipitation, an extension of periods of water shortage against period with excess of water and a significant increase by about 23,9 % for deficit of water that gathers negative differences uncompensated during periods of surplus.

Therefore, the research presented in this article have highlighted significant changes in potential evapotranspiration in relation to climate changes for the 2000 - 2009 studied period, in Sfântu Gheorghe area - Danube Delta, showing an increase of precipitation deficit and an increase of climate aridity .

Indirect method used in this paper work to determine the potential evapotranspiration was based on the values of air temperature and Thornthwaite's diagrams and tables. In this way a general view of a time variation of PET for Sfântu Gheorghe area - Danube Delta, has been created.

The advantages of this indirect method results from the fact that it doesn't require a large number of measured meteorological parameters and that it can be easily applied obtaining good estimates.

In the future it is intended that research should continue in order to see whether the growth trend of a interannual and annual potential evaporation is kept over the period 2000 - 2009.

No doubt that climate change is underway affecting Earth's biodiversity.

Biggest challenge in this respect is related to the marine area, but it is unclear to what extent these changes in climate will affect ecosystems.

What is known is that the temperatures that rise steadily and increasingly frequent extreme weather events are those that have influence on migrating wildlife and also causes invasive species.

Coastal areas offer considerable benefits to society while human activities are exerting considerable pressure on coastal ecosystems. Therefore, these benefits to society are in danger (Nobre, 2009).

6. Acknowledgment

Research carried out were conducted at the Center for Coastal Research and Environmental Protection, Department of Meteorology and Hydrology at the University of Bucharest, Romania.

7. References

Allen, R.G.; Pereira, L.S.; Raes, D. & Smith, M. (1998). Crop Evapotranspiration—Guidelines for Computing Crop Water Requirements. Food and Agriculture Organization of the United Nations. FAO Irrigation and drainage, Rome, ISBN 92-5-104219-4

Andréassian, V.; Perrin, Ch. & Michel, C. (2004). Impact of imperfect potential evapotranspiration knowledge on the efficiency and parameters of watershed models. *Journal of Hydrology*, Vol. 286, pp.19–35, ISSN 0022-1694

Bandoc, G. (2009). Costal phenologic cycles for Sfantu Gheorghe station (Danube Delta). *Journal of Environ. Protection and Ecology*, Vol. 9, No. 4, pp 953-960, ISSN 1311 – 5065

Bandoc, G. & Golumbeanu, M (2010). Climate variability influence to the potential evapotranspiration regime of Sfantu Gheorghe Delta Shore. *Journal of Environmental Protection and Ecology*, Vol. 10, No. 1, pp.172 -181, ISSN 1311 – 5065

Baxter, E.V.; Nadim, S.; Farajalla & Nalneesh, G. (1996). Integrated GIS and distributed storm water runoff modeling. In: Goodchild, et al. (Eds.), *GIS and Environmental Modeling Progress and Research Issues*. Donald F. Hemenway Jr., Fort Collins, pp. 199–204, ISBN 0470-236-779

Berbecel, L.; Socor,O. & Roşca, V. (1970). Current concepts in studying the phenomenon evapotranspiration (in romanian). *Rev. Hidrotehnica*, Vol. 15, No. 5, pp. 265-274

Bouchet, R. J. (1964). Évaporation réelle, évaporation – transpiration potentielle et production agricole, în l'eau et la production végétale, Inst. Nat. *De la Rech. Agr.*, Paris, pp. 151 – 232

Buchmann, N. (2000). Biotic and abiotic factors controlling soil respiration rates in Picea abies stands. *Soil Biol. Biochem*, Vol. 32, pp. 1625–1635, ISSN 0038-0717

Busuioc,A; Caian, M.; Cheval, S.; Bojariu, R.; Boroneant, C.; Baciu, M.; Dumitrescu, A. (2010). *Climate variability and change in Romania*, Ed. ProUniversitaria, pp. 59-72, ISBN 978-973-129-549-7, Bucureşti, România

Casals, P.; Gimeno, C.; Carrara, A.; Lopez-Sangil, L. & Sanz, M. (2009). Soil CO_2 efflux and extractable organic carbon fractions under simulated precipitation events in a Mediterranean Dehesa. *Soil Biol. Biochem*, Vol. 41, pp. 1915-1922, ISSN 0038-0717

Caselles, V.;Artigao, M.M.; Hurtado; E.; Coll, C. & Brasa, A. (1998). Mapping actual evapotranspiration by combining landsat TM and NOAA-AVHRR images: application to the Barrax Area, Albacete, Spain. *Remote Sensing of Environment*, Vol. No. 63, pp. 1-10, ISSN 0034-4257

Chattopadhyay, N. & Hulme, M. (1997). Evaporation and potential evapotranspiration in India under conditions of recent and future climatic change. *Agricultural and Forest Meteorology* , Vol. 87, No. 1, pp. 55-75. ISSN 0168-1923

Chiriţă, C.; Vlad, I.; Păunescu, C.; Pătrăşcoiu, N.; Roşu, C. & Iancu, I. (1977). *Forest sites* (in romanian). Ed. Academiei RSR, Bucureşti, România

Choudhury, B.J. (1997). Global pattern of potential evaporation calculated from the Penman–Monteith equation using satellite and assimilated data. *Remote Sens. Environ*, Vol. 61, pp. 64–81, ISSN 0034-4257

Chen, D.; Gao, G.; Xu, C.-Y. & Ren, G. (2005). Comparison of the Thornthwaite method and pan data with the standard Penman-Monteith estimates of reference evapotranspiration in China. *Climate research*, Vol. 28, pp. 123-132 ISSN 1616-1572

Chuanyana, Z.; Zhongrena, N. & Zhaodonga, F. (2004). GIS-assisted spatially distributed modeling of the potential evapotranspiration in semi-arid climate of the Chinese Loess Plateau. *Journal of Arid Environments*, Vol. 58, pp. 387–403, ISSN 0140-1963

Cleugh, H.A.; Leuning, R.; Mu, Q. & Running, S.W. (2007). Regional evaporation estimates from flux tower and MODIS satellite data. *Remote Sens. Environ*, Vol. 106, pp. 285–304, ISSN 0034-4257

Donciu, C. (1958). Evapotranspiration in the RPR (in romanian). *Rev. Hidrotehnica*, Vol. 3, No. 1, pp. 129-135

Donciu, C. (1983). Evapotranspiration and soil water balance (in romanian), *Memoriile Secţiilor Ştiinţifice*, Seria IV, tom VI, nr. 2, pp. 347-366, Edit. Acad. R.S.R., Bucureşti

Douglas, E. M.; Jacobs, J. M.; Sumner, D, M. & Ray, R. L. (2009). A comparison of models for estimating potential evapotranspiration for Florida land cover types. *Journal of Hydrology*, Vol. 373, pp. 366–376, ISSN 0022-1694

Eagleman, J. R. (1967). Pan evaporation, potential and actual evaporation, *Journal of Applied Meteorology*, Vo. 6, No 3, pp. 482-488, ISSN 1520-0450

Granger, R.J. (1997). Comparison of surface and satellite derived estimates of evapotranspiration using a feedback algorithm. In: Kite, G.W., Pietroniro, A., Pultz, T. (Eds.), Applications of Remote Sensing in Hydrology. Proceedings of the Symposium No. 17 NHRI, Saskatoon, Canada. National Hydrology Research Institute (NHRI), pp. 21–81.

Hargreaves, G.H. & Samani, Z.A. (1982). Estimating potential evapotranspiration (Tech. Note). *Journal of Irrigation and Drainage Engineering*, Vol. 108, No. 3, pp. 225–230, ISSN 0733-9437

Henning, I. & Henning, D. (1981). Potential evapotranspiration in mountain geo – ecosystems of different altitudines and latitudes. *Mountain Research and Development* Vol. 1, pp. 267-274, ISSN 0276-4741

Irmak, A. & Kamble, B. (2009). Evapotranspiration data assimilation with genetic algorithms and SWAP model for on-demand irrigation. Irrigation Science, Vol. 28, No.1, pp.101-112 , ISSN 1432-1319

Köppen, W. (1900). Versuch einer Klassifikation der Klimate, vorzugsweise nach ihren Beziehungen zur Pflanzenwelt. *Geogr. Zeitschr*

Kouzmov, K. (2002). Climatic changes in the region of Vidin and their efect on the agroclimatic resources. *Journal of Environmental Protection and Ecology*, Vol. 3, No.3, pp. 126-131, ISSN 1311 – 5065

Kumar, M.; Raghuwanshi, N.S.; Singh, R.; Wallender, W.W. & Pruitt, W.O. (2002). Estimating evapotranspiration using Artificial Neural Network. *Journal of Irrigation and Drainage Engineering*, ASCE 128, pp. 224–233, ISSN 0733-9437

Li, H.; Yan, J.; Yue, X. & Wang, M. (2008a). Significance of soil temperature and moisture for soil respiration in a Chinese mountain area. *Agric. Forest. Meteorol*, Vol. 148, pp. 490-503, ISSN 0168-1923

Li, Z.; Wang, Y.; Zhou, Q.; Wu, J.; Peng, J. & Chang, H. (2008b). Spatiotemporal variability of land surface moisture based on vegetation and temperature characteristics in Northern Shaanxi Loess plateau, China. *Journal of Arid Environments*, Vol. 72, pp. 974–985, ISSN 0140-1963

Lioubimtseva, E. & Henebry, G.M. (2009). Climate and environmental change in arid Central Asia: Impacts, vulnerability and adaptations. *Journal of Arid Environments*, Vol. 73. pp. 963–977, ISSN 0140-1963

Lu, J.; Sun, G.; McNulty,S.; & Amatya, D. M. (2005). A comparison of six potential evapotranspiration methods for regional use in the Southeastern United States. *Journal of the American Water Resources Association (JAWRA)* Vol. 41 (3), pp. 621-633, ISSN 1752-1688

Monteith, J.L. (1965). Evaporation and environment. *Symposium of the Society for Experimental Biology*, Vol. 19, pp. 205–224

Moore, I.D. (1996). Hydrologic modeling and GIS. In: Goodchild, et al. (Eds.), *GIS and Environmental Modeling Progress and Research Issues*. Donald F. Hemenway Jr., Fort Collins, pp. 143–149, ISBN 0470-236-779

Nobre, A.M.; Ferreira, J.G; Nunes, J.P; Yan, X; Bricker, S.; Corner, R.; Groom,S.; Gu, H.; Hawkins, A.J.S.; Hutson, R.; Dongzhao Lan, D.; Lencart e Silva,J.D.; Pascoe,P.; Telfer, T.; Zhang, X. & Zhu, M. (2010). Assessment of coastal management options by means of multilayered ecosystem models. Estuarine, *Coastal and Shelf Science*, Vol. 87, No. 1, pp. 43-62, ISSN 0272-7714

Oguz, T.; Dippner, J. & Kaymaz, Z. (2006). Climatic regulation of the Black Sea hydro-meteorological and ecological properties at interannual-to-decadal time scales. *Journal of Marine Systems* Vol. 60, No. 3-4, pp. 235–254, ISSN 0924-7963

Oudin, L.; Michel, C. & Anctil, F. (2005a). Which potential evapotranspiration input for a lumped rainfall–runoff model? Part 1 – Can rainfall–runoff models effectively handle detailed potential evapotranspiration inputs?. *Journal of Hydrology*, Vol. 303, pp. 275–289, ISSN 0022- 1694

Oudin, L.; Hervieu, F.; Michel, C.; Perrin, C.; Andreassian, V.; Anctil, F. & Loumagne, C. (2005b). Which potential evapotranspiration input for a lumped rainfall–runoff model? Part 2 – Towards a simple and efficient potential evapotranspiration model for rainfall–runoff modeling. *Journ. of Hydr.*, Vol. 303, pp. 290–306, ISSN 0022-1694

Palutikof, J.P.; Goddes, S.C.M. & Guo, X. (1994). Climate change, potential evapotranspiration and moisture availability in the Mediterranean Basin. *International Journal of Climatology*, Vol. 14 , No. 8, pp. 853-869, ISSN 0899-8418

Penman, H. L. (1946). Natural evaporation from open water, bare soil and grass. *Proceedings of the Royal Society of London. Series A, Mathematical and Physical Sciences*, Vol. 193, No. 1032 (Apr. 22, 1948), pp. 120-145

Ponce, V.M. (1989). Engineering hydrology: principles *and practices*. John Wiley and Sons, New York, pp. 48-51, ISBN: 0471147354

Raich, J.W. & Schlesinger, W.H. (1992). The global carbon dioxide flux in soil respiration and its relationship to vegetation and climate. *Tellus* Vol. 44B, pp. 81–89, ISSN 1600-0899

Smith, B.A.; McClendon, R.W. & Hoogenboom, G. (2006). Improving air temperature prediction with Artificial Neural Networks. *International Journal of Computational Intelligence*, Vol. 3, No. 3, pp. 179–186, ISSN 0883-9514

Srinivasan, R.; Arnold, J.; Rosenthal, W. & Muttiah, R.S. (1996). Hydrologic Modeling of Texas Gulf Basin using GIS. In: Goodchild, et al. (Eds.), *GIS and Environmental Modeling Progress and Research Issues*, Donald F. Hemenway Jr., Fort Collins, pp. 213–219, ISBN 0470-236-779

Stefano, C.D. & Ferro, V. (1997). Estimation of evapotranspiration by Hargreaves formula and remotely sensed data in semi-arid Mediterranean areas. *Journal of Agricultural Engineering Research*, Vol. 68, pp. 189–199, ISSN 0021-8634

Stewarta, J.B.; Watts, C.J.; Rodriguez, J.C.; De Bruin, H.A.R.; van den Berg, A.R. & Garatuza-Payan, J. (1999). Use of satellite data to estimate radiation and evaporation for Northwest Mexico. *Agricultural Water Manag.*, Vol. 38, pp. 181–193, ISSN 0378-3774

Tang, R.L.; Li, Z.L. & Tang, B.H. (2010). An application of the Ts–VI triangle method with enhanced edges determination for evapotranspiration estimation from MODIS data in and semi-arid regions: implementation and validation. *Remote Sensing of Environment*, Vol. 114, No. 3,pp. 540–551, ISSN 0034-4257

Thornthwaite, C.W. (1948). An approach towards a rational classification of climate. *Geographical Revue*,Vol. 38, pp. 55-94

Thomas, A. (2000a). Spatial land temporal characteristics of potential evapotranspiration trends over China. *Inter. Journal of Climatology*, Vol. 20, pp. 381-396, ISSN 0899-8418

Thomas, A. (2000b). Climatic changes in yield index and soil water deficit trends in China. *Agricultural and Forests Meteorology*, Vol. 102, pp. 71-81, ISSN 0168-1923

Torres, A.F., Walker, W.R. & McKee, M. (2011). Forecasting daily potential evapotranspiration using machine learning and limited climatic data. *Agricultural Water Management*, Vol. 98 , pp.553-562, ISSN 0378-3774

Turc, L. (1954). Calcul du bilan de l'eau evaluation en function des precipitations et des temperatures. In *Association International d'Hydrology*, Assemblée Génrale de Rome, Tome III , No.3, pp. 188-202

Vespremeanu, E. (2000). *The Danube Delta tourist map* 1:200 000. Ed Amco Press, Bucureşti, România

Vespremeanu, E. (2004). *Geography of the Black Sea*, Ed. Univ. din Bucureşti, ISBN 973-575-925-X, Bucureşti, România

Walter, H. (1955). Die Klimadiagramme als Mittel zur Beurteilung der Klimaverhältnisse für ökologische, vegetationskundliche und landwirtschaftliche Zwecke. *Berichte der Deutschen Botanischen Gesellschaft* Vol. 68, pp. 331-344

Walter, H., & Lieth, H. (1960). Kimadiagramm-Weltatlas, Fischer-Verlag, Jena

Walter, H. (1999). Vegetation und Klimazonen. Grundriß der globalen Ökologie. Ulmer, ISBN 3-8252-0014-0, Stuttgart, Germania

****Climate of Romania* (2008). National Meteorological Administration. Ed. Academiei Române, Bucureşti, România

3

Evapotranspiration of Partially Vegetated Surfaces

L.O. Lagos[1,2], G. Merino[1], D. Martin[2], S. Verma[2] and A. Suyker[2]
[1]Universidad de Concepción Chile
[2]University of Nebraska-Lincoln
[1]Chile
[2]USA

1. Introduction

Latent heat flux equivalent to Evapotranspiration (ET) is the total amount of water lost via transpiration and evaporation from plant surfaces and the soil in an area where a crop is growing. Since 80-90% of precipitation received in semiarid and subhumid climates is commonly used in evapotranspiration, accurate estimations of ET are very important for hydrologic studies and crop water requirements. ET determination and modelling is not straightforward due to the natural heterogeneity and complexity of agricultural and natural land surfaces. In evapotranspiration modelling it is very common to represent vegetation assuming a single source of energy flux at an effective height within the canopy. However, when crops are sparse, the single source/sink of energy assumption in such models is not entirely satisfied. Improvements using multiple source models have been developed to estimate ET from crop transpiration and soil evaporation. Soil evaporation on partially vegetated surfaces over natural vegetation and orchards includes not only the soil under the canopy but also areas of bare soil between vegetation that contribute to ET. Soil evaporation can account for 25-45% of annual ET in agricultural systems. In irrigated agriculture, partially vegetated surfaces include fruit orchards (i.e. apples, oranges, vineyards, avocados, blueberries, and lemons among others), which cover a significant portion of the total area under irrigation.

In semiarid regions, direct soil evaporation from sparse barley or millet crops can account for 30% to 60% of rainfall (Wallace et al., 1999). On a seasonal basis, sparse canopy soil evaporation can account for half of total rainfall (Lund & Soegaard, 2003). Allen (1990) estimated the soil evaporation under a sparse barley crop in northern Syria and found that about 70% of the total evaporation originated from the soil. Lagos (2008) estimated that under irrigated maize conditions soil evaporation accounted for around 26-36% of annual evapotranspiration. Under rain-fed maize conditions annual evaporation accounted for 36-39% of total ET. Under irrigated soybean the percentage was 41%, and under rainfed soybean conditions annual evaporation accounted for 45-47% of annual ET. Massman (1992) estimated that the soil contribution to total ET was about 30% for a short grass steppe measurement site in northeast Colorado. In a sparse canopy at the middle of the growing season, and after a rain event, more than 50% of the daily ET corresponds to directly soil evaporation (Lund & Soegaard, 2003). Soil evaporation can be maximized under frequent rainfall or irrigation events, common conditions in agricultural systems for orchard with drip or micro sprinklers systems. If some of this unproductive loss of water could be retained in the soil and used as transpiration, yields could be increased without increased rainfall or the use of supplemental irrigation (Wallace et al., 1999). The measurement and modelling of soil evaporation on partially vegetated surfaces is crucial to estimate how much water is lost to the atmosphere via soil evaporation. Consequently, better water management can be proposed for water savings.

Partially vegetated surface accounts for a significant portion of land surface. It occurs seasonally in all agricultural areas and throughout the year in or chard and natural land covers. Predictions of ET for these conditions have not been thoroughly researched. In Chile, agricultural orchards with partially vegetated surfaces include apples, oranges, avocados, cherries, vineyards, blueberries, and berries, among others. According to the agricultural census (INE, 2007) the national orchard surface covers more than 324,000 ha, representing 30% of the total surface under irrigation.

Similar to the Shuttleworth and Wallace (1985), Choudhury and Monteith (1988) and Lagos (2008) models, the modelling of evapotranspiration for partially vegetated surfaces can be accomplished using explicit solutions of the equations that define the conservation of heat and water vapor fluxes for partially vegetated surfaces and soil. Multiple-layer models offer the possibility to represent these conditions to solve the surface energy balance and consequently, estimate evapotranspiration. Modelling is essential to predict long-term trends and to quantify expected outcomes. Since ET is such a large component of the hydrologic cycle in areas with partially vegetated surfaces, small changes in the calculation of ET can result in significant changes in simulated water budgets. Thus, good data and accurate modelling of ET is essential for predicting not only water requirements for agricultural crops but also to predict the significance of irrigation management decisions and land use changes to the entire hydrologic cycle.

Currently, several methods and models exist to predict natural environments under different conditions. More complex models have been developed to account for more variables affecting model performance. However, the applicability of these models has been limited by the difficulties and tedious algorithms needed to complete estimations. Mathematical algorithms used by multiple-layer models can be programmed in a software package to facilitate and optimize ET estimation by any user. User-friendly software facilitates the use of these improved methods; users (i.e. students) can use the computer model to study the behaviour of the system from a set of parameters and initial conditions.

Accordingly, in this chapter, a review of models that estimate ET for partially covered surfaces that occur normally in agricultural systems (i.e. orchards or vineyards) is presented, and the needs for further research are assessed.

2. ET modelling review

Evapotranspiration (ET) is the total amount of water lost via transpiration and evaporation from plant surfaces and the soil in an area where a crop is growing. Traditionally, ET from agricultural fields has been estimated using the two-step approach by multiplying the weather-based reference ET (Jensen et al., 1971; Allen et al., 1998 and ASCE, 2002) by crop coefficients (Kc) to make an approximate allowance for crop differences. Crop coefficients are determined according to the crop type and the crop growth stage (Allen et al., 1998). However, there is typically some question regarding whether the crops grown compare with the conditions represented by the idealized Kc values (Parkes et al., 2005; Rana et al.,

2005; Katerji & Rana, 2006; Flores, 2007). In addition, it is difficult to predict the correct crop growth stage dates for large populations of crops and fields (Allen et al., 2007).

A second method is to make a one-step estimate of ET based on the Penman-Monteith (P-M) equation (Monteith, 1965), with crop-to-crop differences represented by the use of crop-specific values of surface and aerodynamic resistances (Shuttleworth, 2006). ET estimations using the one-step approach with the P-M model have been studied by several authors (Stannard, 1993; Farahani & Bausch, 1995; Rana et al., 1997; Alves & Pereira, 2000; Kjelgaard & Stockle, 2001; Ortega-Farias et al., 2004; Shuttleworth, 2006; Katerji & Rana, 2006; Flores, 2007; Irmak et al., 2008). Although different degrees of success have been achieved, the model has generally performed more satisfactorily when the leaf area index (LAI) is large (LAI>2). Results shows that the "big leaf" assumption used by the P-M model is not satisfied for sparse vegetation and crops with partial canopy cover.

A third approach consists of extending the P-M single-layer model to a multiple-layer model (i.e. two layers in the Shuttleworth-Wallace (S-W) model (Shuttleworth-Wallace, 1985) and four layers in the Choudhury-Monteith (C-M) model (Choudhury & Monteith, 1988). Shuttleworth and Wallace (1985) combined a one-dimensional model of crop transpiration and a one-dimensional model of soil evaporation. Surface resistances regulate the heat and mass transfer in plant and soil surfaces, and aerodynamic resistances regulate fluxes between the surface and the atmospheric boundary layer. Several studies have evaluated the performance of the S-W model to estimate evapotranspiration (Farahani & Baush,1995; Stannard, 1993; Lafleur & Rouse, 1990; Farahani & Ahuja, 1996; Iritz et al. 2001; Tourula & Heikinheimo, 1998; Anadranistakis et al., 2000; Ortega-Farias et al., 2007). Field tests of the model have shown promising results for a wide range of both agricultural and non-agricultural vegetation.

Farahani and Baush (1995) evaluated the performance of the P-M model and the S-W model for irrigated maize. Their main conclusion was that the Penman-Monteith model performed poorly when the leaf area index was less than 2 because soil evaporation was neglected in calculating surface resistance. Results of the S-W model were encouraging as it performed satisfactorily for the entire range of canopy cover. Stannard (1993) compared the P-M, S-W and Priestley-Taylor ET models for sparsely vegetated, semiarid rangeland. The P-M model was not sufficiently accurate (hourly r^2 =0.56, daily r^2=0.60); however, the S-W model performs significantly better for hourly (r^2=0.78) and daily data (r^2=0.85). Lafleur and Rouse (1990) compared the S-W model with evapotranspiration calculated from the Bowen Ratio Energy Balance technique over a range of LAI from non-vegetated to fully vegetated conditions. The results showed that the S-W model was in excellent agreement with the measured evapotranspiration for hourly and day-time totals for all values of LAI. Using the potential of the S-W model to partition transpiration and evaporation, Farahani and Ahuja (1996) extended the model to include the effects of crop residues on soil evaporation by the inclusion of a partially covered soil area and partitioning evaporation between the bare and residue-covered areas. Iritz et al. (2001) applied a modified version of the S-W model to estimate evapotranspiration for a forest. The main modification consisted of a two-layer soil module, which enabled soil surface resistance to be calculated as a function of the wetness of the top soil. They found that the general seasonal dynamics of evaporation were fairly well simulated with the model. Tourula and Heikinheimo (1998) evaluated a modified version of the S-W model in a barley field. A modification of soil surface resistance and aerodynamic resistance, over two growing seasons, produced daily and hourly ET estimates in good agreement with the measured evapotranspiration. The performance of the S-W model was evaluated against two eddy covariance systems by Ortega-Farias et al. (2007) over a Cabernet

Sauvignon vineyard. Model performance was good under arid atmospheric conditions with a correlation coefficient (r^2) of 0.77 and a root mean square error (RMSE) of 29 Wm^{-2}.

Although good results have been found using the Shuttleworth-Wallace approach, the model still needs an estimation or measurement of soil heat flux (G) to estimate ET. Commonly, G is calculated as a fixed percentage of net radiation (Rn). Shuttleworth and Wallace (1985) estimated G as 20% of the net radiation reaching the soil surface. In the FAO56 method, Allen et al. (1998) estimated daily reference ET (ETr and ETo), assuming that the soil heat flux beneath a fully vegetated grass or alfalfa reference surface is small in comparison with Rn (i.e. G=0). For hourly estimations, soil heat flux was estimated as one tenth of the Rn during the daytime and as half of the Rn for the night time when grass was used as the reference surface. Similarly, G was assumed to be 0.04xRn for the daytime and 0.2xRn during the night time for an alfalfa reference surface. A more complete surface energy balance was presented by Choudhury and Monteith (1988). The proposed method developed a four-layer model for the heat budget of homogeneous land surfaces. The model is an explicit solution of the equations which define the conservation of heat and water vapor in a system consisting of uniform vegetation and soil. An important feature was the interaction of evaporation from the soil and transpiration from the canopy expressed by changes in the vapor pressure deficit of the air in the canopy. A second feature was the ability of the model to partition the available energy into sensible heat, latent heat, and soil heat flux for the canopy/soil system.

Similar to Shuttleworth-Wallace (1985), the Choudhury-Monteith model included a soil surface resistance to regulate the heat and mass transfer at the soil surface. However, residue effects on the surface energy balance are not included in the model. Crop residue generally increases infiltration and reduces soil evaporation. Surface residue affects many of the variables that determine the evaporation rate. These variables include Rn, G, aerodynamic resistance and surface resistances to transport of heat and water vapor fluxes (Steiner, 1994; Horton et al., 1996; Steiner et al., 2000).

Caprio et al. (1985) compared evaporation from three mini-lysimeters installed in bare soil and in a 14 and 28 cm tall standing wheat stubble. After nine days of measurements, evaporation from the lysimeter with stubble was 60% of the evaporation measured from bare soil. Enz et al. (1988) evaluated daily evaporation for bare soil and stubble-covered soil surfaces. Evaporation was always greater from the bare soil surface until it was dry, then evaporation was greater from the stubble covered-surface because more water was available. Evaporation from a bare soil surface has been described in three stages. An initial, energy-limited stage occurs when enough soil water is available to satisfy the potential evaporation rates. A second, falling rate stage is limited by water flow to the soil surface, while the third stage has a very low, nearly constant evaporative rate from very dry soil (Jalota & Prihar, 1998). Steiner (1989) evaluated the effect of residue (from cotton, sorghum and wheat) on the initial, energy-limited rate of evaporation. The evaporation rate relative to bare soil evaporation was described by a logarithmic relationship. Increasing the amount of residue on the soil surface reduced the relative evaporation rate during the initial stage. Bristow et al. (1986) developed a model to predict soil heat and water budgets in a soil-residue-atmosphere system. Results from application of the model indicate that surface residues decreased evaporation by roughly 36% compared with simulations from bare soil.

With the recognition of the potential of multiple-layer models to estimate ET, a modified surface energy balance model (SEB) was developed by Lagos (2008) and Lagos et al. (2009) to include the effect of crop residue on evapotranspiration. The model relies mainly on the Schuttleworth-Wallace (1985) and Choudhury and Monteith (1988) approaches and has the potential to predict evapotranspiration for varying soil cover ranging from partially residue-covered soil to closed canopy surfaces. Improvements to aerodynamic resistance, surface canopy resistance and soil resistances for the transport of heat and water vapor were also suggested.

2.1 The SEB model

The modified surface energy balance (SEB) model has four layers (Figure 1), the first extended from the reference height above the vegetation and the sink for momentum within the canopy, a second layer between the canopy level and the soil surface, a third layer corresponding to the top soil layer and a lower soil layer where the soil atmosphere is saturated with water vapor. The soil temperature at the bottom of the lower level was held constant for at least a 24h period.

The SEB model distributes net radiation (Rn), sensible heat (H), latent heat (λE), and soil heat fluxes (G) through the soil/residue/canopy system. Horizontal gradients of the potentials are assumed to be small enough for lateral fluxes to be ignored, and physical and biochemical energy storage terms in the canopy/residue/soil system are assumed to be negligible. The evaporation of water on plant leaves due to rain, irrigation or dew is also ignored.

The SEB model distributes net radiation (Rn) into sensible heat (H), latent heat (λE), and soil heat fluxes (G) through the soil-canopy system (Figure 2). Total latent heat (λE) is the sum of latent heat from the canopy (λEc), latent heat from the soil (λEs) and latent heat from the residue-covered soil (λEr). Similarly, sensible heat is calculated as the sum of sensible heat from the canopy (Hc), sensible heat from the soil (Hs) and sensible heat from the residue covered soil (Hr).

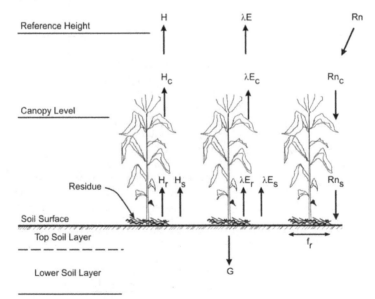

Fig. 1. Fluxes of the surface energy balance model (SEB).

The total net radiation is divided into that absorbed by the canopy (Rnc) and the soil (Rns) and is given by Rn = Rnc + Rns. The net radiation absorbed by the canopy is divided into latent heat and sensible heat fluxes as Rnc = λEc +Hc. Similarly, for the soil Rns = Gos + Hs, where Gos is a conduction term downwards from the soil surface and is expressed as Gos = λEs + Gs, where Gs is the soil heat flux for bare soil. Similarly, for the residue-covered soil Rns = Gor + Hr where Gor is the conduction downwards from the soil covered by residue. The conduction is given by Gor = λEr + Gr where Gr is the soil heat flux for residue-covered soil.

Total latent heat flux from the canopy/residue/soil system is the sum of the latent heat from the canopy (transpiration), latent heat from the soil and latent heat from the residue-covered soil (evaporation), calculated as:

$$\lambda E = \lambda E_c + (1 - fr) \cdot \lambda E_s + fr \cdot \lambda E_r \tag{1}$$

where fr is the fraction of the soil affected by residue. Similarly, the total sensible heat is given by:

$$H = H_c + (1 - fr) \cdot H_s + fr \cdot H_r \tag{2}$$

The differences in vapor pressure and temperature between levels can be expressed with an Ohm's law analogy using appropriate resistance and flux terms (Figure 2). The sensible and latent heat fluxes from the canopy, from bare soil and soil covered by residue are expressed by (Shuttleworth & Wallace, 1985):

$$H_c = \frac{\rho \cdot c_p \cdot (T_1 - T_b)}{r_1} \quad \text{and} \quad \lambda E_c = \frac{\rho \cdot C_p \cdot (e_1^* - e_b)}{\gamma \cdot (r_1 + r_c)} \tag{3}$$

$$H_s = \frac{\rho \cdot C_p \cdot (T_2 - T_b)}{r_2} \quad \text{and} \quad \lambda E_s = \frac{\rho \cdot C_p \cdot (e_L^* - e_b)}{\gamma \cdot (r_2 + r_s)} \tag{4}$$

$$H_r = \frac{\rho \cdot C_p \cdot (T_{2r} - T_b)}{r_2 + r_{rh}} \quad \text{and} \quad \lambda E_r = \frac{\rho \cdot C_p \cdot (e_{Lr}^* - e_b)}{\gamma \cdot (r_2 + r_s + r_r)} \tag{5}$$

where, ρ is the density of moist air, C_p is the specific heat of air, γ is the psychrometric constant, T_1 is the mean canopy temperature, T_2 is the temperature at the soil surface, T_b is the air temperature within the canopy, T_{2r} is the temperature of the soil covered by residue, r_1 is an aerodynamic resistance between the canopy and the air, r_c is the surface canopy resistance, r_2 is the aerodynamic resistance between the soil and the canopy, r_s is the resistance to the diffusion of water vapor at the top soil layer, r_{rh} is the residue resistance to transfer of heat, r_r is the residue resistance to the transfer of vapor acting in series with the soil resistance r_s, e_b is the vapor pressure of the atmosphere at the canopy level, e_1^* is the saturation vapor pressure in the canopy, e_L^* is the saturation vapor pressure at the top of the wet layer, and e_{Lr}^* is the saturation vapor pressure at the top of the wet layer for the soil covered by residue.

Conduction of heat for the bare-soil and residue-covered surfaces are given by:

$$G_{os} = \frac{\rho \cdot C_p \cdot (T_2 - T_L)}{r_u} \quad \text{and} \quad G_s = \frac{\rho \cdot C_p \cdot (T_L - T_m)}{r_L} \tag{6}$$

$$G_{or} = \frac{\rho \cdot C_p \cdot (T_{2r} - T_{Lr})}{r_u} \quad \text{and} \quad G_r = \frac{\rho \cdot C_p \cdot (T_{Lr} - T_m)}{r_L} \tag{7}$$

where; r_u and r_L are resistance to the transport of heat for the upper and lower soil layers, respectively, T_L and T_{Lr} are the temperatures at the interface between the upper and lower layers for the bare soil and the residue-covered soil, and T_m is the temperature at the bottom of the lower layer which was assumed to be constant on a daily basis.

Choudhury and Monteith (1988) expressed differences in saturation vapor pressure between points in the system as linear functions of the corresponding temperature differences. They found that a single value of the slope of the saturation vapor pressure, Δ, when evaluated at the air temperature, T_a, gave acceptable results for the components of the heat balance. The vapor pressure differences were given by:

$$e_1^* - e_b^* = \Delta \cdot (T_1 - T_b) e_L^* - e_b^* = \Delta \cdot (T_L - T_b) e_b^* - e_a^* = \Delta \cdot (T_b - T_a) \tag{8}$$

$$\text{and} \qquad e_{Lr}^* - e_b^* = \Delta \cdot (T_{Lr} - T_b)$$

The above equations were combined and solved to estimate fluxes. Details are provided by Lagos (2008). The solution gives the latent and sensible heat fluxes from the canopy as:

$$\lambda E_c = \frac{\Delta \cdot r_1 \cdot Rn_c + \rho \cdot C_p \cdot (e_b^* - e_b)}{\Delta \cdot r_1 + \gamma \cdot (r_1 + r_c)} \quad \text{and} \quad H_c = \frac{\gamma \cdot (r_1 - r_c) \cdot Rn_c - \rho \cdot C_p \cdot (e_b^* - e_b)}{\Delta \cdot r_1 + \gamma \cdot (r_1 + r_c)} \tag{9}$$

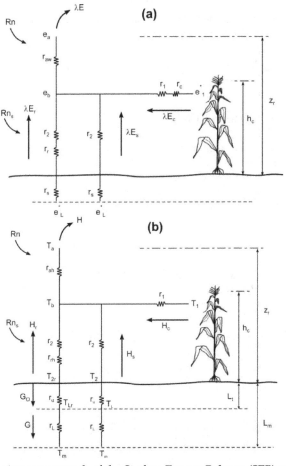

Fig. 2. Schematic resistance network of the Surface Energy Balance (SEB) model a) Latent heat flux and b) Sensible heat flux.

Similarly, latent and sensible heat fluxes from bare soil surfaces are estimated by:

$$\lambda E_s = \frac{Rn_s \cdot \Delta \cdot r_2 \cdot r_L + \rho \cdot C_p \cdot [(e_b^* - e_b) \cdot (r_u + r_L + r_2) + (T_m - T_b) \cdot \Delta \cdot (r_u + r_2)]}{\gamma \cdot (r_2 + r_s) \cdot (r_u + r_L + r_2) + \Delta \cdot r_L \cdot (r_u + r_2)} \quad (10)$$

$$H_s = \frac{Rn_s \cdot r_L \cdot \Delta - \lambda E_s \cdot [r_L \cdot \Delta + \gamma \cdot (r_2 + r_s)] + \rho \cdot C_p \cdot (e_b^* - e_b) - \rho \cdot C_p \cdot \Delta \cdot (T_b - T_m)}{r_L \cdot \Delta} \quad (11)$$

The latent and sensible heat fluxes from the residue-covered soil are simulated with:

$$\lambda E_r = \frac{Rn_s \cdot \Delta \cdot (r_2 + r_{rh}) \cdot r_L + \rho \cdot C_p \cdot [(e_b^* - e_b) \cdot (r_u + r_L + r_2 + r_{rh}) + (T_m - T_b) \cdot \Delta \cdot (r_u + r_2 + r_r)]}{\gamma \cdot (r_2 + r_s + r_r) \cdot (r_u + r_L + r_2 + r_{rh}) + \Delta \cdot r_L \cdot (r_u + r_2 + r_{rh})} \quad (12)$$

$$H_r = \frac{Rn_s \cdot r_L \cdot \Delta - \lambda E_r \cdot [r_L \cdot \Delta + \gamma \cdot (r_2 + r_s + r_r)] + \rho \cdot C_p \cdot (e_b^* - e_b) - \rho \cdot C_p \cdot \Delta \cdot (T_b - T_m)}{r_L \cdot \Delta} \quad (13)$$

Values for T_b and e_b are necessary to estimate latent heat and sensible heat fluxes. The values of the parameters can be expressed as:

$$e_b = \left(T_b \cdot (\Delta \cdot A_2 - A_3) + \frac{A_1}{\rho \cdot C_p} - \Delta \cdot A_2 \cdot T_a + A_2 \cdot e_a^* + T_m \cdot A_3 + \frac{e_a}{\gamma \cdot r_{aw}}\right) \cdot \left(\frac{\gamma \cdot r_{aw}}{1 + A_2 \cdot \gamma \cdot r_{aw}}\right) \quad (14)$$

$$T_b = \left[\frac{B_1}{\rho \cdot C_p} + T_a \cdot \left(\frac{1}{r_{ah}} - \Delta \cdot B_2\right) + (e_a^* - e_b) \cdot B_2 + T_m \cdot B_3\right] \cdot \left(\frac{r_{ah}}{1 - \Delta \cdot B_2 \cdot r_{ah} + B_3 \cdot r_{ah}}\right) \quad (15)$$

where, r_{ah} is the aerodynamic resistance for heat transport, r_{aw} is the aerodynamic resistance for water vapor transport, e_a is the vapor pressure at the reference height, and e_a^* is the saturated vapor pressure at the reference height. Six coefficients (A_1, A_2, A_3 and B_1, B_2 and B_3) are involved in these expressions. These coefficients depend on environmental conditions and other parameters. The expressions to compute the coefficients are given by (Lagos, 2008):

$$A_1 = \frac{\Delta \cdot r_1 \cdot Rn_c}{\Delta \cdot r_1 + \gamma \cdot (r_1 + r_c)} + (1 - f_r) \cdot \frac{Rn_s \cdot \Delta \cdot r_2 \cdot r_L}{\gamma \cdot (r_2 + r_s) \cdot (r_u + r_L + r_2) + \Delta \cdot r_L \cdot (r_u + r_2)} +$$
$$f_r \cdot \frac{Rn_s \cdot \Delta \cdot (r_2 + r_{rh}) \cdot r_L}{\gamma \cdot (r_2 + r_s + r_r) \cdot (r_u + r_L + r_2 + r_{rh}) + \Delta \cdot r_L \cdot (r_u + r_2 + r_{rh})} \quad (16)$$

$$A_2 = \frac{1}{\Delta \cdot r_1 + \gamma \cdot (r_1 + r_c)} + (1 - f_r) \cdot \frac{(r_u + r_L + r_2)}{\gamma \cdot (r_2 + r_s) \cdot (r_u + r_L + r_2) + \Delta \cdot r_L \cdot (r_u + r_2)} +$$
$$f_r \cdot \frac{(r_u + r_L + r_2 + r_{rh})}{\gamma \cdot (r_2 + r_s + r_r) \cdot (r_u + r_L + r_2 + r_{rh}) + \Delta \cdot r_L \cdot (r_u + r_2 + r_{rh})} \quad (17)$$

$$A_3 = \left[(1 - f_r) \cdot \frac{\Delta \cdot (r_u + r_2)}{\gamma \cdot (r_2 + r_s) \cdot (r_u + r_L + r_2) + \Delta \cdot r_L \cdot (r_u + r_2)}\right.$$
$$\left. + f_r \frac{\Delta \cdot (r_u + r_2 + r_{rh})}{\gamma \cdot (r_2 + r_s + r_r) \cdot (r_u + r_L + r_2 + r_{rh}) + \Delta \cdot r_L \cdot (r_u + r_2 + r_{rh})}\right] \quad (18)$$

$$B_1 = \left[Rn_c \cdot \frac{\gamma \cdot (r_1 + r_c)}{\Delta \cdot r_1 + \gamma \cdot (r_1 + r_c)} + Rn_s \cdot \binom{(1 - f_r) \cdot (1 - \Delta \cdot r_2 \cdot r_L \cdot X_s) +}{f_r \cdot (1 - \Delta \cdot (r_2 + r_{rh}) \cdot r_L \cdot X_r)} \right] \qquad (19)$$

$$B_2 = \frac{-1}{\Delta \cdot r_1 + \gamma \cdot (r_1 + r_c)} + (1 - f_r) \cdot \left(\frac{1}{r_L \Delta} - (r_u + r_L + r_2) \cdot X_s \right)$$

$$+ f_r \cdot \left(\frac{1}{r_L \Delta} - (r_u + r_L + r_2 + r_{rh}) \cdot X_r \right) \qquad (20)$$

$$B_3 = \left[(1 - f_r) \cdot \left(\frac{1}{r_L} - \Delta \cdot (r_u + r_2) \cdot X_s \right) + f_r \cdot \left(\frac{1}{r_L} - \Delta \cdot (r_u + r_2 + r_{rh}) \cdot X_r \right) \right] \qquad (21)$$

$$X_s = \left(\frac{1}{\gamma \cdot (r_2 + r_s) \cdot (r_u + r_L + r_2) + \Delta \cdot r_L \cdot (r_u + r_2)} \right) \left(\frac{(r_L \cdot \Delta + \gamma \cdot (r_2 + r_s))}{r_L \cdot \Delta} \right) \text{ and}$$

$$X_r = \left(\frac{1}{\gamma \cdot (r_2 + r_s + r_r) \cdot (r_u + r_L + r_2 + r_{rh}) + \Delta \cdot r_L \cdot (r_u + r_2 + r_{rh})} \right) \left(\frac{(r_L \cdot \Delta + \gamma \cdot (r_2 + r_s + r_r))}{r_L \cdot \Delta} \right) \qquad (22)$$

These relationships define the surface energy balance model which is applicable to conditions ranging from closed canopies to surfaces with bare soil or those partially covered with residue. Without residue, the model is similar to that by Choudhury and Monteith (1988).

2.1.1 Determination of the SEB model parameters

In the following sections, the procedures to compute parameter values for the model are detailed. The parameters are as important as the formulation of the energy balance equations.

2.1.1.1 Aerodynamic resistances

Thom (1972) stated that heat and mass transfer encounter greater aerodynamic resistance than the transfer of momentum. Accordingly, aerodynamic resistances to heat (r_{ah}) and water vapor transfer (r_{aw}) can be estimated as:

$$r_{ah} = r_{am} + r_{bh} \qquad \text{and} \qquad r_{aw} = r_{am} + r_{bw} \qquad (23)$$

where r_{am} is the aerodynamic resistance to momentum transfer, and r_{bh} and r_{bw} are excess resistance terms for heat and water vapor transfer.

Shuttleworth and Gurney (1990) built on the work of Choudhury and Monteith (1988) to estimate r_{am} by integrating the eddy diffusion coefficient over the sink of momentum in the canopy to a reference height z_r above the canopy, giving the following relationship for r_{am}:

$$r_{am} = \frac{1}{k \cdot u^*} \cdot Ln \left(\frac{z_r - d}{h - d} \right) + \frac{h}{\alpha \cdot K_h} \cdot \left[\exp \left(\alpha \cdot \left(1 - \frac{z_0 + d}{h} \right) \right) - 1 \right] \qquad (24)$$

where k is the von Karman constant, u^* is the friction velocity, z_0 is the surface roughness, d is the zero-plane displacement height, K_h is the value of eddy diffusion coefficient at the top of the canopy, h is the height of vegetation, and α is the attenuation coefficient. A value of α = 2.5, which is typical for agricultural crops, was recommended by Shuttleworth and Wallace (1985) and Shuttleworth and Gurney (1990).

Verma (1989) expressed the excess resistance for heat transfer as:

$$r_{bh} = \frac{k \cdot B^{-1}}{k \cdot u^*} \tag{25}$$

where B^{-1} represents a dimensionless bulk parameter. Thom (1972) suggests that the product kB^{-1} equal approximately 2 for most arable crops.

Excess resistance was derived primarily from heat transfer observations (Weseley & Hicks 1977). Aerodynamic resistance to water vapor was modified by the ratio of thermal and water vapor diffusivity:

$$r_{bw} = \frac{k \cdot B^{-1}}{k \cdot u^*} \left(\frac{k_1}{D_v}\right)^{2/3} \tag{26}$$

where, k_1 is the thermal diffusivity and D_v is the molecular diffusivity of water vapor in air. Similarly, Shuttleworth and Gurney (1990) expressed the aerodynamic resistance (r_2) by integrating the eddy diffusion coefficient between the soil surface and the sink of momentum in the canopy to yield:

$$r_2 = \frac{h \cdot \exp(\alpha)}{\alpha \cdot K_h} \cdot \left[\exp\left(\frac{-\alpha \cdot z_o'}{h}\right) - \exp\left(\frac{-\alpha \cdot (d + z_o)}{h}\right) \right] \tag{27}$$

where z_o' is the roughness length of the soil surface. Values of surface roughness (z_o) and displacement height (d) are functions of leaf area index (LAI) and can be estimated using the expressions given by Shaw and Pereira (1982).

The diffusion coefficients between the soil surface and the canopy, and therefore the resistance for momentum, heat, and vapor transport are assumed equal although it is recognized that this is a weakness in the use of the K theory to describe through-canopy transfer (Shuttleworth & Gurney, 1990). Stability is not considered.

2.1.1.2 Canopy resistances

The mean boundary layer resistance of the canopy r_1, for latent and sensible heat flux, is influenced by the surface area of vegetation (Shuttleworth & Wallace, 1985):

$$r_1 = \frac{r_b}{2 \cdot LAI} \tag{28}$$

where r_b is the resistance of the leaf boundary layer, which is proportional to the temperature difference between the leaf and surrounding air divided by the associated flux (Choudhury & Monteith, 1988). Shuttleworth and Wallace (1985) noted that resistance r_b exhibits some dependence on in-canopy wind speed, with typical values of 25 s m^{-1}. Shuttleworth and Gurney (1990) represented r_b as:

$$r_b = \frac{100}{\alpha} \cdot \left(\frac{w}{u_h}\right)^{1/2} \cdot \left(1 - \exp\left(\frac{-\alpha}{2}\right)\right)^{-1} \tag{29}$$

where w is the representative leaf width and u_h is the wind speed at the top of the canopy. This resistance is only significant when acting in combination with a much larger canopy surface resistance, and Shuttleworth and Gurney (1990) suggest that r_1 could be neglected for foliage completely covering the ground. Using $r_b = 25$ s m^{-1} with an LAI = 4, the corresponding canopy boundary layer resistance is $r_1 = 3$ s m^{-1}.

Canopy surface resistance, r_c, can be calculated by dividing the minimum surface resistance for a single leaf (r_l) by the effective canopy leaf area index (LAI). Five environmental factors have been found to affect stomata resistance: solar radiation, air temperature, humidity, CO_2 concentration and soil water potential (Yu et al., 2004). Several models have been developed to estimate stomata conductance and canopy resistance. Stannard (1993) estimated r_c as a function of vapor pressure deficit, leaf area index, and solar radiation as:

$$r_c = \left[C_1 \cdot \frac{LAI}{LAI_{max}} \cdot \frac{C_2}{C_2 + VPD_a} \cdot \frac{Rad \cdot (Rad_{max} + C_3)}{Rad_{max} \cdot (Rad + C_3)} \right]^{-1} \tag{30}$$

where LAI_{max} is the maximum value of leaf area index, VPD_a is vapor pressure deficit, Rad is solar radiation, Rad_{max} is the maximum value of solar radiation (estimated at 1000 W m^{-2}) and C_1, C_2 and C_3 are regression coefficients. Canopy resistance does not account for soil water stress effects.

2.1.1.3 Soil resistances

Farahani and Bausch (1995), Anadranistakis et al. (2000) and Lindburg (2002) found that soil resistance (r_s) can be related to volumetric soil water content in the top soil layer. Farahani and Ahuja (1996) found that the ratio of soil resistance when the surface layer is wet relative to its upper limit depends on the degree of saturation (θ/θ_s) and can be described by an exponential function as:

$$r_s = r_{so} \cdot \exp\left(-\beta \cdot \frac{\theta}{\theta_s} \right) \qquad \text{and} \qquad r_{so} = \frac{L_t \cdot \tau_s}{D_v \cdot \emptyset} \tag{31}$$

where L_t is the thickness of the surface soil layer, τ_s is a soil tortuosity factor, D_v is the water vapor diffusion coefficient and $\emptyset$ is soil porosity, θ is the average volumetric water content in the surface layer, θ_s is the saturation water content, and β is a fitting parameter. Measurements of θ from the top 0.05 m soil layer were more effective in modeling r_s than θ for thinner layers.

Choudhury and Monteith (1988) expressed the soil resistance for heat flux (r_L) in the soil layer extending from depth L_t to L_m as:

$$r_L = \frac{\rho \cdot C_p \cdot (L_m - L_t)}{K} \tag{32}$$

where K is the thermal conductivity of the soil. Similarly, the corresponding resistance for the upper layer (r_u) of depth L_t and conductivity K' as:

$$r_u = \frac{\rho \cdot C_p \cdot L_t}{K'} \tag{33}$$

2.1.1.4 Residue resistances

Surface residue is an integral part of many cropping systems. Bristow and Horton (1996) showed that partial surface mulch cover can have dramatic effects on the soil physical environment. The vapor conductance through residue has been described as a linear function of wind speed. Farahani and Ahuja (1996) used results from Tanner and Shen (1990) to develop the resistance of surface residue (r_r) as:

$$r_r = \frac{L_r \cdot \tau_r}{D_v \cdot \emptyset_r} (1 + 0.7 \cdot u_2)^{-1} \qquad (34)$$

where L_r is residue thickness, τ_r is residue tortuosity, D_v is vapor diffusivity in still air, $\emptyset_r$ is residue porosity and u_2 is wind speed measured two meters above the surface. Due to the porous nature of field crop residue layers, the ratio $\tau_r/\emptyset_r$ is about one (Farahani & Ahuja, 1996).

Similar to the soil resistance, Bristow and Horton (1996) and Horton et al. (1996) expressed the resistance of residue for heat transfer, r_{rh}, as:

$$r_{rh} = \frac{\rho \cdot C_p \cdot L_r}{K_r} \qquad (35)$$

where K_r is the residue thermal conductivity.

The fraction of the soil covered by residue (f_r) can be estimated using the amount and type of residue (Steiner et al., 2000). The soil covered by residue and the residue thickness are estimated using the expressions developed by Gregory (1982).

2.1.2 SEB model inputs

Inputs required to solve multiple layer models (i.e. Shuttleworth and Wallace (1985), Choudhury and Monteith (1988) and Lagos (2008) models) are net radiation, solar radiation, air temperature, relative humidity, wind speed, LAI, crop height, soil texture, soil temperature, soil water content, residue type, and residue amount. In particular, net radiation, leaf area index, soil temperatures and residue amount are variables rarely measured in the field, other than at research sites. Net radiation and soil temperature models can be incorporated into surface energy balance models to predict evapotranspiration from environmental variables typically measured by automatic weather stations.

Similar to the Shuttleworth and Wallace (1985) and Choudhury and Monteith (1988) models, measurements of net radiation and estimations of net radiation absorbed by the canopy are necessary for the SEB model. Beer's law is used to estimate the penetration of radiation through the canopy and estimates the net radiation reaching the surface (Rn_s) as:

$$Rn_s = Rn \cdot \exp(-C_{ext} \cdot LAI) \qquad (36)$$

where Cext is the extinction coefficient of the crop for net radiation. Consequently, net radiation absorbed by the canopy (Rnc) can be estimated as Rnc = Rn – Rn_s.

2.1.3 SEB model evaluation

An irrigated maize field site located at the University of Nebraska Agricultural Research and Development Center near Mead, NE (41°09′53.5″N, 96°28′12.3″W, elevation 362 m) was used for model evaluation. This site is a 49 ha production field that provides sufficient upwind fetch of uniform cover required for adequately measuring mass and energy fluxes using eddy covariance systems. The area has a humid continental climate and the soil corresponds to a deep silty clay loam (Suyker & Verma, 2009). The field has not been tilled since 2001. Detailed information about planting densities and crop management is provided by Verma et al. (2005) and Suyker and Verma (2009).

Soil water content was measured continuously at four depths (0.10, 0.25, 0.5 and 1.0 m) with Theta probes (Delta-T Device, Cambridge, UK). Destructive green leaf area index and biomass measurements were taken bi-monthly during the growing season. The eddy covariance measurements of latent heat, sensible heat and momentum fluxes were made using an omnidirectional three dimensional sonic anemometer (Model R3, Gill Instruments Ltd., Lymington, UK) and an open-path infrared CO_2/H_2O gas analyzer system (Model LI7500, Li-Cor Inc, Lincoln, NE). Fluxes were corrected for sensor frequency response and variations in air density. More details of measurements and calculations are given in Verma et al. (2005). Air temperature and humidity were measured at 3 and 6 meters (Humitter 50Y, Vaisala, Helsinki, Finland), net radiation at 5.5 m (CNR1, Kipp and Zonen, Delft, NLD) and soil heat flux at 0.06 m (Radiation and Energy Balance Systems Inc, Seattle, WA). Soil temperature was measured at 0.06, 0.1, 0.2 and 0.5 m depths (Platinum RTD, Omega Engineering, Stamford, CT). More details are given in Verma et al. (2005) and Suyker and Verma (2009).

Evapotranspiration predictions from the SEB model were compared with eddy covariance flux measurements during 2003 for an irrigated maize field. To evaluate the energy balance closure of eddy covariance measurements, net radiation was compared against the sum of latent heat, sensible heat, soil heat flux and storage terms. Storage terms include soil heat storage, canopy heat storage, and energy used in photosynthesis. Storage terms were calculated by Suyker and Verma (2009) following Meyers and Hollinger (2004). During these days, the regression slope for energy balance closure was 0.89 with a correlation coefficient of $r^2 = 0.98$.

For model evaluation, 15 days under different LAI conditions were selected to initially test the model, however further work is needed to test the model for entire growing seasons and during longer periods. Hourly data for three 5-day periods with varying LAI conditions (LAI = 0, 1.5 and 5.4) were used to compare measured ET to model predictions. Input data of the model included hourly values for: net radiation, air temperature, relative humidity, soil temperature at 50 cm, wind speed, solar radiation and soil water content. During the first 5-day period, which was prior to germination, the maximum net radiation ranged from 240 to 720 W m-2, air temperature ranged from 10 to 30°C, soil temperature was fairly constant at 16°C and wind speed ranged from 1 to 9 m s-1 but was generally less than 6 m s-1 (Figure 3). Soil water content in the evaporation zone averaged 0.34 m3 m-3and the residue density was 12.5 ton/ha on June 6, 2003. Precipitation occurred on the second and fifth days, totaling 17 mm.

Evapotranspiration estimated with the SEB model and measured using the eddy covariance system is given in Figure 4. ET fluxes were the highest at midday on June 6, reaching approximately 350 W m-2. The lowest ET rates occurred on the second day. Estimated ET tracked measured latent heat fluxes reasonably well. Estimates were better for days without precipitation than for days when rainfall occurred. The effect of crop residue on evaporation from the soil is shown in Figure 4 for this period. Residue reduced cumulative evaporation by approximately 17% during this five-day period. Evaporation estimated with the SEB model on June 6 and 9 was approximately 3.5 mm/day, totaling approximately half of the total evaporation for the five days.

During the second five-day period, when plants partially shaded the soil surface (LAI = 1.5), the maximum net radiation ranged from 350 to 720 W m-2 and air temperature ranged from 10 to 33°C (Figure 5). The soil temperature was nearly constant at 20°C. Wind speed ranged from 0.3 to 8 m s-1 but was generally less than 6 m s-1. The soil water content was about 0.31

m^3 m^{-3} and the residue density was 12.2 ton/ha on June 24, 2003. Precipitation totaling 3mm occurred on the fifth day. The predicted rate of ET estimated with the SEB model was close to the observed data (Figure 6). Estimates were smaller than measured values for June 24, which was the hottest and windiest day of the period. The ability of the model to partition ET into evaporation and transpiration for partial canopy conditions is also illustrated in Figure 6. Evaporation from the soil represented the majority of the water used during the night, and early or late in the day. During the middle of the day transpiration represented approximately half of the hourly ET flux.

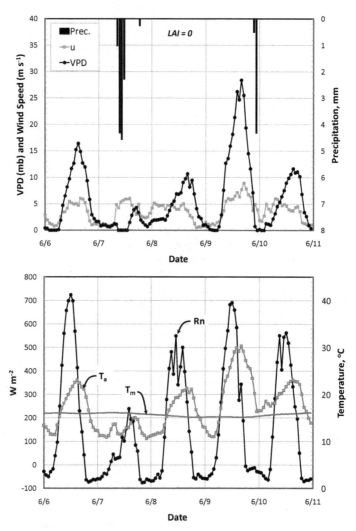

Fig. 3. Environmental conditions during a five-day period without canopy cover for net radiation (Rn), air temperature (T$_a$), soil temperature (T$_m$), precipitation (Prec.), vapor pressure deficit (VPD), and wind speed (u).

The last period represents a fully developed maize canopy that completely shaded the soil surface. The crop height was 2.3 m and the LAI was 5.4. Environmental conditions for the period are given in Figure 7. The maximum net radiation ranged from 700 to 740 W m^{-2} and air temperature ranged from 15 to 36 °C during the period. Soil temperature was fairly constant during the five days at 21.5°C and wind speed ranged from 0.3 to 4 m s^{-1}. The soil water content was about 0.25 m^3 m^{-3} and the residue density was 11.8 ton/ha on July 16, 2003. Precipitation totaling 29 mm occurred on the third day. Observed and predicted ET fluxes agreed for most days with some differences early in the morning during the first day and during the middle of several days (Figure 8). Transpiration simulated with the SEB model was nearly equal to the simulated ET for the period as evaporation rates from the soil was very small.

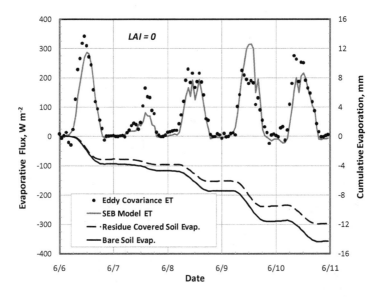

Fig. 4. Evapotranspiration estimated by the Surface Energy Balance (SEB) model and measured by an eddy covariance system and simulated cumulative evaporation from bare and residue-covered soil for a period without plant canopy cover.

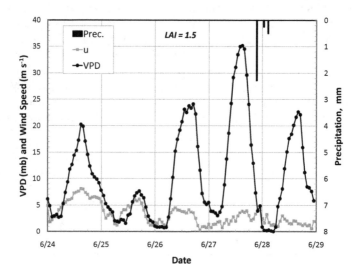

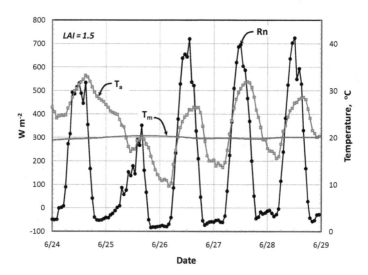

Fig. 5. Environmental conditions for a five-day period with partial crop cover for net radiation (Rn), air temperature (Ta), soil temperature (Tm), precipitation (Prec), vapor pressure deficit (VPD), and wind speed (u).

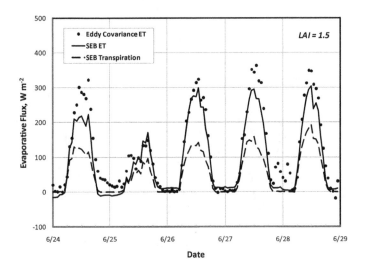

Fig. 6. Evapotranspiration and transpiration estimated by the Surface Energy Balance (SEB) model and ET measured by an eddy covariance system for a 5-day period with partial canopy cover.

Hourly measurements and SEB predictions for the three five-day periods were combined to evaluate the overall performance of the model (Figure 9). Results show variation about the 1:1 line; however, there is a strong correlation and the data are reasonably well distributed about the line. Modeled ET is less than measured for latent heat fluxes above 450 W m^{-2}. The model underestimates ET during hours with high values of vapor pressure deficit (Figure 6 and 8), this suggests that the linear effect of vapor pressure deficit in canopy resistance estimated with equation (30) produce a reduction on ET estimations. Further work is required to evaluate and explore if different canopy resistance models improve the performance of ET predictions under these conditions. Various statistical techniques were used to evaluate the performance of the model. The coefficient of determination, Nash-Sutcliffe coefficient, index of agreement, root mean square error and the mean absolute error were used for model evaluation (Legates & McCabe 1999; Krause et al., 2005; Moriasi et al., 2007; Coffey et al. 2004). The coefficient of determination was 0.92 with a slope of 0.90 over the range of hourly ET values. The root mean square error was 41.4 W m^{-2}, the mean absolute error was 29.9 W m^{-2}, the Nash-Sutcliffe coefficient was 0.92 and the index of agreement was 0.97. The statistical parameters show that the model represents field measurements reasonably well. Similar performance was obtained for daily ET estimations (Table 1). Analysis is underway to evaluate the model for more conditions and longer periods. Simulations reported here relied on literature-reported parameter values. We are also exploring calibration methods to improve model performance.

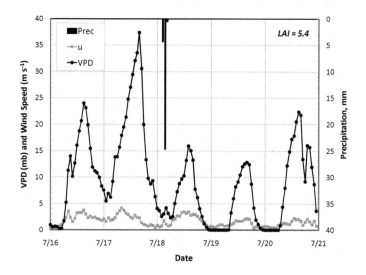

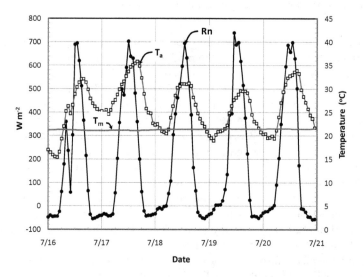

Fig. 7. Environmental conditions for 5-day period with full canopy cover for net radiation (Rn), air temperature (Ta), soil temperature (Tm), precipitation (Prec), vapor pressure deficit (VPD) and wind speed (u).

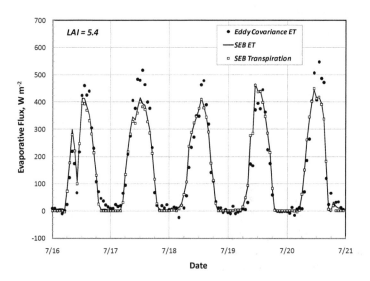

Fig. 8. Evapotranspiration and transpiration estimated by the Surface Energy Balance (SEB) model and ET measured by an eddy covariance system during a period with full canopy cover.

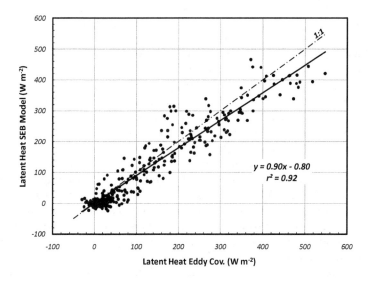

Fig. 9. Measured versus modeled hourly latent heat fluxes.

Date	LAI $m^2 m^{-2}$	Evapotranspiration (mm day^{-1})	
		SEB	EC
6-Jun	0	3.2	3.7
7-Jun	0	0.7	1.4
8-Jun	0	2.3	3.2
9-Jun	0	3.5	2.7
10-Jun	0	2.4	3.5
24-Jun	1.5	2.9	4.4
25-Jun	1.5	1.7	2.1
26-Jun	1.5	4.1	4.3
27-Jun	1.5	4.0	5.0
28-Jun	1.5	3.8	4.7
16-Jul	5.4	5.1	5.1
17-Jul	5.4	5.8	6.8
18-Jul	5.4	5.2	5.0
19-Jul	5.4	5.0	4.1
20-Jul	5.4	5.1	5.4

Table 1. Daily evapotranspiration estimated with the Surface Energy Balance (SEB) model and measured from the Eddy Covariance (EC) system.

2.2 The modified SEB model for Partially Vegetated surfaces (SEB-PV)

Although good performance of multiple-layer models has been recognized, multiple-layer models estimate more accurate ET values under high LAI conditions. Lagos (2008) evaluated the SEB model for maize and soybean under rainfed and irrigated conditions; results indicate that during the growing season, the model more accurately predicted ET after canopy closure (after LAI=4) than for low LAI conditions. The SEB model, similar to S-W and C-M models, is based on homogeneous land surfaces. Under low LAI conditions, the land surface is partially covered by the canopy and soil evaporation takes place from soil below the canopy and areas of bare soil directly exposed to net radiation. However, in multiple-layer models, evaporation from the soil has been only considered below the canopy and hourly variations in the partitioning of net radiation between the canopy and the soil is often disregarded. Soil evaporation on partially vegetated surfaces & inorchards and natural vegetation include not only soil evaporation beneath the canopy but also evaporation from areas of bare soil that contribute directly to total ET.

Recognizing the need to separate vegetation from soil and considering the effect of residue on evaporation, we extended the SEB model to represent those common conditions. The modified model, hereafter the SEB-PV model, distributes net radiation (Rn), sensible heat (H), latent heat (λE), and soil heat fluxes (G) through the soil/residue/canopy system. Similar to the SEB model, horizontal gradients of the potentials are assumed to be small enough for lateral fluxes to be ignored, and physical and biochemical energy storage terms in the canopy/residue/soil system are assumed to be negligible. The evaporation of water on plant leaves due to rain, irrigation or dew is also ignored.

The SEB-PV model has the same four layers described previously for SEB (Figure 10):the first extended from the reference height above the vegetation and the sink for momentum within the canopy, a second layer between the canopy level and the soil surface, a third layer corresponding to the top soil layer and a lower soil layer where the soil atmosphere is saturated with water vapor.

Total latent heat (λE) is the sum of latent heat from the canopy (λEc), latent heat from the soil (λEs) beneath the canopy, latent heat from the residue-covered soil (λEr) beneath the canopy, latent heat from the soil (λEbs) directly exposed to net radiation and latent heat from the residue-covered soil (λEbr) directly exposed to net radiation.

$$\lambda E = [\lambda E_c + \lambda E_s(1 - f_r) + \lambda E_r f_r]F_V + [\lambda E_{bs}(1 - f_r)](1 - F_V) \tag{37}$$

Where fr is the fraction of the soil affected by residue and Fv is the fraction of the soil covered by vegetation. Similarly, sensible heat is calculated as the sum of sensible heat from the canopy (Hc), sensible heat from the soil (Hs) and sensible heat from the residue covered soil (Hr), sensible heat from the soil (Hbs) directly exposed to net radiation and latent heat from the residue-covered soil (Hbr) directly exposed to net radiation.

$$H = [Hc + Hs(1 - fr) + Hr\, fr]\, Fv + [Hbs(1 - fr) + Hbr\, fr](1 - Fv) \tag{38}$$

For the fraction of the soil covered by vegetation, the total net radiation is divided into that absorbed by the canopy (Rnc) and the soil beneath the canopy (Rns) and is given by Rn = Rnc + Rns. The net radiation absorbed by the canopy is divided into latent heat and sensible heat fluxes as Rnc = λEc + Hc. Similarly, for the soil Rns = Gos + Hs, where Gos is a conduction term downwards from the soil surface and is expressed as Gos = λEs + Gs, where Gs is the soil heat flux for bare soil. Similarly, for the residue covered soil Rns = Gor + Hr where Gor is the conduction downwards from the soil covered by residue. The conduction is given by Gor = λEr + Gr where Gr is the soil heat flux for residue-covered soil. For the area without vegetation, total net radiation is divided into latent and sensible heat fluxes as Rn = λEbs + λEbr + Hbs + Hbr.

The differences in vapor pressure and temperature between levels can be expressed with an Ohm's law analogy using appropriate resistance and flux terms (Figure 10). Latent and sensible flux terms with in the resistance network were combined and solved to estimate total fluxes. The solution gives the latent and sensible heat fluxes from the canopy, the soil beneath the canopy and the soil covered by residue beneath the canopy similar to equations (9), (10), (11), (12) and (13).

The new expressions for latent heat flux of bare soil and soil covered by residue, both directly exposed to net radiation are:

For bare soil:

$$\lambda E_{bs} = \frac{(R_n \cdot \Delta \cdot (r_{2b}) \cdot r_L + \rho \cdot C_p \cdot ((e_b^* - e_b) \cdot r_u + r_L + r_{2b}) + (T_m - T_b) \cdot \Delta \cdot (r_u + r_{2b}))}{\gamma \cdot (r_{2b} + r_s) \cdot (r_u + r_L + r_{2b}) + \Delta \cdot r_L \cdot (r_u + r_{2b})} \tag{39}$$

For residue covered soil:

$$\lambda E_{br} = \frac{R_n \cdot \Delta \cdot (r_{2b} + r_{rh}) \cdot r_L + \rho \cdot C_p \cdot ((e_b^* - e_b) \cdot (r_u + r_L + r_{2b} + r_{rh}) + (T_m - T_b) \cdot \Delta \cdot (r_u + r_{2b} + r_r))}{\gamma \cdot (r_{2b} + r_s + r_r) \cdot (r_u + r_L + r_{2b} + r_{rh}) + \Delta \cdot r_L \cdot (r_u + r_{2b} + r_{rh})} \tag{40}$$

These relationships define the surface energy balance model, which is applicable to conditions ranging from closed canopies to surfaces partially covered by vegetation. If Fv = 1 the model SEB-PV is similar to the original SEB model and with Fv=1 without residue, the model is similar to that by Choudhury and Monteith (1988).

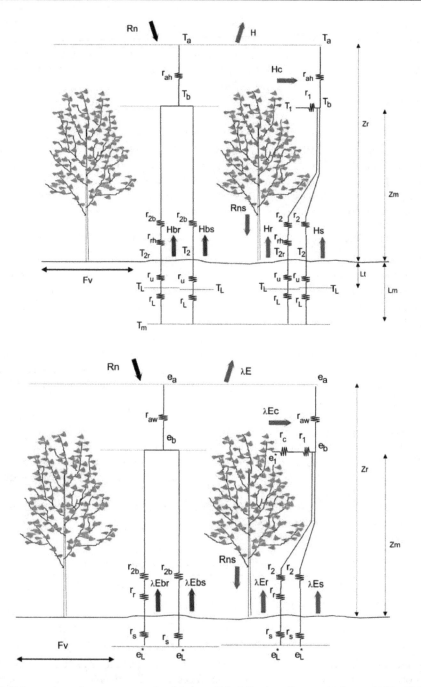

Fig. 10. Schematic resistance network of the modified Surface Energy Balance (SEB - PV) model for partially vegetated surfaces a) Sensible heat flux and b) Latent heat flux.

2.2.1 Model resistances
Model resistances are similar to those described by the SEB model; however, a new aerodynamic resistance (r_{2b}) for the transfer of heat and water flux is required for the surface without vegetation.

The aerodynamic resistance between the soil surface and Zm (r_{2b}) could be calculated by assuming that the soil directly exposed to net radiation is totally unaffected by adjacent vegetation as:

$$r_{as} = \frac{\ln\left(\frac{z_m}{z_0}\right)^2}{k^2 u} \tag{41}$$

According to Brenner and Incoll (1997), actual aerodynamic resistance (r_{2b}) will vary between r_{as} for Fv=0 and r_2 when the fractional vegetative cover Fv=1. The form of the functional relationship of this change is not known, r_{2b} was varied linearly between r_{as} and r_2 as:

$$r_{2b} = FV(r_2) + (1 - FV)(r_{as}) \tag{42}$$

2.2.2 Model inputs
The proposed SEB-PV model requires the same inputs of the SEB model plus the fraction of the surface covered by vegetation (Fv).

2.3 Sensitivity analysis
A sensitivity analysis was performed to evaluate the response of the SEB model to changes in resistances and model parameters. Meteorological conditions, crop characteristics and soil/residue characteristics used in these calculations are given in Table 2. Such conditions are typical for midday during the growing season of maize in southeastern Nebraska. The sensitivity of total latent heat from the system was explored when model resistances and model parameters were changed under different LAI conditions. The effect of the changes in model parameters and resistances were expressed as changes in total ET (λE) and changes in the crop transpiration ratio. The transpiration ratio is the ratio between crop transpiration (λEc) over total ET (transpiration ratio= λEc / λE).

The response of the SEB model was evaluated for three values of the extinction coefficient (Cext = 0.4, 0.6 and 0.8), three conditions of vapor pressure deficit (VPDa = 0.5 kPa, 0.1 kPa and 0.25 kPa) three soil temperatures (T_m=21°C, $0.8 \times T_m$=16.8 °C and $1.2 \times T_m$=25.2 °C) (Figure 11), changes in the parameterization of aerodynamic resistances (the attenuation coefficient, α= 1, 2.5 and 3.5), the mean boundary layer resistance, r_b (±40%) the crop height, h (±30%)), selected conditions for the soil surface resistance, r_s (0, 227, and 1500 s m^{-1}) (Figure 12), four values for residue resistance, r_r (0, 400, 1000, and 2500 s m^{-1}), and changes of ±30% in surface canopy resistance, r_c (Figure 13).

In general, the sensitivity analysis of model resistances showed that simulated ET was most sensitive to changes in surface canopy resistance for LAI > 0.5 values, and soil surface resistance and residue surface resistance for small LAI values (LAI < ~3). The model was less sensitive to changes in the other parameters evaluated.

Variable	Symbol	Value	Unit
Net Radiation	Rn	500	W m^{-2}
Air temperature	Ta	25	^{o}C
Relative humidity	RH	68	%
Wind speed	U	2	m s^{-1}
Soil Temperature at 0.5 m	Tm	21	^{o}C
Solar radiation	Rad	700	W m^{-2}
Canopy resistance coeff.	C1, C2, C3	5, 0.005, 300	
Maximum leaf area index	LAImax	6	m^{2} m^{-2}
Soil water content	Θ	0.25	m^{3} m^{-3}
Saturation soil water content	Θs	0.5	m^{3} m^{-3}
Soil porosity	ϕ	0.5	m^{3} m^{-3}
Soil tortuosity	τs	1.5	
Residue fraction	Fr	0.5	
Thickness of the residue layer	Lr	0.02	m
Residue tortuosity	τr	1	
Residue porosity	ϕ r	1	
Upper layer thickness	Lt	0.05	m
Lower layer depth	Lm	0.5	m
Soil roughness length	Zo$'$	0.01	m
Drag coefficient	Cd	0.07	
Reference height	Z	3	m
Attenuation coefficient	α	2.5	
Maximum solar radiation	Radmax	1000	W m^{-2}
Extinction coefficient	Cext	0.6	
Mean leaf width	W	0.08	m
Water vapor diffusion coefficient	Dv	2.56x10^{-5}	m^{2} s^{-1}
Fitting parameter	β	6.5	
Soil thermal conductivity, upper layer	K	2.8	W m$^{-1 o}$C^{-1}
Soil thermal conductivity, lower layer	K$'$	3.8	W m$^{-1 o}$C^{-1}

Table 2. Predefined conditions for the sensitivity analysis.

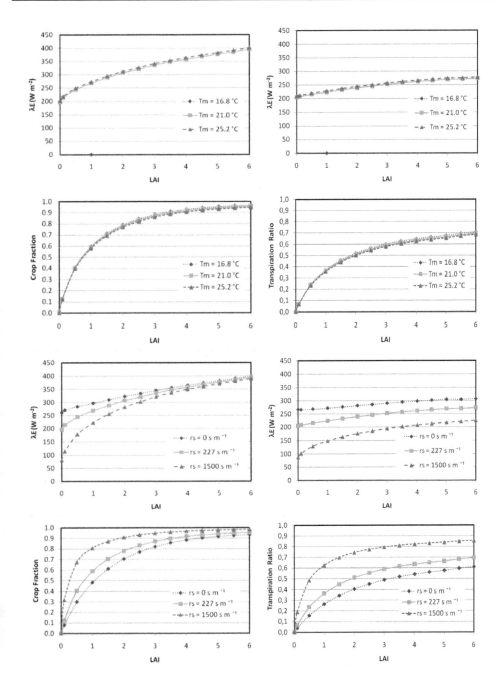

Fig. 11. Sensitivity analysis of the SEB-PV model for Fv=1 (left) and Fv=0,5 (right) under different soil temperatures Tm, and soil resistance conditions.

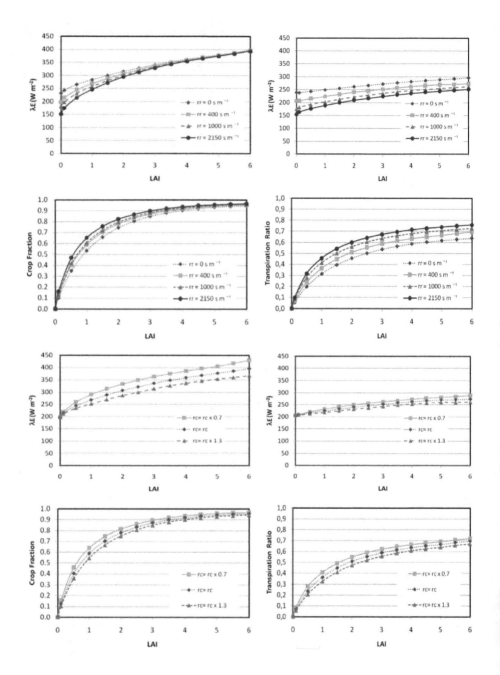

Fig. 12. Sensitivity analysis of the SEB-PV model for Fv=1 (left) and Fv=0,5 (right) under different residue and canopy conditions.

3. Conclusions

A surface energy balance model (SEB) based on the Shuttleworth-Wallace and Choudhury-Monteith models was developed to account for the effect of residue, soil evaporation and canopy transpiration on ET. The model describes the energy balance of vegetated and residue-covered surfaces in terms of driving potential and resistances to flux. Improvements in the SEB model were the incorporation of residue into the energy balance and modification of aerodynamic resistances for heat and water transfer, canopy resistance for water flux, residue resistance for heat and water flux, and soil resistance for water transfer. The model requires hourly data for net radiation, solar radiation, air temperature, relative humidity, and wind speed. Leaf area index and crop height plus soil texture, temperature and water content as well as the type and amount of crop residue are also required. An important feature of the model is the ability to estimate latent, sensible and soil heat fluxes. The model provides a method for partitioning ET into soil/residue evaporation and plant transpiration, and a tool to estimate the effect of residue ET on water balance studies. Comparison between estimated ET and measurements from an irrigated maize field provide support for the validity of the surface energy balance model. Further evaluation of the model is underway for agricultural and natural ecosystems during growing seasons and dormant periods. We are developing calibration procedures to refine parameters and improve model results.

The SEB model was modified for modeling evapotranspiration of partially vegetated surfaces given place to the SEB-PV model. The SEB-PV model can be used for partitioning total ET on canopy transpiration and soil evaporation beneath the canopy and soil directly exposed to net radiation. The model can be used for partitioning net radiation into not only latent heat fluxes but also sensible heat fluxes from each surface. A preliminary sensitivity analysis shows that similar to the SEB model, the proposed modification was sensitive to soil surface resistance, residue resistance, canopy resistance and vapor pressure deficit. Further model evaluation is needed to test this approach. A model to estimate Rn and a model to estimate soil temperature T_m from air temperature and soil conditions are also required to reduce the required inputs of the model.

4. List of variables

Rn   Net Radiation (W m-2).
Rn_c Net Radiation absorbed by the canopy (W m-2).
Rn_s Net Radiation absorbed by the soil (W m-2).
λE Total latent heat flux (W m-2).
λE_c Latent heat flux from the canopy (W m-2).
λE_s Latent heat flux from the soil (W m-2).
λE_r Latent heat flux from the residue-covered soil (W m-2).
λE_{bs} Latent heat from the soil directly exposed to net radiation (W m-2).
λE_{br} Latent heat from the residue-covered soil directly exposed to net radiation (W m-2).
H   Total Sensible heat flux (W m-2).
H_c Sensible heat flux from the canopy (W m-2).
H_s Sensible heat flux from the soil (W m-2).
H_r Sensible heat flux from the residue-covered soil (W m-2).
G_{os} Conduction flux from the soil surface (W m-2).
G_{or} Conduction flux from the residue-covered soil surface (W m-2).
G_s Soil heat flux for bare soil (W m-2).

G_r Soil heat flux for residue-covered soil (W m^{-2}).

f_r Fraction of the soil covered by residue (0-1).

ρ Density of moist air (Kg m^{-3}).

C_p Specific heat of air (J Kg^{-1} ^{o}C^{-1}).

γ Psychrometric constant (Kpa ^{o}C^{-1}).

T_a Air temperature (^{o}C).

T_b Air temperature at canopy height (^{o}C).

T_1 Canopy temperature (^{o}C).

T_2 Soil surface temperature (^{o}C).

T_{2r} Soil surface temperature below the residue (^{o}C).

T_L Soil temperature at the interface between the upper and lower layers for bare soil (^{o}C).

T_{Lr} Soil temperature at the interface between the upper and lower layers for residue-covered soil (^{o}C).

T_m Soil temperature at the bottom of the lower layer (^{o}C).

e_a Vapor pressure of the air (mb).

e_b Vapor pressure of the air at the canopy level (mb).

e_1^* Saturated vapor pressure at the canopy (mb).

e_L^* Saturated vapor pressure at the top of the wet layer (mb).

e_b^* Saturated vapor pressure at the canopy level (mb).

e_a^* Saturated vapor pressure of the air (mb).

e_{Lr}^* Saturated vapor pressure at the top of the wet layer for the residue-covered soil (mb).

r_{am} Aerodynamic resistance for momentum transfer (s m^{-1}).

r_{ah} Aerodynamic resistance for heat transfer (s m^{-1}).

r_{aw} Aerodynamic resistance for water vapor (s m^{-1}).

r_{bh} Excess resistance term for heat transfer (s m^{-1}).

r_{bw} Excess resistance term for water vapor (s m^{-1}).

r_1 Aerodynamic resistance between the canopy and the air at the canopy level (s m^{-1}).

r_b Boundary layer resistance (s m^{-1}).

r_2 Aerodynamic resistance between the soil and the air at the canopy level (s m^{-1}).

r_{2b} Actual aerodynamic resistance between the soil surface and Zm (s m^{-1}).

r_{as} Aerodynamic resistance between the soil surface and Zm totally unaffected by adjacent vegetation (s m^{-1}).

r_c Surface canopy resistance (s m^{-1}).

r_r Residue resistance for water vapor flux (s m^{-1}).

r_s Soil surface resistance for water vapor flux (s m^{-1}).

r_{rh} Residue resistance to transfer of heat (s m^{-1}).

r_r Residue resistance for heat flux (s m^{-1}).

r_u Soil heat flux resistance for the upper layer (s m^{-1}).

r_L Soil heat flux resistance for the lower layer (s m^{-1}).

Δ Slope of the saturation vapor pressure (mb ^{o}C^{-1}).

h   Vegetation height (m).

LAI   Leaf area index (m^2 m^{-2}).

LAI$_{max}$ Maximum value of leaf area index (m^2 m^{-2}).

d   Zero plane displacement (m).

z_r Reference height above the canopy (m).

Z_m Reference height (m).

z_o Surface roughness length (m).

z_o' Roughness length of the soil surface (m).

k   Von-Karman Constant.
k_h Diffusion coefficient at the top of the canopy (m^2 s^{-1}).
u^* Friction velocity (m s^{-1}).
α   Attenuation coefficient for eddy diffusion coefficient within the canopy.
B^{-1} Dimensionless bulk parameter.
VPD_a Vapor pressure deficit (mb).
Rad   Solar radiation (W m^{-2}).
Rad_{max} Maximum value of solar radiation (W m^{-2}).
w   Mean leaf width (m).
u_h Wind speed at the top of the canopy (m s^{-1}).
L_t Thickness of the surface soil layer (m).
L_m Thickness of the surface and bottom soil layers (m)
r_{so} Soil surface resistance to the vapor flux for a dry layer (m s^{-1}).
τ_s Soil tortuosity.
D_v Water vapor diffusion coefficient (m^2 s^{-1}).
k_1 Thermal diffusivity (m^2 s^{-1}).
φ   Soil porosity.
β   Fitting parameter.
θ   Volumetric soil water content (m^3 m^{-3}).
θ_s Saturation water content of the soil (m^3 m^{-3}).
L_r Residue thickness (m).
τ_r Residue tortuosity.
ϕ_r Residue porosity.
u_2 Wind speed at two meters above the surface (m s^{-1}).
K   Thermal conductivity of the soil, upper layer (W m^{-1} $^oC^{-1}$).
K'   Thermal conductivity of the soil, lower layer (W m^{-1} $^oC^{-1}$).
K_r Thermal conductivity of the residue layer (W m^{-1} $^oC^{-1}$).
C_{ext} Extinction coefficient.
Fv   Fraction of the soil covered by vegetation.
H_{bs} Sensible heat from the soil (W m^{-2}).
H_{br} Latent heat from the residue-covered soil (W m^{-2}).

5. Acknowledgments

We thank the University of Nebraska Program of Excellence, the University of Nebraska-Lincoln Institute of Agriculture and Natural Resources, Fondo Nacional de Desarrollo Cientifico y Tecnologico (FONDECYT 11100083) and Fondo de Fomento al Desarrollo Cientifico y Tecnologico (FONDEF D09I1146) Their support is gratefully recognized.

6. References

Allen, S.J., (1990). Measurement and estimation of evaporation from soil under sparse barley crops in northern Syria. Agric. For. Meteorology, 49: 291-309.

Allen R G, Pereira LS, Raes D and Smith M (1998) Crop Evapotranspiration: Guidelines for computing crop requirement. (Irrigation and Drainage Paper No 56) FAO, Rome, Italy.

Allen, R.G., Tasumi M., and Trezza, R., (2007). Satellite-based energy balance for mapping evapotranspiration with internalized calibration (METRIC)-model. Journal of Irrigation and Drainage Engineering, 133 (4): 380-394.

Alves I., and Pereira, L. S., (2000). Modeling surface resistance from climatic variables? Agricultural Water Management, 42, 371-385.

Anadranistakis M, Liakatas A, Kerkides P, Rizos S, Gavanosis J, and Poulovassilis A (2000) Crop water requirements model tested for crops grown in Greece. Agric Water Manage 45:297-316.

ASCE (2002) The ASCE Standardized equation for calculating reference evapotranspiration, Task Committee Report.Environment and Water Resources Institute of ASCE, New York.

Brenner AJ and Incoll LD (1996) The effect of clumping and stomatal response on evaporation from sparsely vegetated shrublands. Agric For Meteorol 84:187-205.

Bristow KL, Campbell GS, Papendick RI and Elliot LF (1986) Simulation of heat and moisture transfer through a surface residue-soil system. Agric For Meteorol 36:193-214.

Bristow KL and Horton R (1996) Modeling the impact of partial surface mulch on soil heat and water flow. Theor Appl Clim 56(1-2):85-98.

Caprio J, Grunwald G and Snyder R (1985) Effect of standing stubble on soil water loss by evaporation. Agric For Meteorol 34:129-144.

Choudhury BJ and Monteith JL (1988) A four layer model for the heat budget of homogeneous land surfaces. Quarterly J Royal Meteorol Soc 114:373-398.

Coffey ME, Workman SR, Taraba JL and Fogle AW (2004) Statistical procedures for evaluating daily and monthly hydrologic model predictions. Trans ASAE 47:59-68.

Enz J, Brun L and Larsen J (1988) Evaporation and energy balance for bare soil and stubble covered soil. Agric For Meteorol 43:59-70.

Farahani HJ and Ahuja L R (1996) Evapotranspiration modeling of partial canopy/residue covered fields. Trans ASAE 39:2051-2064.

Farahani HJ and Bausch W (1995) Performance of evapotranspiration models for maize - bare soil to closed canopy. Trans ASAE 38:1049-1059.

Flores H (2007) Penman-Monteith formulation for direct estimation of maize evapotranspiration in well watered conditions with full canopy. PhD Dissertation, University of Nebraska-Lincoln, Lincoln, Nebraska.

Gregory JM (1982) Soil cover prediction with various amounts and types of crop residue. Trans ASAE 25:1333-1337.

Horton R, Bristow KL, Kluitenberg GJ and Sauer TJ (1996) Crop residue effects on surface radiation and energy balance-review.Theor Appl Clim 54:27-37.

INE, 2007. Instituto Nacional de Estadisticas.Censo agropecuario. www.censoagropecuario.cl.

Irmak, S., Mutiibwa D., Irmak A., Arkebauer T., Weiss A., Martin D., and Eisenhauer D., (2008). On the scaling up leaf stomatal resistance to canopy resistance using photosynthetic photon flux density. Agricultural and Forest Meteorology, 148 1034-1044.

Iritz Z, Tourula T, Lindroth A and Heikinheimo M (2001) Simulation of willow short-rotation forest evaporation using a modified Shuttleworth-Wallace approach. Hydrolog Processes 15:97-113.

Jalota SK and Prihar SS (1998) Reducing soil water evaporation with tillage and straw mulching. Iowa State University, Ames, Iowa.

Jensen ME, Burman RD and Allen RG (1990) Evapotranspiration and irrigation water requirements. ASCE Manuals and Reports on Engineering Practice No. 70 332 pp.

Jensen, J.M., Wright, J.L., and Pratt , B.J., 1971. Estimating soil moisture depletion from climate, crop and soil data. Transactions of the ASAE, 14 (6): 954-959.

Katerji, N. and Rana, G. (2006). Modeling evapotranspiration of six irrigated crops under Mediterranean climate conditions. Agricultural and Forest Meteorology, 138 142-155.

Kjelgaard JF and CO Stockle (2001) Evaluating surface resistance for estimating corn and potato evapotranspiration with the Penman-Monteith model. Trans ASAE 44:797-805.

Krause P, Boyle DP and Base F (2005) Comparison of different efficiency criteria for hydrological model assessment. Adv Geosciences 5:89-97.

Lafleur P and Rouse W (1990) Application of an energy combination model for evaporation from sparse canopies. Agric For Meteorol 49:135-153.

Lagos LO (2008) A modified surface energy balance to model evapotranspiration and surface canopy resistance. PhD Dissertation University of Nebraska-Lincoln Lincoln, Nebraska.

Lagos L.O. Martin D.L., Verma S. Suyker A. and Irmak S. (2009). Surface Energy Balance Model of Transpiration from Variable Canopy Cover and Evaporation from Residue Covered or Bare Soil Systems. Irrigation Science. (28)1:51-64.

Legates DR and McCabe GJ (1999) Evaluating the use of goodness of fit measures in hydrologic and hydroclimatic model validation. Water Res Res 35(1): 233-241.

Lindburg M (2002) A soil surface resistance equation for estimating soil water evaporation with a crop coefficient based model. M.Sc. Thesis University of Nebraska-Lincoln Lincoln, Nebraska.

Lund M.R. and Soegaard H. (2003). Modelling of evaporation in a sparse millet crop using a two-source model including sensible heat advection within the canopy. Journal of Hydrology, 280: 124-144.

Massman, W.J.(1992). A Surface energy balance method for partitioning evapotranspiration data into plant and soil components for a surface with partial canopy cover. Water Resources Research. 28:1723-1732.

Meyers TP and Hollinger SE (2004) An assessment of storage terms in the surface energy balance of maize and soybean. Agric For Meteorol 125:105-115.

Monteith JL (1965) Evaporation and the environment. Proc Symposium Soc Expl Biol 19:205-234.

Monteith J.L., and Unsworth M.H. (2008). Principles of environmental physics.Academic Press, Burlington, MA USA.

Moriasi DN, Arnold J G, Van Liew MW, Bingner RL, Harmel RD and Veith TL (2007) Model evaluation guidelines for systematic quantification of accuracy in watershed simulations. Trans ASAE 50:885-900.

Ortega-Farias S, Carrasco M and Olioso A (2007) Latent heat flux over Cabernet Sauvignon vineyard using the Shuttleworth and Wallace model. Irrig Sci 25:161-170.

Ortega-Farias S, Olioso A and Antonioletti R (2004) Evaluation of the Penman-Monteith model for estimating soybean evapotranspiration. Irrig Sci 23:1-9.

Parkes M., Jiang W., and Knowles R. (2005). Peak crop coefficient values for shaanxi north-west China. Agricultural Water Management, 73 149-168.

Penman H L (1948) Natural evaporation from open water, bare soil and grass. Proc Royal Soc London , Series A, 193:120-146.

Rana G, Katerji N, Mastrorilli M, El Moujabber M and Brisson N (1997) Validation of a model of actual evapotranspiration for water stressed soybeans. Agric For Meteorol 86:215-224.

Rana, G., Katerji, N., and De Lorenzi F. (2005). Measurement and modeling of evapotranspiration of irrigated citrus orchard under Mediterranean conditions. Agricultural and Forest Meteorology, 128 199-209.

Shaw RH and Pereira AR (1982) Aerodynamic roughness of a plant canopy: A numerical experiment. Agric Meteorol 26(1):51-65.

Shuttleworth WJ (2006) Towards one-step estimation of crop water requirements. Trans ASAE 49: 925-935.

Shuttleworth WJ and Gurney R (1990) The theoretical relationship between foliage temperature and canopy resistance in sparse crops. Quarterly J Royal Meteorol Soc 116:497-519.

Shuttleworth WJ and Wallace JS (1985) Evaporation from sparse crops-an energy combination theory. Quarterly J Royal Meteorol Soc 111:839-855.

Stannard DI (1993) Comparison of Penman-Monteith, Shuttleworth-Wallace, and modified Priestley-Taylor evapotranspiration models for wildland vegetation in semiarid rangeland. Water Res Res 29(5):1379-1392.

Steiner J (1989) Tillage and surface residue effects on evaporation from soils. Soil Sci Soc Am J 53:911-916.

Steiner J (1994) Crop residue effects on water conservation. Managing agricultural residues, Unger P ed, Lewis Publisher, Boca Raton, Florida, 41-76.

Steiner J, Schomberg H, Unger P and Cresap J (2000) Biomass and residue cover relationships of fresh and decomposing small grain residue. Soil Sci Soc Am J 64:2109-2114.

Suyker A and Verma S (2009) Evapotranspiration of irrigated and rainfed maize-soybean cropping systems. Agric For Meteorol 149:443-452.

Tanner B and Shen Y (1990) Water vapor transport through a flail-chopped corn residue. Soil Sci Soc Am J 54(4):945-951.

Thom AS (1972) Momentum, mass and heat exchange of vegetation. Quarterly J Royal Meteorol Soc 98:124-134.

Todd RW, Klocke NL, Hergert GW and Parkhurst AM (1991) Evaporation from soil influenced by crop shading, crop residue, and wetting regime. Trans ASAE 34:461-466.

Tourula T and Heikinheimo M (1998) Modeling evapotranspiration from a barley field over the growing season.Agric For Meteorol 91:237-250.

Verma S (1989) Aerodynamic resistances to transfer of heat, mass and momentum. Proc (Estimation of Areal Evapotranspiration) Vancouver BC Canada IAHS Publ #177:13-20.

Verma SB, Dobermann A, Cassman KG, Walters DT, Knops JM, Arkebauer TJ, Suyker AE, Burba GG, Amos B, Yang H, Ginting D, Hubbard KG, Gitelson AA, and Walter-Shea EA (2005) Annual carbon dioxide exchange in irrigated and rainfed maize-based agroecosystems. Agric For Meteorol 131:77-96.

Wallace J.S., Jackson N.A. and C.K. Ong. (1999). Modelling soil evaporation in an agroforestry system in Kenya. Agricultural and Forest Meteorology 94: 189-202.

Weseley ML and Hicks BB (1977) Some factors that affect the deposition rates of sulfur dioxide and similar gases on vegetation. J Air Pollution Control Assoc 27(11):1110-1116.

Yu Q, Zhnag Y, Liu Y and Shi P (2004) Simulation of the stomatal conductance of winter wheat in response to light, temperature and CO2 changes. Annals of Bot 93:435-441.

Evapotranspiration – A Driving Force in Landscape Sustainability

Martina Eiseltová[1,2], Jan Pokorný[3], Petra Hesslerová[3,4] and Wilhelm Ripl[5]
[1]Crop Research Institute
[2]Environment and Wetland Centre
[3]Enki, o.p.s.
[4]Faculty of Environmental Sciences,
Czech University of Life Sciences, Prague
[5]Aquaterra System Institute
[1,2,3,4]Czech Republic
[5]Germany

1. Introduction

It is clear from the ever-growing evidence that human interference with vegetation cover and water flows have considerably impacted water circulation in the landscape and resulted in major changes in temperature distribution. Human changes in land use – extensive river channelization, forest clearance and land drainage – have greatly altered patterns of evapotranspiration over the landscape. To comprehend how the changes in evapotranspiration impact landscape sustainability it is necessary to take a holistic view of landscape functioning and gain understanding of the underlying natural processes.

The Earth's surface has been shaped by water - in interaction with geological processes - for billions of years. Water and the water cycle - along with living organisms - have been instrumental in the development of the Earth's atmosphere; free oxygen in the atmosphere is the result of the activity of autotrophic, photosynthetic organisms (stromatolites) that evolved in seawater some 3.5 billions years ago. This was the beginning of aerobic metabolism and enabled the evolution of higher organisms, including higher plants.

The emergence of terrestrial plants some 400 million years ago has played a major role in the amelioration of the climate. The process of evapotranspiration – evaporation from surfaces and transpiration by plants - is instrumental in temperature and water distribution in time and space. Whilst evaporation is a passive process driven solely by solar energy input, transpiration involves an active movement of water through the body of plants - transferring water from the soil to the atmosphere. The process of transpiration is also driven by solar energy but plants have the ability to control the rate of transpiration through their stomata and have developed many adaptations to conserve water when water is scarce.

Water vapour is the main greenhouse gas playing a protective role against heat loss from the Earth's surface; on average the earth is about 33°C warmer than it otherwise would be without water vapour and the other greenhouse gases in the atmosphere (water vapour's contribution being about 60 % on average, Schlesinger 1997). Water, thanks to its high heat-carrying capacity, is able to redistribute much of the solar heat energy received by the Earth through the water cycle: by evapotranspiration and condensation. Water evapotranspiration and condensation therefore plays an instrumental role in climate control with regard to temperature distribution in time and space, i.e. reducing the peaks and modulating the amplitudes of high and low temperatures on the land surface - making conditions on Earth suitable for life.

The natural vegetation cover that has developed over the Earth throughout millennia is best suited to utilize and dissipate the incoming solar energy, and to use the available water and matter in the most energy-efficient way. There is ample evidence for this. Since the time that human civilization begun greatly interfering with the landscape's natural vegetation cover - clearing forests, ploughing savannas and draining wetlands for agricultural use and urban settlements - many environmental problems have started to appear. More recently environmentally sustainable management systems have been sought - with various degrees of effort and understanding of the underlying problems.

In this chapter we will provide evidence of the role of water and vegetation in shaping the climate. Using data and observations from a virgin forest in Austria we will present and discuss the play rules of nature and offer a definition of landscape sustainability. We will present a living example of reduced precipitation over an area of 4000 square kilometres following the partial clearance of the Mau Forest in western Kenya and describe the situation in the de-watered landscape of the open-cast mining area of North-West Bohemia, Czech Republic. The connection between the disturbed water cycle and matter losses in the predominantly-agricultural Stör River catchment in Germany will be demonstrated and the role of evapotranspiration in maintaining landscape sustainability discussed.

2. The play rules of nature in search of sustainability

2.1 The energy-dissipative properties of water

Life on Earth depends on energy, water and a few basic elements (mainly C, H, O, N, P, S and about 20 others) that constitute living tissue. The biogeochemical cycles - the continuous cycles of matter and water - are essential for life to be sustained. The cycles are primarily powered by the energy received from the Sun. Driven by the sun's radiation water is cycled continuously: playing an instrumental role in energy dissipation and the cycling of matter. The dissipation of solar energy at the Earth's surface – i.e. the distribution of energy in time and space - creates suitable thermal conditions for natural processes and life on Earth.

To understand how the natural processes involved in energy dissipation are inter-related Ripl (1992, 1995) proposed a conceptual model based on the energy dissipative properties of water. In his Energy-Transport-Reaction Model (ETR Model), Ripl considered three essential processes (Fig. 1) that control the dissipation of energy:

- the process of water evaporation and condensation;
- the process of dissolution and precipitation of salts; and
- the process of disintegration and recombination of the water molecule within the biological cell

With water's high capacity for carrying energy in the form of latent heat, most energy is dissipated by the physical processor property of evaporation and condensation, making water a very efficient cooler or heater. When water changes from a liquid to its gaseous phase - as in evapotranspiration - energy is stored in the water vapour in the form of latent

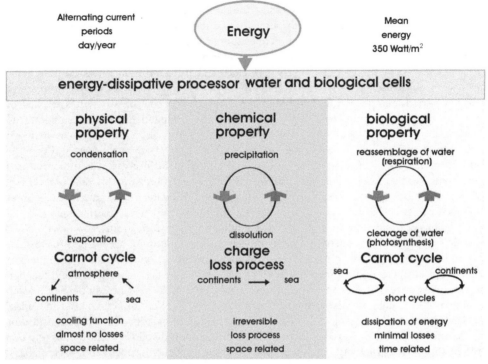

Fig. 1. Three processor properties of water

heat and the local area is cooled down. At night or early morning when water condenses on cooler surfaces, energy in the form of latent heat is released and the local area is warmed up. Without water, the energy of the incoming radiation is transformed into sensible heat and the local area becomes overheated during the day and likewise far cooler at night (as is well known from desert areas, with differences between day and night temperatures typically exceeding 50°C). Water-saturated landscapes provide much more stable environments than do dry terrestrial systems. In landscapes with water - abundant aquatic ecosystems, wetlands and soils with high water retention capacity - about 80 % of incoming solar energy is stored as latent heat of water vapour via evapotranspiration, whilst in de-watered landscapes (with a low-water retention capacity) the vast majority of solar energy is transformed into sensible heat (Pokorný et al. 2010b). In exceptional cases when, for example, hot air of low relative humidity moves across a wetland surrounded by dry areas, even more than equivalent of 100% of solar radiation can be stored safely in latent heat (Monteith 1975, Ryszkowski & Kedziora 1987, Kučerová et al. 2001). Below in Sections 3 and 4 we will show the high temperature differences measured between de-watered areas and sites with a good supply of water and high evapotranspiration.

Water has another important natural property - the ability to separate the charges in a given amount of molecules into protons and electrons. This chemical processor property of water is responsible for the dissolution of salts - using up the water's heat energy in the formation of ionic solutions – and then if concentrated by subsequent evaporation of the water crystals can be precipitated from the solute, releasing the same amount of energy as was required by the dissolution process. However, through dissolution and precipitation a much smaller fraction of energy is dissipated compared to evaporation and condensation.

In pure distilled water at 20°C, 10^{-7} moles of water are dissociated into protons (H$^+$) and electron-charged hydroxyl ions (OH$^-$). These electric charges represent chemical potentials, i.e. energy with the potential to be converted into chemical reactions. The number of charged parts (ions) per volume of water constitutes the concept of reactivity (pH, law of mass action). Importantly, reactivity is to a large part dependent on the temperature-, concentration- and pH gradients existing at various interfaces. Such interfaces between solid, liquid and gaseous phases are of special interest in all energy processes and provide sites for steady rates of change. Being essential tools for life processes, nature produces membranes and surfaces where life's important reactions can most readily take place. Even without there being differences in temperature at a liquid- (water-) solid interface, chemical reactions can still readily take place due to the singularity of charge distributions and the modulations of thermal motion (the thermal 'jiggling' of molecules / ions).

Kinetic energy ($mv^2/2$) consists of the frequency and amplitude of accelerated masses. At the interfaces between two phases (e.g. liquid-solid) a modulation of the mass movement of ions (molecules) can occur, especially in amplitude; reactivity is thus enhanced and reaction probabilities increased (in conditions of decreased pH and elevated proton density). As an example of this, take the distribution of highly-diluted, colloidal organic matter in a glass beaker of water. The organic colloids are coagulated at the glass wall, attracted and thus concentrated by the lowered pH conditions at the liquid-solid interface; this enables potential bacterial activity such as, for example, quicker growth of bacteria and decomposition of organic matter. Such phenomena are ubiquitous in nature: always occurring, for example, between the root membranes of plants and the interstitial water of the soil. Evapotranspiration by the leaves of plants lowers the water content in the capillary network of the soil interstitium, giving access to the oxygen of the air and thus exerting a positive feedback on root activity. If the 'water pump' of a productive growing plant should for some reason stop, then electron density (i.e. low redox conditions) will rise and decomposition processes will be severely retarded. Thus the activity of evapotranspiration – the switching on or off of the plant's water pump – controls soil bacterial activity and mineralization processes. In this way highly-efficient processes – control mechanisms closely connecting functioning plant systems and soil - are able to maintain loss-free conditions in the soil. Minerals and nutrients become 'available' only when the plant is actively growing and thus are readily 'used up'. The losses induced by the percolation of 'free' nutrients and minerals released by mineralization through to rivers via sub-surface groundwater flow are thus minimized. Such a mechanism is steadily optimizing the sustainable development of vegetation cover over the landscape by minimizing the irreversible losses from land sites to the sea (Ripl 2010).

Water is also the most important agent in the biological processes of production (photosynthesis) and decomposition (respiration) of organic matter. During photosynthesis water is split into reactive 2H and O. Oxygen is released to the atmosphere and hydrogen is used for the reduction of carbon dioxide to carbohydrates - organic compounds including sugars, starch and cellulose. The solar energy bound in organic matter is released again during mineralization (decomposition) when oxygen is used up to split sugars back into

CO_2 and H_2O. As the production and breakdown of organic matter generally occur within the same site, the biological process can be considered cyclic just like the physical dissipative process. However, considerably less incoming solar energy (about 1 - 2%) is bound by photosynthesis compared to that of water evaporation; the net efficiency of solar energy conversion into plant biomass is usually between 0.5 and a few percent of the incident radiation (for more details see, for example, Blankenship 2002).

The theories and ideas associated with dissipative structures, open dynamic systems operating far from equilibrium, and self-organization (Prigogine & Glansdorff 1971; Prigogine 1980; Prigogine & Stengers 1984) have given us a clearer understanding of how living organisms utilize a throughput of external energy to create new order and structures of increased complexity (Capra 1996). These theories cast light on how ecosystems have organized themselves during evolution: maximizing their sustainability through cycling water and matter and dissipating energy. The dissipation of energy takes place at various scales - from the micro-scale within cells to ecosystems and landscapes (Schneider & Sagan 2005). At the landscape level, evapotranspiration plays an essential role in energy dissipation and as such is highly dependent on the vegetation cover and water availability.

2.2 Plants and water availability

Water is supplied to the land and its vegetation through precipitation. The various sources of water contributing to precipitation differ in different regions of the Earth. In maritime regions, water derived from evaporation from the sea prevails whilst further inland precipitation may be derived equally from long-distance atmospheric transport of water from the sea and from evapotranspiration from within the basin itself (Schlesinger 1997). Availability of water is one of the most important factors determining the growth of plants: hence the distribution of plants on Earth coincides with the availability of water. Deserts are typically short of water and thus the vegetation is rather scarce or non-existent. Nevertheless, plants have developed a number of different strategies during evolution to cope with both conditions of water abundance on the one hand and water scarcity on the other. For the purpose of this chapter we will focus on mechanisms that plants use to control the local water cycle and why it is important.

There are several mechanisms that plants use to control the loss of water from their tissues. One of these is the operation of stomata, their intricate structure, position on plants, their size and numbers. Stomata are found in the leaf and stem epidermis of plants; they facilitate gas exchange and the passage of water from the leaf or stem tissues to the surrounding air by controlling the rate of transpiration. Stomata consist of a pair of guard cells, the opening between them providing the connection between the external air and the system of intercellular spaces. Plants adapted to dry conditions mostly have small stomata immersed within the epidermis. Numbers of stomata differ from about 50 to 1000 stomata per mm^2. Stomata respond to the amount of water in the leaf tissue and to air humidity: closing when the water content in leaf tissue is low and when ambient air humidity declines. In such cases only a small amount of water is transpired through the cuticle (a wax layer on the epidermis). In plants with a thin cuticle – most wetland plants (hygrophytes) belong to this category – the cuticle transpiration may amount to a considerable percentage of total transpiration. However, cuticle transpiration usually amounts to only a few percent of the water released by stomatal transpiration. The effectiveness of the cuticle in reducing loss of water is well seen in fruits, such as apples and pears, or potato tubers: if unpeeled they can stay many weeks without any great water loss (Harder et al. 1965).

Transpiration by plants can be seen as a water loss in such cases as water scarcity; managers of water reservoirs that supply drinking water would usually see it as a loss. For a plant, however, transpiration is a necessity by which a plant maintains its inner environment within the limit of optimal temperatures. And at the level of landscape, evapotranspiration is the most efficient air conditioning system developed by nature.

In addition to optimising temperature, through evapotranspiration plants control the optimum water balance in their root zone. The activity of plant roots in respect to water uptake regulates the redox conditions in the root zone, thus regulating the rate of organic matter decomposition that makes nutrients available for plants growth. It is therefore most likely that, through evapotranspiration, the vegetation cover controls the irreversible losses of matter: an efficient system where only so much organic matter is decomposed such that those mineral nutrients freed from organic bonds are rapidly taken up by plants for their nutrition.

In dry environments, plants have developed ways to attract water condensation. As water condensation takes place on surfaces, plants growing under the conditions of water-scarcity typically have a high surface-volume ratio. Spines and hairs on plants have developed to increase the plants' surface-volume ratio - thus providing more surfaces for water condensation (Fig. 2). Given the complex role of vegetation in maintaining a water balance, smooth temperature gradients and a control of matter cycles in the landscape, any potential economic profits expected from the destruction of natural vegetation cover need to be carefully weighed against the loss of the functioning role of vegetation.

Fig. 2. Spines and hairs on cacti enhance water condensation in arid environments (Photo: M. Marečková)

2.3 Water dynamics and matter losses

It is generally accepted that water is the most important transport and reaction medium – many chemical reactions can only take place in the presence of water and matter is transported mainly with water flow. Matter that is transported via rivers to the sea – both in a dissolved or particulate form – has to be seen as an irreversible matter loss for continents and their vegetation as it takes millions of years before the sea floor is lifted up to form a new continent. Equally, matter that is leached through the soil to the permanent groundwater is further unavailable for nutrition of the vegetation cover on land. Ripl (1992) used data from palaeolimnological studies of lake sediments in southern Sweden (Digerfeldt 1972) to demonstrate the role of vegetation cover in matter and water flows. Vegetation cover reconstruction and sediment dating has made it possible to document four distinctive stages in landscape and vegetation development in postglacial North European catchments and the relevant matter losses at each stage. During the first stage, the bare soils or soils with scarce pioneer vegetation that occurred after the retrieval of glaciers were prone to elevated soil erosion and high transport of dissolved matter. This was measured as a relatively high rate of matter deposition in lake sediments; analysis showed that sediment deposition rates were highly correlated with the deposition rates of base minerals, nutrients and organic material. When climax vegetation became established within catchments, rates of sediment deposition diminished some ten fold. With a fully developed vegetation cover in catchments, low deposition rates of approximately 0.1 to 0.2 mm per year remained rather constant right through until the second half of the 19th century. Since then increasing rates of sewage discharge to lakes, clearance of forest and intensification of agriculture have led to deposition rates increasing nearly a hundred fold to present levels of 8 to 10 mm per year.

The reduction in matter losses from catchments covered by climax vegetation is ascribed to the increased system efficiency of water and matter recycling. In catchments with a well-developed vegetation cover, water and matter are bound to short-circuited cycles and losses are minimal. In contrast, the increased clearance of forest, exposure of bare land, and drainage of agricultural land have accelerated matter losses from catchments. The lowering of the water table by humans has increased the rate of mineralization of organic matter and also enhanced water percolation through soils that carries away the dissolved mineral ions and nutrients. The increased inputs of nutrients to water bodies were documented by the much higher deposition rates of sediments – the beginning of eutrophication (Digerfeldt 1972, Björk 1988, Björk et al. 1972, 2010).

Ripl et al. (1995) confirmed by a laboratory lysimeter experiment that the water dynamics in a soil substrate has a major impact on the rate of organic matter decomposition; under the conditions of intermittent wet and dry phases more organic matter was mineralized and higher amounts of mineral ions leached through the soil than from the control soil substrate with a continuous water flow. The significance of interchanging dry/wet phases and its decisive role in matter losses can be documented also by many examples of drained lowland fens in northern Europe, where increased matter losses have been observed following fen drainage. The mineralization of organic matter accumulated throughout centuries has been of such dimensions that soil subsidence, for example in the fenland of Cambridgeshire, England, has amounted to more than 4.5 metres following the drainage that took place there in the 1650s (Purseglove 1989). By contrast, permanently moist soils slowly accumulate organic matter and matter losses are minimal.

2.4 Specific features of energy fluxes in wetland ecosystems – primary production and decomposition of organic matter

Wetlands which are eutrophic, i.e., well supplied with plant mineral nutrients, are highly productive because they do not suffer from water shortages. Individual types of wetlands differ significantly - not only in their production of plant biomass but also in their capability of long-term accumulation of dead organic matter (as detritus or peat). This capability depends on the ratio between average rates of primary production and decomposition. For example, bogs are distinguished by their low annual primary production of biomass (usually only 100 to 250 g m^{-2} of dry mass). Nonetheless, the strongly suppressed decomposition of organic matter that is produced in bogs results in a net annual accumulation of dead plant biomass that is eventually transformed into peat. As the peat layer grows upwards, the bog vegetation loses contact with the groundwater rich in minerals and its biomass production slows down. In contrast, though eutrophic fishponds have a typical primary production one order of magnitude higher than in bogs, they often hardly accumulate any dead biomass as the annual decomposition approaches or equals annual net primary production. In fishponds, however, like in other wetlands, the production to decomposition ratio depends on the supply of nutrients (especially P and N), i.e., on the trophic status of the water (Pokorný et al. 2010b). Thus any lake or fishpond, if oversupplied with nutrients, can accumulate a nutrient-rich organic sediment if the decomposition rate cannot keep pace with the extremely high primary production. Eventually, the fishpond becomes a source of nutrients; when oxygen gets depleted and anaerobic conditions at the sediment-water interface occur, phosphorus is released from the sediment enhancing the primary production even further.

2.5 Landscape sustainability
2.5.1 The dissipative-ecological-unit

The Earth's atmosphere has been described by Lovelock (1990) as an open system, far from equilibrium, characterized by a constant flow of energy and matter. Equally, living organisms are open systems with respect to continual flows of energy and matter. However, at a higher organisational level – such as an ecosystem – matter is continually recycled, i.e., what is a waste for one organism becomes a resource for another. Ripl & Hildmann (2000) termed the smallest functional unit that is capable of forming internalized cycles of matter and water while dissipating energy - the *dissipative-ecological unit* (DEU). The steadily increasing resource stability of DEUs is achieved by their reduction of water percolation through soils to the groundwater and instead their increase in local, short-circuited water cycling within ecosystems by enhancing their evapotranspiration.

The concept of the dissipative-ecological-unit is used to demonstrate how nature, when not disturbed by sudden changes in climatic conditions, tends to close cycles of matter, i.e. run an efficient local resource economy and maintain relatively even temperatures and moisture conditions.

2.5.2 Evapotranspiration and landscape sustainability

Results from a detailed study conducted in a predominantly agricultural catchment of the River Stör in NW Germany demonstrated how the destruction of natural vegetation cover over large areas has led to the opening up of cycles due to the disturbance of natural water flow dynamics (Ripl et al. 1995, Ripl & Eiseltová 2010). Water and matter no longer cycle within localized, short-circuited cycles; instead, reduced evapotranspiration has resulted in increased water percolation through the soil accompanied by increased losses of matter. The average losses of dissolved mineral ions measured within the Stör River catchment were alarmingly high, about 1,050 kg of mineral salts per ha and year (excluding NaCl). A detailed description of the measurements performed and methods used can be found in Ripl and Hildmann (2000). Such land management systems are unsustainable in the long-term as soil fertility will inevitably be gradually reduced.

A rather different situation can be observed in an undisturbed ecosystem, such as the rather unique virgin forest of Rothwald in Austria. Here the feedback control mechanism of this complex mature forest ecosystem is functioning according to the rules of nature. It is the interlinked vegetation cover that is in control of the processes. In this dolomitic bedrock area groundwater is very scarce - being present only in minor crevices. Oscillations of the water table within the thick debris layer are mainly controlled by the plants through their evapotranspiration. Despite the relatively high precipitation – over 1,000 mm a year – the run off from the virgin forest remains very low and is restricted mainly to the period of snow melt above frozen ground (February till May). The site does not suffer from shortage of water as can be deduced from the highly damped temperature distribution; the temperature amplitudes between day and night almost never exceed 8-9°C during summer (Ripl et al. 2004). The organic matter decomposition is rather slow due to the water-saturated conditions and the debris layer is rather high. The debris layer was 2-4 times higher in the Rothwald virgin forest that in the large areas of neighbouring managed forest (Splechtna, pers. comm., 2000). Water analyses of melted snow samples showed extremely low conductivity values (Table 1). This indicates that there is a much quicker turnover of water evaporated from the virgin forest in relation to precipitation brought from long distances away, as such precipitation water would have about 10 times higher conductivity. It is estimated that very short water cycles with a frequency of one day or less must be prevalent.

	Conductivity at 20° C mS m^{-1}	Alkalinity mmol l^{-1}	pH
Max	1.45	0.09	7.22
Min	0.26	0.00	4.73
Median	0.60	0.01	6.27
MW	0.72	0.03	6.49
no. of sites	17	16	16

Table 1. Conductivity, alkalinity and pH measured in melted snow from Rothwald virgin forest.

Based on the findings described above we can define landscape sustainability as the efficiency of the landscape to recycle water and matter, and to dissipate the incoming solar energy. We have provided evidence that matter losses increase with increased water percolation through soil – as a result of reduced evapotranspiration due to natural vegetation clearance. In the following sections we provide data from a thermal camera and satellite images. These data give supporting evidence that evapotranspiration plays a major role in the dissipation of the incoming solar energy and dampening temperature amplitudes.

3. Evapotranspiration as seen by thermal camera

Pictures of the landscape using a thermal camera show distinct differences in the temperatures of forest, grassland, bare soil and buildings. Even over relatively small areas of a few square metres, temperature differences can be over 20° C. Dry surfaces, such as concrete, when exposed to sunshine are the warmest, despite their higher albedo (higher reflection of solar radiation). This demonstrates that the surface temperatures in the landscape are controlled mainly by the process of water evapotranspiration while the albedo plays a less important role.

On a sunny day, dry surfaces such as the road show the highest temperature (up to 45° C), whilst meadows and forests have lower temperatures as they are cooled by evapotranspiration (Fig. 3). The cooling efficiency depends on water availability and vegetation type. The maize field (Fig. 4) shows a higher temperature over the bare soil (up to 47 °C) than on the top of the stand (32° C). Air heated by a warm soil ascends upwards and takes away water vapour. In hot air crops lose a high amount of water in the form of water vapour.

Fig. 3. Surface temperature of a drained meadow, road and forest as seen by thermovision camera on 17 July 2009 at 9.40 GMT+1 near the town of Třeboň, Czech Republic.

Fig. 4. Maize field and its surface temperature as seen by thermovision camera on 16 July 2010 at 14.19 GMT+1 in the vicinity of the town of Třeboň, Czech Republic.

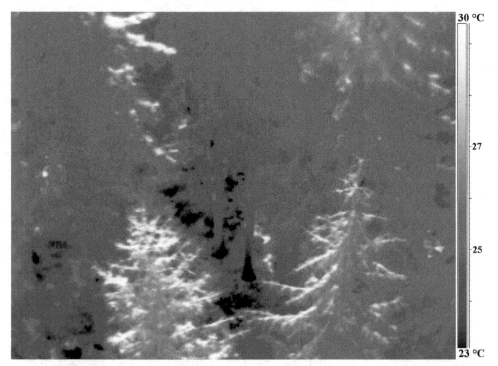

Fig. 5. Surface temperature of forest canopy as seen by thermovision camera on 13 July 2010 at 14.15 GMT+1 in Novohradské hory, Czech Republic.

The vertical distribution of temperature in a forest canopy is opposite to that observed in a maize field. During a sunny day in a forest, a temperature inversion – a lower temperature at ground level in the shrub layer (23 - 26° C) than on tree crowns in the forest canopy (29.5° C) - has been observed (Fig. 5). A heavier cold air stays at the ground and hence the water vapour may condense on herb and shrub vegetation even during a sunny day. When temperatures go down at night the air becomes more saturated and condensation occurs above the tree canopy. Makarieva and Gorshkov (2010) have shown that intensive condensation is associated with the high evaporation from natural forest cover that is able to maintain regions of low atmospheric pressure on land – i.e. forests constitute acceptor regions for water condensation and precipitation.

4. Use of satellite images to assess cooling efficiency of vegetation cover

Satellite remote sensing data from the Landsat thermal infrared channel provide a suitable tool to evaluate the spatial and temporal distribution of land surface temperatures. This can be used to assess and compare the cooling efficiency of different vegetation cover types or land use. Two model sites (in central Europe and eastern Africa) were selected to demonstrate the role that a functioning vegetation cover plays in energy dissipation compared to the situation of bare or sparsely vegetated land characterized by highly reduced evapotranspiration and shortage of available water.

4.1 North-west Bohemia
4.1.1 Site description and methods

Landsat multispectral satellite data were used to analyze the effects of different land cover types on surface temperature. Two scenes - from 1 July 1995 and 10 August 2004 - were used to compare, firstly, the long-time change in vegetation cover and its effect on temperature distribution, and, secondly, the effect of seasonality of farming land. The selected model area of North-West Bohemia, Czech Republic and Saxony, Germany covers 8,722 km² (102 x 85 km). The site was selected to include different landscape types with a heterogeneous mix of land use – highly intensive agriculture, industrial areas and an open-cast brown coal mining area, small-scale farming lands, and broad-leaf and coniferous forests.

Supervised classification methods (Mather & Tso, 2009) were used to classify the land cover; five categories were defined – bare grounds, water, forest, non-forest vegetation and clouds. Surface radiation temperatures were calculated from the standard mono-window algorithm (Sobrino et al. 2004), using a conversion of thermal radiance values from the Landsat thermal infrared channel. As the satellite images selected differ in year and season, the temperature data were standardized by the normalization method of z-scores using the following equation:

$$Z = \frac{x_i - \overline{x}}{\sigma}$$

where x_i is the temperature value of a pixel, $\overline{x}$ is the mean average temperature, and σ is standard deviation.

Date	Min. temperature °C	Max. temperature °C	Mean average temperature °C	Standard deviation
1 July 1995	9.6	46	23.1	2.9
10 August 2004	12.3	46	24.1	4.8

Table 2. Temperature values calculated from Landsat thermal channel.

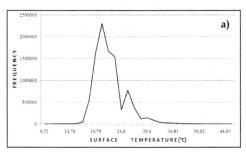
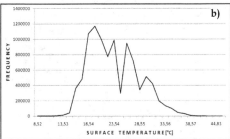

Fig. 6. Histograms showing frequencies of temperature distribution in satellite images of a) 1995 and b) 2004.

The result of normalization is a temperature image with a relative scale showing a range from minimum to maximum temperature. The real temperature values are given (Table 2) and their frequency of distribution displayed by histograms (Fig. 6).

4.1.2 Satellite data interpretation
The relationship between different land-cover types and their relative surface temperature
distributions is shown in Figures 7 and 8. Surface temperature is an indicator of the system's

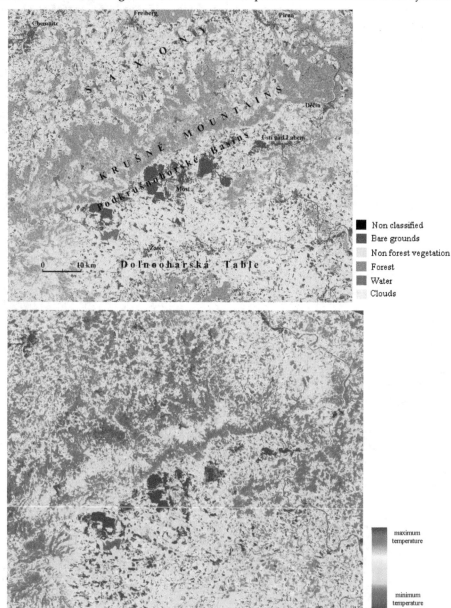

Fig. 7. Land cover (upper image) and temperature distribution (bottom image) over the
model area of North-West Bohemia and Saxony obtained on 1 July 1995 at 9.40 GMT+1. The
surface temperature data were obtained from Landsat thermal channel TM 6.

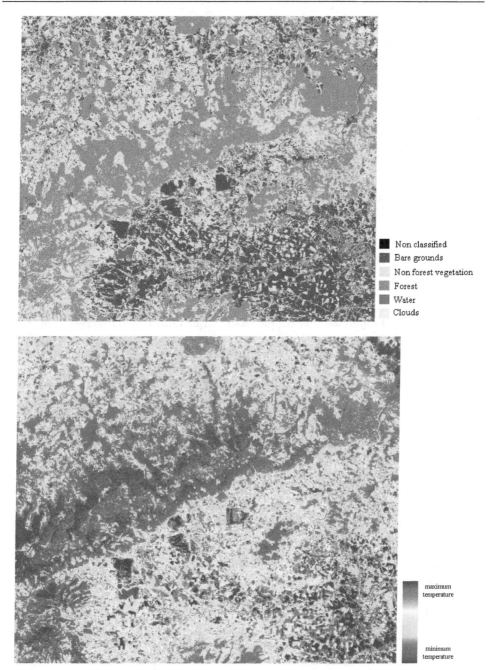

Fig. 8. Land cover (upper image) and temperature distribution (bottom image) over the model area of North-West Bohemia and Saxony obtained on 10 August 2004 at 9.40 GMT+1. The surface temperature data were obtained from Landsat thermal channel TM 6.

ability to convert (dissipate) the incoming solar energy; loss of vegetation is accompanied by changes in the distribution of solar radiation, resulting in a temperature rise. The lowest temperature range (20 - 22 °C) in both satellite scenes was obtained for a deciduous forest, followed by a coniferous forest (17 - 23°C) on the scene from the year 2004. In the satellite scene from 1995, the coniferous (spruce) forest across the top of the Krusne Hory mountains shows remarkably higher temperatures. This is explained by its lower cooling capacity, i.e. lower evapotranspiration rates, as, at that time, the forest was dying off due to the extreme depositions of sulphur dioxide emitted mainly in the 1980s. By 2004, the forest had mostly recovered and this can be seen through the recovery of the forest's cooling function and enhanced temperature damping. The highest temperatures (ranging between 31 - 45 °C) were found at the sites of open-cast brown coal mines and spoil heap tailings, but also on areas of arable fields after crop harvest (August 2004). Furthermore, this category of the highest temperatures, classified as bare grounds, also included urban areas, industrial and commercial zones, communication infrastructures (roads), as well as relatively natural surfaces such as rock outcrops, and peat bogs affected by drainage and/or peat mining. The temperature range of 23 - 32°C characterized the non-forest vegetation - a rather heterogeneous category of land cover that included meadows, areas with sparse vegetation, peatbogs and arable land in 1995 when this was still covered by green crops.

The temperature distribution over the prevailing different land uses depends on the water availability and rate of evapotranspiration. It illustrates the seasonal variability in the damping of temperatures over arable land: early in the season (satellite scene 1 July 1995; Fig. 7) when the crops are still green, the arable land belongs to the lower temperature class compared to the August 2004 scene (Fig. 8) when the crops had already been harvested and the sites fell into the highest temperature class, i.e. the sites were greatly overheated. Furthermore, we can observe the negative impact of the large arable fields of former state farms in the Czech Republic compared to smaller field sizes in Saxony (Germany) which show better temperature damping. The lack of functioning vegetation can be also seen in the shape of the temperature histograms (Fig. 6) - the histogram of 2004 is wider reflecting a higher spread of temperatures whilst the histogram of 1995 (with fields still mainly covered by crops) is narrower and shifted to lower values, reflecting the better cooling by vegetation. Sites with bare ground undoubtedly belong to the warmest places in the landscape; due to the lack of water evapotranspiration, more solar energy is transformed into sensible heat (raising the site's temperature) than into latent heat of water vapour. The higher albedo of bare ground (concrete, etc.) and the lower albedo of forests does not play such an important role when compared to the cooling effect of evapotranspiration (Pokorný et al. 2010a).

4.2 Mau Forest in Western Kenya

The Mau Forest complex, located about 150 km northwest of Nairobi, at an altitude between 1200 - 2600 m, is referred to as one of the largest remaining continuous blocks of indigenous forest in eastern Africa. With a high annual precipitation (reaching about 1000 mm on eastern slopes and more than 2000 mm on western ones) it is an area which includes the headwaters of many rivers feeding into the Rift Valley lakes (Lake Natron, Turkana, Victoria, Nakuru, Naivasha, Elmenteita and Baringo). In the last 25 years the site has been subject to extensive deforestation: forest cover of 5200 km² in 1986 was reduced to a mere 3400 km² in 2009 (for details, see Fig. 9). The availability of satellite images since the 1980s has enabled us to demonstrate the effect the forest clearance has had on temperature distribution over the whole area. Extreme rises in temperature (by more than 20° C; see Fig. 10) can be observed on sites of deforestation. Its consequences are also evident in the Rift Valley region, between lakes Nakuru and Naivasha. Areas that have been converted into fast-growing plantation forest show the opposite trend, i.e., temperature damping.

Fig. 9. Changes in the extent of the Mau Forest, East Africa, between the years 1986 and 2009.

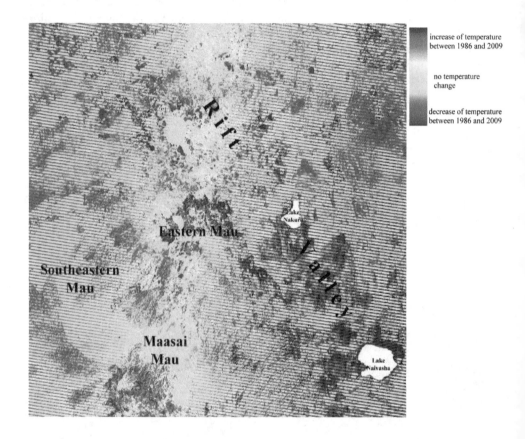

Fig. 10. Changes of temperature between the years 1986 and 2009 in the Mau Forest complex, East Africa, obtained as a difference of the standardized temperatures. The surface temperature data were obtained from Landsat thermal channel TM 6.

5. Discussion and conclusions

Deforestation and land drainage for agriculture or urbanisation has led to accelerated water discharge from catchments. From the self-regulating dissipative structures described earlier, loss of vegetation along with the water shortages has caused a shift to highly negative circumstances, with such consequences as temperature swings of increased amplitude and frequency leading to turbulent motion in warm dry air. The loss of functioning vegetation has been extensive for some time, as observed and reported in the Millennium Environmental Assessment (2005): every year some 60,000 km^2 of badly managed land is becoming a desert and about 200,000 km^2 of land loses agricultural productivity. The lack of water and ecosystem functionality now affects 30–40% of our global landmass.

Life on land is only possible when soil contains enough moisture for green plants to grow. The continuous runoff of river water discharge from land to ocean needs to be compensated by the opposite transport of water vapour from ocean to continent. Makarieva and Gorshkov (2010) have shown the role of the forest cover in a condensation-induced water cycle that maintains a flow of moist air from ocean to continent. They evaluated precipitation measured along transects from ocean to land on different continents; they revealed that along transects with continuous forest cover precipitation reaches as far as 1000 km inland (in hardly diminished amounts), whilst above deforested landscape precipitation rapidly diminishes when distance from the ocean exceeds 600 km. They identified two principles as to how the forest attracts and retains water. Firstly, the forest vertical architecture induces a temperature inversion: during the day, temperatures in the forest understorey are lower than that of the forest crown; in this way losses of water vapour to the atmosphere are reduced. Secondly, at night, water vapour condenses above the forest canopy causing a decrease in air pressure above the canopy; this 'sucks in' air horizontally and can bring moist air from the ocean enhancing further water condensation above the forest. Induced by living organisms, this atmospheric circulation maintaining the hydrological cycle on land has been termed the 'biotic pump' of atmospheric moisture (Makarieva & Gorshkov 2007).

In addition to the long water cycle (ocean to land), the role of the short water cycle has been emphasized by Ripl (1995, 2010) as playing an important role in climate amelioration and landscape sustainability. Figures 7, 8 and 10 clearly demonstrate the role of a healthy forest in modulating surface temperatures contributing to climate amelioration. In contrast, the bare or scarcely-vegetated land with low water supplies is often overheated as a result of the reduced evapotranspiration. Furthermore, matter cycles are dependent on water circulation; water being the most important transport and reaction medium. Not only does vegetation largely control water runoff and precipitation over land - with the help of evapotranspiration – but it also controls matter flows. By controlling soil moisture, vegetation governs the process of organic matter decomposition and hence the circulation of matter. Over millions of years of evolution biological communities have optimized their self-organization to run a highly efficient resource economy and hence maintain their sustainability.

We have provided evidence that the sustainability of river catchments has been seriously impaired by large-scale deforestation and drainage. The accelerated discharge of water via rivers to the sea - caused by extensive deforestation, land drainage and hence the landscape's reduced capacity for water retention – brings about overly high matter losses. Data provided for Germany have shown that areal matter losses have reached between 1 and 1.5 tons of dissolved matter per ha per year on average (Ripl & Eiseltová 2009). Areas of high matter losses correlate with areas of reduced evapotranspiration, and hence landscape overheating, as shown in the detailed study of the Stör River catchment in Germany (Ripl & Hildmann 2000) and three small sub-montane catchments in the Czech Republic (Procházka et al. 2001). The deforestation of large areas in tropical regions has resulted in a temperature increase of about 20°C (Hesslerová & Pokorný, 2010).

In arid zones, where irrigation is often a necessity for crop production, the situation is not better. Farmers may try to minimize the use of water for irrigation as water is scarce and costly; they will kill weeds by herbicides to reduce unwanted water losses through evapotranspiration. The ground thus remains void of an understorey or ground layer that would protect the soil from overheating, and rising hot air takes water vapour away. Even large irrigated areas, such as the cotton fields in Central Asia or irrigated farmland in Australia (e.g. the Murray-Darling Basin) do not achieve closed water cycles. Instead the excessive use of water for irrigation from rivers have had detrimental effects, such as the drying out of the Aral Sea due to water withdrawal from Amu-Darya and Syr-Darya (Central Asia) or the degradation of wetlands in the mouth of the Murray River in Australia. An additional problem is an increasing soil salinity in irrigated areas. There is an urgent need that agricultural research focuses on how to close water cycles in the landscape and the development of farming systems with a more vertically-layered vegetation structure keeping water and lower temperatures during a sunny day.

Observations of nature and studies of natural processes have offered us some understanding as to how nature tends to close the cycles of water and matter so that losses – water discharge and transport of matter via rivers to the sea – are kept to a minimum. We have provided evidence that evapotranspiration plays far the most important role in damping temperature amplitudes and helping to prevent large-scale overheating of land and atmosphere. Hence it is the vegetation cover that ameliorates the climate and can mitigate climate change. Studying the natural processes in a virgin forest in Austria has revealed how natural vegetation cover closes the cycles of water and matter and efficiently dissipates excesses in solar energy. An important question remains to be answered. How can we achieve such an efficient resource economy in a human-managed landscape - efficient water and matter recycling and energy dissipation as achieved by any undisturbed fully-functioning natural ecosystem? Below we offer some thoughts that we think are worth considering if society seriously wants to address landscape sustainability:

- An assembly of organisms that is ecologically-optimized will show the best local resource utilization in a given space; this ensemble will thus be the one that is able to grow and to expand over the area of that site. That is, at least, until some shift in the surrounding conditions immediately outside the given area should occur - and then another organism ensemble becomes the most efficient with respect to the available resources. According to the direct experience of farmers, the two mostly limiting factors for growth and expansion in our landscape are usually water and nitrogen.

- Farmers should be seen by society as the 'managers of our landscape': only their experience 'in tune' with their local environment - in direct feedback mode with the properties and harmonic patterns of their own locality - can rescue society's life-giving 'hardware', the land, and provide a sustainable management. The short water cycle -

with its inherent 'loss-free' matter flow - controlled by the land manager would appear to be the only way to a sustainable society. However, if intelligent land management is to be successful it has to be paid for: through appropriate rewards according to a land manager's achievements towards sustainability - such as low matter losses and efficient solar energy dissipation.

- All other conceptions of nature protection and conservation - that attempt some 'esoteric' protection of the landscape, with farmers trying to do 'nature conservation' by preserving structures in time and space - are in the long run deemed to fail. 'Fixed' structures cannot be sustainable within ecosystems that are living on dynamic changes towards keeping matter (biomass and soil) in place. Land management, as practiced today, that follows 'one rule fits all' centralized planning at the EU-level, must be seen as mismanagement; such planning results in an ever-growing disturbance of vegetation, climate, cooling and soil fertility, leading to steadily-growing desertification and loss of water, climate instabilities and increased food insecurities.

- There is not the slightest evidence for the belief that chasing the most necessary-for-life gas CO_2 through the trading of 'indulgence' certificates - and burying it deep down into what is mostly water-saturated zones - will change back a distorted climate. Neither has it been proved that, in an open atmosphere, CO_2 is acting as the driving greenhouse gas in the atmosphere as much as the far more dynamic water vapour under the aerodynamic conditions driving an ever-increasing number of wind-mills. To establish increasing areas of water evapotranspiration as the most desirable cooling mechanism, and dew formation as the most important process controlling air pressure in interaction with the vegetation cover of landscapes, would seem to be a far better strategy.

The water cycle is akin to the 'bloodstream' of the biosphere. Returning water to the landscape and restoring more natural vegetation cover is the only way to restore landscape sustainability. More attention in present-day science needs to be devoted to the study of the role of vegetation in the water cycle and climate amelioration. Restoration of a more natural vegetation cover over the landscape seems to be the only way forward.

Based on our current scientific knowledge, we can propose two criteria for assessing sustainable land management. These criteria are: the efficiency of an ecosystem to recycle water and matter; and its efficiency to dissipate solar energy. It is land managers that can substantially contribute to the restoration of the water cycle, climate amelioration and reduction of irreversible matter losses with river water flows to the sea. It is in the interest of society as a whole that land managers (farmers, foresters) be rewarded for their actions towards sustainable management of their land. Suitable tools to assess the achievements of individual land managers with respect to sustainable management of their land are: (1) continuous monitoring of conductivity – a measure of dissolved load - and flow rates in streams in order to estimate matter losses; and (2) the regular evaluation of satellite thermal channel images to assess temperature damping, i.e. the effectiveness of land use to dissipate solar energy. Restoration of natural 'cooling structures' – vegetation with its evapotranspiration and condensation-induced water circulation – is essential to renew landscape sustainability.

6. Acknowledgements

We would like to thank Mr. Steve Ridgill for improving the English text.

7. References

Björk, S. (1988). Redevelopment of lake ecosystems - a case study approach. *Ambio*, 17, 90-98.

Björk, S. et al. 1972. Ecosystem studies in connection with the restoration of lakes. *Verh. Internat. Verein. Limnol.* 18: 379-387.

Björk, S., Pokorný, J. & Hauser, V. (2010). Restoration of Lakes Through Sediment Removal, with Case Studies from Lakes Trummen, Sweden and Vajgar, Czech Republic. In: Eiseltová, M. (ed). *Restoration of Lakes, Streams, Floodplains, and Bogs in Europe: Principles and Case Studies.* Springer, Dordrecht, pp. 101-122.

Blankenship R.E. (2002). *Molecular Mechanisms of Phostosynthesis.* Blackwell Science, 336pp.

Capra F. (1996). *The Web of Life: A New Synthesis of Mind and Matter.* Harper Collins Publishers, New York, 320 pp.

Digerfeldt, G. (1972). The post-glacial development of lake Trumen. Regional vegetation history, water level changes and paleolimnology. *Folia Limnologica Scandinavica*, 16: 104.

Harder, R., Schumacher, W., Firbas, F. & von Denffer, D. (1965). *Strasburger's Textbook of Botany.* English translation from ed. 28 (1962) by Bell, P. and Coombe, D. Longmans. Green and Co. London.

Hesslerová, P. & Pokorný, J. (2010). Forest clearing, water loss and land surface heating as development costs. *Int. J. Water*, Vol. 5, No. 4, pp. 401 – 418.

Kučerová, A., Pokorný, J., Radoux, M., Němcová, M., Cadelli, D. & Dušek, J. (2001). Evapotranspiration of small-scale constructed wetlands planted with ligneous species. In: Vymazal, J. (Ed.): *Transformations of Nutrients in Natural and Constructed Wetlands*, Backhuys, Leiden, pp. 413-427.

Lovelock J. (1990): *The Ages of Gaia – A biography of Our Living Earth.* Oxford University Press, Oxford, 252 pp.

Makarieva, A. M. & Gorshkov, V. G. (2007). Biotic pump of atmospheric moisture as driver of the hydrological cycle on land. *Hydrology and Earth System Sciences*, Vol. 11, No. 2, pp. 1013-1033.

Makarieva, A. M. & Gorshkov, V. G. (2010). The Biotic Pump: Condensation, atmospheric dynamics and climate. *Int. J. Water*, Vol. 5, No. 4, pp. 365-385.

Mather, P.M. & Tso, B. (2009). *Classification Methods for Remotely Sensed Data.* CRC Press, Boca Raton, 332 pp.

Millennium Environmental Assessment (2005) *Ecosystems and Human Well-being: Desertification Synthesis*, World Resources Institute, Washington DC.

Monteith, J. L. (1975). *Vegetation and Atmosphere*, Academic Press, London.

Pokorný, J., Brom, J., Čermák, J. Hesslerová, P., Huryna, H., Nadyezhdina, N. & Rejšková, A. (2010a). Solar energy dissipation and temperature control by water and plants. *Int. J. Water*, Vol. 5, No. 4, pp. 311 – 336.

Pokorný, J., Květ, J., Rejšková, A. & Brom, J. (2010b). Wetlands as energy-dissipating systems. *J. Ind. Microbiol. Biotechol.*, Vol. 37, No. 12, pp. 1299 – 1305.

Prigogine, I. (1980). *From Being To Becoming: Time and Complexity in the Physical Sciences.* Freeman, San Francisco, 272 pp.

Prigogine, I. & Glansdorff, P. (1971). *Thermodynamic Theory of Structure, Stability and Fluctuations*, Wiley, New York.

Prigogine, I., Stengers, I. (1984). Order Out of Chaos. Bantam Books, New York, 349.

Procházka, J., Hakrová, P., Pokorný, J., Pecharová, E., Hezina, T., Wotavová, K., Šíma, M. & Pechar, L. (2001). Effect of different management practices on vegetation development, losses of soluble matter and solar energy dissipation in three small sub-mountain catchments. In: Vymazal, J. (ed.). *Transformations of nutrients in natural and constructed wetlands.* Backhuys, Leiden, pp. 143–175.

Purseglove, J. (1989). *Taming the Flood.* Oxford University Press, Oxford, 307pp.

Ripl, W. (1992). Management of Water Cycle: An Approach to Urban Ecology. *Water Pollution Journal Canada*, Vol. 27, No. 2, pp. 221-237.

Ripl, W. (1995). Management of water cycle and energy flow for ecosystem control: the energy-transport-reaction (ETR) model. *Ecological Modelling*, Vol. 78, No. 1-2, pp. 61-76.

Ripl, W. (2010). Losing fertile matter to the sea: How landscape entropy affects climate. *Int. J. Water*, Vol. 5, No. 4, pp. 353-364.

Ripl, W. & Eiseltová, M. (2009). Sustainable land management by restoration of shor water cycles and prevention of irreversible matter losses from topsoils. *Plant Soil Environ.*, Vol. 55, No. 9, pp. 404-410.

Ripl, W. & Eiseltová, M. (2010). Criteria for Sustainable Restoration of the Landscape. In: Eiseltová, M. (ed). *Restoration of Lakes, Streams, Floodplains, and Bogs in Europe: Principles and Case Studies.* Springer, Dordrecht, pp. 1-24.

Ripl, W. & Hildmann, C. (2000). Dissolved load transported by rivers as an indicator of landscape sustainability. *Ecological Engineering*, 14: 373–387.

Ripl, W., Hildmann, C., Janssen, T., Gerlach, I., Heller, S. & Ridgill, S. (1995). Sustainable redevelopment of a river and its catchment: the Stör River project. In: Eiseltová, M., Biggs, J. (eds.). *Restoration of Stream Ecosystems – An Integrated Catchment Approach.* IWRB Publishing, Slimbridge, Publ. No. 37: 76–112.

Ripl, W., Splechtna, K., Brande, A., Wolter, K.D., Janssen T., Ripl, W. jun. & Ohmeyer, C. (2004). *Funktionale Landschaftsanalyse im Albert Rothschild Wildnisgebiet Rothwald.* Im Auftrag von LIL (Verein zur Förderung der Landentwicklung und intakter Lebensräume) NÖ Landesregierung, Österreich, Final Report, 154 pp.

Ryszkowski, L. & Kedziora, A. (1987). Impact of agricultural landscape structure on energy flow and water cycling. *Landscape Ecology*, Vol. 1, No. 2, pp. 85-94.

Schlesinger, W. H. (1997). *Biochemistry: An Analysis of Global Change.* 2nd ed., Academic Press, San Diego, 588 pp.

Schneider, E. D. & Sagan, D. (2005). *Into the Cool: Energy Flow, Thermodynamics, and Life.* University of Chicago Press, Chicago.

Sobrino, J.A., Jiménez-Muñoz, J.C. & Paolini, L. (2004). Land surface temperature retrieval from LANDSAT TM 5, *Remote Sensing of Environment*, Vol. 90, pp. 434–440.

5

Development of Hybrid Method for the Modeling of Evaporation and Evapotranspiration

Sungwon Kim
Dongyang University,
Republic of Korea

1. Introduction

Evaporation is the process whereby liquid water is converted to water vapor and removed from the evaporating surface. In hydrological practice, the estimation of evaporation can be achieved by direct or indirect methods. Direct method is based on the field measurements. Evaporation pan have also been used and compared to estimate evaporation by researchers (Choudhury, 1999; McKenzie and Craig, 2001; Vallet-Coulomb et al., 2001). The Class A evaporation pan is one of the most widely used instruments for the measurements of evaporation from a free water surface. The pan evaporation (PE) is widely used to estimate the evaporation of lakes and reservoirs (Finch, 2001). Many researchers have tried to estimate the evaporation through the indirect methods using the climatic variables, but some of these techniques require the data which cannot be easily obtained (Rosenberry et al., 2007).

Evapotranspiration (ET) is the sum of volume of water used by vegetation, evaporated from the soil, and intercepted precipitation (Singh, 1988). ET plays an important role in our environment at global, regional, and local scales. ET is observed using a lysimeter directly or can be calculated using the water balance method or the climatic variables indirectly. Because the measurements of ET using a lysimeter directly, however, requires much unnecessary time and needs correct and careful experience, it is not always possible in field measurements. Thus, an empirical approach based on the climatic variables is generally used to calculate ET (Penman, 1948; Allen et al., 1989). In the early 1970s, the Food and Agricultural Organization of the United Nations (FAO), Rome, developed practical procedures to calculate the crop water requirements (Doorenbos & Pruitt, 1977), which have become the widely accepted standard for irrigation studies. A common practice for estimating ET from a well-watered agricultural crop is to calculate the reference crop ET such as the grass reference ET (ET_0) or the alfalfa reference ET (ET_r) from a standard surface and to apply an appropriate empirical crop coefficient, which accounts for the difference between the standard surface and the crop ET.

Recently, the outstanding results using the neural networks model in the fields of PE and ET modeling have been obtained (Bruton et al., 2000; Sudheer et al., 2003; Trajkovic et al., 2003; Trajkovic, 2005; Keskin and Terzi, 2006; Kisi, 2006; Kisi, 2007; Zanetti et al., 2007; Jain et al., 2008; Kumar et al., 2008; Landeras et al., 2008; Kisi, 2009; Kumar et al., 2009; Tabari et al.,

2009; Chang et al., 2010; Guven and Kisi, 2011). Sudheer et al. (2002) investigated the prediction of Class A PE using the neural networks model. They used the neural networks model for the evaporation process using proper combinations of the observed climate variables such as temperature, relative humidity, sunshine duration, and wind speed for the neural networks model. Shiri and Kisi (2011) investigated the ability of genetic programming (GP) to improve the accuracy of daily evaporation estimation. They used proper combinations of air temperature, sunshine hours, wind speed, relative humidity, and solar radiation for GP. Kumar et al. (2002) developed the neural networks models to calculate the daily ET. They used proper combinations of the observed climatic variables such as solar radiation, temperature, relative humidity, and wind speed for the neural networks models. Kisi & Ozturk (2007) used the neuro-fuzzy models to calculate FAO-56 PM ET_o using the observed climatic variables. They used proper combinations of the observed climatic variables such as air temperature, solar radiation, wind speed, and relative humidity for the neuro-fuzzy models. Kim & Kim (2008) developed the neural networks model embedding the genetic algorithm for the modeling of the daily PE and ET_r simultaneously, and constructed the optimal neural networks model using the uncertainty analysis of the input layer nodes/variables. Furthermore, they suggested the 2-dimensional and 3-dimensional maps for PE and ET_r to provide the reference data for irrigation and drainage system, Republic of Korea. And, the recent researches combining the stochastic models and the neural networks models in the fields of hydrology and water resources have been accomplished. Mishra et al. (2007) developed a hybrid model, which combined a linear stochastic model and a nonlinear neural networks model, for drought forecasting. Kim (2011) investigated the modeling of the monthly PE and ET_r simultaneously using the specific method, which combined the stochastic model with the neural networks models.

The purpose of this study is to develop the hybrid method for the modeling of the monthly PE and FAO-56 PM ET_o simultaneously. The hybrid method represents the combination of Univariate Seasonal periodic autoregressive moving average (PARMA) model and support vector machine neural networks model (SVM-NNM). For this research, first, the stochastic model, Univariate Seasonal PARMA(1,1) model, is used for the generation of the reliable data, which are considered as the training dataset. Therefore, the observed data are considered as the testing dataset. Second, the neural networks model, SVM-NNM, is used for the modeling of the monthly PE and FAO-56 PM ET_o simultaneously. Homogeneity evaluation using the One-way ANOVA and Mann-Whitney U test, furthermore, is carried out for the observed and calculated PE and FAO-56 PM ET_o data. And, the correlation relationship between the observed PE and FAO-56 PM ET_o data can be derived using the bivariate linear regression analysis model (BLRAM), respectively.

2. Calculation of FAO-56 PM ET_o

Penman (1948) combination method links evaporation dynamics with the flux of net radiation and aerodynamic transport characteristics of the natural surface. Based on the observations that latent heat transfer in plant stem is influenced not only by these abiotic factors, Monteith (1965) introduced a surface conductance term that accounted for the response of leaf stomata to its hydrologic environment. This modified form of the Penman-Monteith (PM) ET_o model. Jensen et al. (1990) measured ET_o using the lysimeters at 11 stations located in the different climatic zones of various regions around the world. They compared the results of the lysimeters with those of 20 different empirical equations and methodologies for ET_0 measurements. It was found that PM ET_0 model showed the optimal results over all the climatic zones. If the observed/measured data for ET_0 does not exist, therefore, PM ET_0 model can be considered as a standard methodology to calculate ET_0. In Gwangju and Haenam stations which were selected for this study, there are no observed data for ET_0 using a lysimeter. The data calculated using PM ET_0 model can be assumed as the observed ET_0, whose reliability was verified by many previous studies. All calculation procedures as used in PM ET_0 model are based on the FAO guidelines as laid down in the publication No. 56 of the Irrigation and Drainage Series of FAO "Crop Evapotranspiration-Guidelines for Computing Crop Water Requirements" (1998). Therefore, FAO-56 PM ET_0 equation means PM ET_0 equation suggested by the Irrigation and Drainage Paper No. 56, FAO. FAO-56 PM ET_0 equation is given by Allen et al. (1998) and can be shown as the following equation (1).

$$\text{FAO-56 PM } ET_0 = \frac{0.408\Delta(R_n - G) + \gamma(900/(T + 273))u_2(e_s - e_a)}{\Delta + \gamma(1 + 0.34u_2)} \qquad (1)$$

where FAO-56 PM ET_0 = the grass reference evapotranspiration (mm/day); R_n = the net radiation at the crop surface (MJ/m² day); G = the soil heat flux density (MJ/m² day); T = the mean air temperature at 2m height (°C); u_2 = the wind speed at 2m height (m/sec); e_s = the saturation vapor pressure (kPa); e_a = the actual vapor pressure (kPa); $e_s - e_a$ = the saturation vapor pressure deficit (kPa); Δ = the slope vapor pressure curve (kPa/°C); and γ = the psychometric constant (kPa/°C). FAO CROPWAT 8.0 computer program has been used to calculate FAO-56 PM ET_0 and extraterrestrial radiation (R_a). FAO CROPWAT 8.0 computer program allows the user to enter the climatic data available including maximum temperature (T_{max}), minimum temperature (T_{min}), mean relative humidity (RH_{mean}), mean wind speed (WS_{mean}), and sunshine duration (SD) for calculating FAO-56 PM ET_0. On the base of climatic data available, FAO CROPWAT 8.0 computer program calculates the solar radiation reaching soil surface. Fig. 1(a)-(b) show the calculation of FAO-56 PM ET_0 using FAO CROPWAT 8.0 computer program in Gwangju and Haenam stations, respectively.

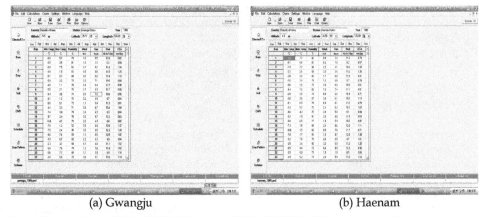

(a) Gwangju   (b) Haenam

Fig. 1. Calculation of FAO-56 PM ET_0 using FAO CROPWAT 8.0 Computer Program

3. Stochastic model

3.1 Univariate seasonal periodic autoregressive moving average model

Stationary ARMA models have been widely applied in stochastic hydrology for modeling of annual time series where the mean, variance, and the correlation structure do not depend on time. For seasonal hydrologic time series, such as monthly series, seasonal statistics including the mean and standard deviation may be reproduced by a stationary ARMA model by means of standardizing the underlying seasonal series. Hydrologic time series such as monthly streamflows, PE, and FAO-56 PM ET_0 are usually characterized by different dependence structure depending on the season (Salas, 1993). One may extend Univariate Seasonal periodic autoregressive (PAR) model to include periodic moving average (MA) parameters. Such a model is Univariate Seasonal periodic autoregressive moving average (PARMA) model and is expressed as Univariate Seasonal PARMA(p,q) model. The stochastic models are generally simple to use. When the errors, however, happen in model identification and parameter estimation, the generated data using the stochastic models cannot reconstruct the statistical properties of the observed data exactly. Furthermore, the high-order PARMA(p,q) models have generally many parameters, and the calculation process is much complex (Salas et al., 1980). In this study, the author determined in advance 4 kinds of Univariate Seasonal PARMA(p,q) models including PARMA(1,1), PARMA(1,2), PARMA (2,1), and PARMA(2,2), which are the low-order models and contain the seasonal properties. In general, the low-order Univariate Seasonal PARMA(p,q) models are useful for the periodic hydrologic time series modeling (Salas et al., 1982). Furthermore, the author generated 100 years data in advance using each Univariate Seasonal PARMA(p,q) model for the climatic variables of the neural networks models, respectively. As a result, the author selected Univariate Seasonal PARMA(1,1) model, which shows the best statistical properties and is simple in parameter estimation. Univariate Seasonal PARMA(1,1) model has been applied to monthly streamflow time series from the previous studies (Tao and Delleur, 1976; Hirsch, 1979), and is shown as the following equation (2).

$$y_{v,\tau} = \mu_\tau + \phi_{1,\tau}(y_{v,\tau-1} - \mu_{\tau-1}) + \varepsilon_{v,\tau} - \theta_{1,\tau}\varepsilon_{v,\tau-1} \qquad (2)$$

where $y_{v,\tau} / y_{v,\tau-1}$ = the monthly PE and FAO-56 PM ET_0 for year= v and month= $\tau / \tau - 1$; $\mu_\tau / \mu_{\tau-1}$ = the means for month= $\tau / \tau - 1$; $\phi_{1,\tau}$ = the seasonal autoregressive parameter for month= τ ; $\theta_{1,\tau}$ = the seasonal moving average parameter for month= τ ; $\varepsilon_{v,\tau} / \varepsilon_{v,\tau-1}$ = uncorrelated noise terms; v = year; τ = month $(1,2,...,\omega)$; and ω =12. Furthermore, Univariate Seasonal PARMA(2,1), PARMA(2,2) models, and more complex multiplicative PARMA(p,q) models may be needed for hydrologic modeling and simulation when the preservation of both the seasonal and the annual statistics is desired (Salas and Abdelmohsen, 1993).

3.2 Construction of Univariate Seasonal PARMA(p,q) model

The author used Univariate Seasonal PARMA(1,1) model to generate the sufficient training dataset, and obtained two generated samples. They included the input nodes/variables including mean temperature (T_{mean}), maximum temperature (T_{max}), minimum temperature (T_{min}), mean dew point temperature (DP_{mean}), minimum relative humidity (RH_{min}), mean relative humidity (RH_{mean}), mean wind speed (WS_{mean}), maximum wind speed (WS_{max}), and sunshine duration (SD) in mean values and the output nodes/variables including PE and

FAO-56 PM ET_0 in total values, respectively. Furthermore, they were generated for 100 years (Short-term), 500 years (Mid-term), and 1000 years (Long-term), respectively. The author selected the second generated sample, and the first 50 years of the 100, 500, and 1000 years was abandoned to eliminate the biases, respectively. The parameters of Univariate Seasonal PARMA(1,1) model were determined using the method of approximate least square, respectively.

4. Support Vector Machine Neural Networks Model (SVM-NNM)

SVM-NNM has found wide application in several areas including pattern recognition, regression, multimedia, bio-informatics and artificial intelligence. Very recently, SVM-NNM is gaining recognition in hydrology (Dibike et al., 2001; Khadam & Kaluarachchi, 2004). SVM-NNM implements the structural risk minimization principle which attempts to minimize an upper bound on the generalization error by striking a right balance between the training performance error and the capacity of machine. The solution of traditional neural networks models such as MLP-NNM may tend to fall into a local optimal solution, whereas global optimum solution is guaranteed for SVM-NNM (Haykin, 2009). SVM-NNM is a new kind of classifier that is motivated by two concepts. First, transforming data into a high-dimensional space can transform complex problems into simpler problems that can use linear discriminant functions. Second, SVM-NNM is motivated by the concept of training and using only those inputs that are near the decision surface since they provide the most information about the classification. The first step in SVM-NNM is transforming the data into a high-dimensional space. This is done using radial basis function (RBF) that places a Gaussian at each sample data. Thus, the feature space becomes as large as the number of sample data. RBF uses backpropagation to train a linear combination of the gaussians to produce the final result. SVM-NNM, however, uses the idea of large margin classifiers for the training performance. This decouples the capacity of the classifier from the input space and at the same time provides good generalization. This is an ideal combination for classification (Vapnik, 1992, 2000; Principe et al., 2000; Tripathi et al., 2006).

In this study, the basic ideas of SVM-NNM are reviewed. Consider the finite training pattern (x_i, y_i). where $x_i \in \Re^n$ = a sample value of the input vector x considering of N training patterns; and $y_i \in \Re^n$ = the corresponding value of the desired model output. A nonlinear transformation function $\phi(\cdot)$ is defined to map the input space to a higher dimension feature space, $\Re^{n_h}$. According to Cover's theorem (Cover, 1965), a linear function, $f(\cdot)$, could be formulated in the high dimensional feature space to look for a nonlinear relationship between inputs and outputs in the original input space. It can be written as the following equation (3).

$$\bar{y} = f(x) = w^T \phi(x) + b \tag{3}$$

where $\bar{y}$ = the actual model output. The coefficient w and b are adjustable model parameters. In SVM-NNM, we aim at minimizing the empirical risk. It can be written as the following equation (4).

$$R_{emp} = \frac{1}{N} \sum_{i=1}^{N} \left| y_i - \bar{y}_i \right|_\varepsilon \tag{4}$$

where R_{emp} = the empirical risk; and $\left|y_i - \overline{y}_i\right|_\varepsilon$ = the Vapnik's ε-insensitive loss function. Following regularization theory (Haykin, 2009), the parameters w and b are calculated by minimizing the cost function. It can be written as the following equation (5).

$$\psi_\varepsilon(w,\xi,\xi^*) = \frac{1}{2}w^Tw + C\sum_{i=1}^{N}(\xi_i + \xi_i^*) \tag{5}$$

subject to the constraints: 1) $y_i - \overline{y}_i \leq \varepsilon + \xi_i$ i = 1, 2, ... , N, 2) $-y_i + \overline{y}_i \leq \varepsilon + \xi_i^*$ i = 1, 2, ... , N, and 3) $\xi_i, \xi_i^* \geq 0$ i = 1, 2, ... , N. where $\psi_\varepsilon(w,\xi,\xi^*)$ = the cost function; ξ_i, ξ_i^* = positive slack variables; and C = the cost constant. The first term of the cost function, which represents weight decay, is used to regularize weight sizes and to penalize large weights. This helps in improving generalization performance (Hush and Horne, 1993). The second term of the cost function, which represents penalty function, penalizes deviations of $\overline{y}$ from y larger than $\pm\varepsilon$ using Vapnik's ε-insensitive loss function. The cost constant C determines the amount up to which deviations from ε are tolerated. Deviations above ε are denoted by ξ_i, whereas deviations below – ε are denoted by ξ_i^*. The constrained quadratic optimization problem can be solved using the method of Lagrangian multipliers (Haykin, 2009). From this solution, the coefficient w can be written as the following equation (6).

$$w = \sum_{i=1}^{N}(\alpha_i - \alpha_i^*)\phi(x_i) \tag{6}$$

where α_i, α_i^* = the Lagrange multipliers, which are positive real constants. The data points corresponding to non-zero values for $(\alpha_i - \alpha_i^*)$ are called support vectors. In SVM-NNM to calculate PE and FAO-56 PM ET_o, there are several possibilities for the choice of kernel function, including linear, polynomial, sigmoid, splines and RBF. In this study, RBF is used to map the input data into higher dimensional feature space. RBF can be written as the following equation (7).

$$k(x,x_j) = \Phi_1 = \exp(-B_1R_j^2) = \exp\left(-\frac{\left\|x_i - x_j\right\|^2}{2\sigma^2}\right) \tag{7}$$

where i, j = the input layer and the hidden layer; $K(x,x_j) = \Phi_1$ = the inner product kernel function; $B_1 = \frac{1}{2\sigma^2}$, and has a constant value; and σ = the width/spread of RBF, which can be adjusted to control the expressivity of RBF. The function for the single node of the output layer which receives the calculated results of RBF can be written as the following equation (8).

$$G_k = [\sum_{j=1}^{N}(\alpha_j - \alpha_j^*)\cdot K(x,x_j)] + B \tag{8}$$

where k = the output layer; G_k = the calculated value of the single output node; and B = the bias in the output layer. Equation (8), finally, takes the form of equation (9) and (10), which represents SVM-NNM for the modeling of PE and FAO-56 PM ET_o.

$$PE = W_1 \cdot \Phi_2(G_k) = W_1 \cdot \Phi_2[[\sum_{j=1}^{N}(\alpha_j - \alpha_j^*) \cdot K(x,x_j)] + B] \tag{9}$$

$$FAO\text{-}56\ PM\ ET_o = W_2 \cdot \Phi_2(G_k) = W_2 \cdot \Phi_2[[\sum_{j=1}^{N}(\alpha_j - \alpha_j^*) \cdot K(x,x_j)] + B] \tag{10}$$

where $\Phi_2(\cdot)$ = the linear sigmoid transfer function; W_1 = the specific weights connected to the output variable of PE; and W_2 = the specific weights connected to the output variable of FAO-56 PM ET_o. A number of SVM-NNM computer programs are now available for the modeling of PE and FAO-56 PM ET_o. NeuroSolutions 5.0 computer program was used to develop SVM-NNM structure. Fig. 2 shows the developed structure of SVM-NNM. From the Fig. 2, the input nodes/variables of climatic data are mean temperature (T_{mean}), maximum temperature (T_{max}), minimum temperature (T_{min}), mean dew point temperature (DP_{mean}), minimum relative humidity (RH_{min}), mean relative humidity (RH_{mean}), mean wind speed (WS_{mean}), maximum wind speed (WS_{max}), and sunshine duration (SD) in mean values (01/1985-12/1990). The output nodes/variables of climatic data are PE and FAO-56 PM ET_o in total values (01/1985-12/1990).

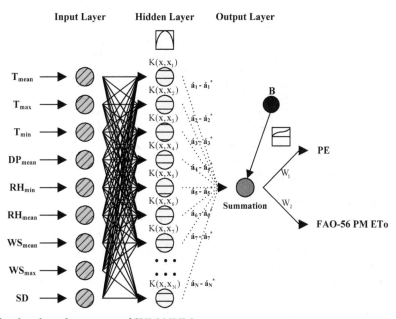

Fig. 2. The developed structure of SVM-NNM

5. Study scope and data

In this study, Gwangju and Haenam stations from the Yeongsan River catchment are selected among the 71 weather stations including Jeju-do under the control of the Korea meteorological administration (KMA). They have possessed long-term climatic data dating back over at least 30 years. The Yeongsan River catchment covers an area of 3455 km², and lies between latitudes 34.4°N and 35.2°N, and between longitudes 126.2°E and 127.0°E. Fig. 3 shows the Yeongsan River catchment including Gwangju and Haenam stations. The climatic data, which was necessary for the modeling of PE and FAO-56 PM ET$_o$, were collected from the Internet homepage of water management information system (www.wamis.go.kr) and the Korea meteorological administration (www.kma.go.kr).

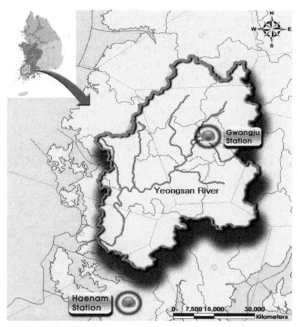

Fig. 3. The Yeongsan River Catchment including Gwangju and Haenam stations

6. SVM-NNM performance

6.1 Performance statistics

The performance of SVM-NNM to account for calculating the monthly PE and FAO-56 PM ET$_o$ was evaluated using a wide variety of standard statistics index. A total of 3 different standard statistics indexes were employed; the coefficient of correlation (CC), root mean square error (RMSE), and Nash-Sutcliffe coefficient (R²) (Nash & Sutcliffe, 1970; ASCE, 1993). Table 1 shows summary of the statistics index in this study. where $\overline{y}_i(x) =$ the calculated PE and FAO-56 PM ET$_o$ (mm/month); $y_i(x) =$ the observed PE and FAO-56 PM ET$_o$ (mm/month); $\overline{u}_y =$ mean of the calculated PE and FAO-56 PM ET$_o$ (mm/month); $u_y =$ mean of the observed PE and FAO-56 PM ET$_o$ (mm/month); and n = total number of the monthly PE and FAO-56 PM ET$_o$ considered. A model which is effective in the modeling of PE and FAO-56 PM ET$_o$ accurately, and efficient in capturing the complex relationship among the various inputs and output variables involved in a particular problem, is considered the best. CC, RMSE, and R² statistics quantify the efficiency of SVM-NNM in capturing the extremely complex, dynamic, and nonlinear relationships (Kim, 2011).

Statistics Indexes	Equation
CC	$$\dfrac{\dfrac{1}{n}\sum\limits_{i=1}^{n}[y_i(x)-u_y][\bar{y}_i(x)-\bar{u}_y]}{\sqrt{\dfrac{1}{n}\sum\limits_{i=1}^{n}[y_i(x)-u_y]^2}\sqrt{\dfrac{1}{n}\sum\limits_{i=1}^{n}[\bar{y}_i(x)-\bar{u}_y]^2}}$$
RMSE	$$\sqrt{\dfrac{1}{n}\sum\limits_{i=1}^{n}[\bar{y}_i(x)-y_i(x)]^2}$$
R^2	$$1-\dfrac{\sum\limits_{i=1}^{n}[y_i(x)-\bar{y}_i(x)]^2}{\sum\limits_{i=1}^{n}[y_i(x)-u_y]^2}$$

Table 1. Summary of statistics indexes

6.2 Data normalization

The climatic data used in this study including mean temperature (T_{mean}), maximum temperature (T_{max}), minimum temperature (T_{min}), mean dew point temperature (DP_{mean}), minimum relative humidity (RH_{min}), mean relative humidity (RH_{mean}), mean wind speed (WS_{mean}), maximum wind speed (WS_{max}), and sunshine duration (SD) were normalized for preventing and overcoming problem associated with the extreme values. An important reason for the normalization of input nodes is that each of input nodes represents an observed value in a different unit. Such input nodes are normalized, and the input nodes in dimensionless unit are relocated. The similarity effect of input nodes is thus eliminated (Kim et al., 2009). According to Zanetti et al. (2007), by grouping the daily values into averages, ET$_o$ may be calculated due to their highest stabilization. For data normalization, the data of input and output nodes were scaled in the range of [0 1] using the following equation (11).

$$Y_{norm} = \frac{Y_i - Y_{min}}{Y_{max} - Y_{min}} \tag{11}$$

where Y_{norm} = the normalized dimensionless data of the specific input node/variable; Y_i = the observed data of the specific input node/variable; Y_{min} = the minimum data of the specific input node/variable; and Y_{max} = the maximum data of the specific input node/variable.

6.3 Training performance

The method for calculating parameters is generally called the training performance in the neural networks model category. The training performance of neural networks model is iterated until the training error is reached to the training tolerance. Iteration means one completely pass through a set of inputs and target patterns or data. In general, it is assumed that the neural networks model does not have any prior knowledge about the example problem before it is trained (Kim, 2004). A difficult task with the neural networks model is to choose the number of hidden nodes. The network geometry is problem dependent. This study adopted one hidden layer for the construction of SVM-NNM since it is well known that one hidden layer is enough to represent PE and FAO-56 PM ET_0 nonlinear complex relationship (Kumar et al., 2002; Zanetti et al., 2007). The testing performance in the modeling of PE and FAO-56 PM ET_0, therefore, is carried out using the optimal parameters, which are calculated during the training performance.

The hybrid method, which was developed in this study, consisted of the following training patterns. First, the stochastic model was selected. As explained previously, Univariate Seasonal PARMA(1,1) model, which consisted of 1 pattern only, was used to generate the training dataset. Second, the data, which were generated by Univariate Seasonal PARMA(1,1) model, consisted of 3 patterns including 100 years (Short-term), 500 years (Mid-term), and 1000 years (Long-term), respectively. Finally, the neural networks model, which consisted of 1 pattern only including SVM-NNM, was used for the training and testing performances, respectively. Therefore, the hybrid method consisted of 3 training patterns including 100/PARMA(1,1)/SVM-NNM, 500/PARMA(1,1)/SVM-NNM, and 1000 /PARMA(1,1)/SVM-NNM, respectively. For Gwangju and Haenam stations, the training dataset including the climatic, PE, and FAO-56 PM ET_0 data were generated by Univariate Seasonal PARMA(1,1) model using observed data (01/1985-12/1990) for 100 years (Short-term), 500 years (Mid-term), and 1000 years (Long-term), respectively. After the first 50 years of the generated data for 100, 500, and 1000 years was abandoned to eliminate the biases, the training performance should be carried out using SVM-NNM. Therefore, the total amount of data used for the training performance consisted of 600, 5400, and 11400, respectively. For the training performance of SVM-NNM, NeuroSolutions 5.0 computer program was used to carry out the training performance. Fig. 4(a)-(b) show SVM-NNM training performance using NeuroSolutions 5.0 computer program.

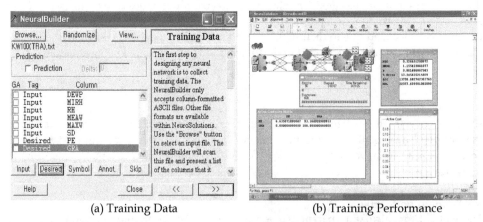

(a) Training Data   (b) Training Performance

Fig. 4. SVM-NNM training performance using NeuroSolution 5.0 Computer Program

Table 2 shows the statistics results of the training performances for 3 training patterns of PE modeling. In PE of Gwangju station, from the Table 2, 100/PARMA(1,1)/SVM-NNM training pattern produced the statistics results with CC value of 0.929, RMSE value of 16.360 mm and R^2 value of 0.858. 500/PARMA(1,1)/SVM-NNM training pattern produced the statistics results with CC value of 0.919, RMSE value of 17.942 mm and R^2 value of 0.833. 1000/PARMA(1,1)/SVM-NNM training pattern produced the statistics results with CC value of 0.905, RMSE value of 20.489 mm and R^2 value of 0.781, respectively. In PE of Haenam station, from the Table 2, 100/PARMA(1,1)/SVM-NNM training pattern produced the statistics results with CC value of 0.965, RMSE value of 9.643 mm and R^2 value of 0.930. 500/PARMA(1,1)/SVM-NNM training pattern produced the statistics results with CC value of 0.958, RMSE value of 10.673 mm and R^2 value of 0.914. 1000/PARMA(1,1)/SVM-NNM training pattern produced the statistics results with CC value of 0.962, RMSE value of 10.105 mm and R^2 value of 0.922, respectively. From the above results, the statistics results of the training performance for 100/PARMA(1,1)/SVM-NNM training pattern were better than those of the training performances for 500/PARMA(1,1)/SVM-NNM and 1000/PARMA(1,1)/SVM-NNM training patterns for PE of Gwangju and Haenam stations, respectively.

Station	Statistics Indexes	100/PARMA(1,1)/ SVM-NNM	500/PARMA(1,1)/ SVM-NNM	1000/PARMA(1,1) SVM-NNM
Gwangju	CC	0.929	0.919	0.905
	RMSE (mm)	16.360	17.942	20.489
	R^2	0.858	0.833	0.781
Haenam	CC	0.965	0.958	0.962
	RMSE (mm)	9.643	10.673	10.105
	R^2	0.930	0.914	0.922

Table 2. Statistics results of the training performances (PE)

Table 3 shows the statistics results of the training performances for 3 training patterns of FAO-56 PM ET_0 modeling. In FAO-56 PM ET_0 of Gwangju station, from the Table 3, 100/PARMA(1,1)/SVM-NNM training pattern produced the statistics results with CC value of 0.975, RMSE value of 8.517 mm and R^2 value of 0.948. 500/PARMA(1,1)/SVM-NNM training pattern produced the statistics results with CC value of 0.966, RMSE value of 10.237 mm and R^2 value of 0.924. 1000/PARMA(1,1)/SVM-NNM training pattern produced the statistics results with CC value of 0.963, RMSE value of 11.061 mm and R^2 value of 0.911, respectively. In FAO-56 PM ET_0 of Haenam station, from the Table 3, 100/PARMA(1,1) /SVM-NNM training pattern produced the statistics results with CC value of 0.965, RMSE value of 9.935 mm and R^2 value of 0.926. 500/PARMA(1,1)/SVM-NNM training pattern produced the statistics results with CC value of 0.956, RMSE value of 10.822 mm and R^2 value of 0.912. 1000/PARMA(1,1)/SVM-NNM training pattern produced the statistics results with CC value of 0.962, RMSE value of 10.014 mm and R^2 value of 0.925, respectively. From the above results, the statistics results of the training performance for 100/ PARMA(1,1)/SVM-NNM training pattern were better than those of the training performances for 500/PARMA(1,1)/SVM-NNM and 1000/PARMA(1,1)/SVM-NNM training patterns for FAO-56 PM ET_0 of Gwangju and Haenam stations, respectively. Therefore, the data length has less effect on the training performance of PE and FAO-56 PM ET_0 in this study. Tokar and Johnson (1999) suggested that the data length has less effect on the neural networks model performance than the data quality. Sivakumar et al. (2002) suggested that it is imperative to select a good training dataset from the available data series. They indicated that the best way to achieve a good training performance seems to be to include most of the extreme events such as very high and very low values in the training dataset. Furthermore, Kim (2011) did not carry out the statistics analysis of the training performance since the training dataset of 6 training patterns consisted of the generated (not observed) data only.

Station	Statistics Indexes	100/PARMA(1,1)/ SVM-NNM	500/PARMA(1,1)/ SVM-NNM	1000/PARMA(1,1) SVM-NNM
Gwangju	CC	0.975	0.966	0.963
	RMSE (mm)	8.517	10.237	11.061
	R^2	0.948	0.924	0.911
Haenam	CC	0.965	0.956	0.962
	RMSE (mm)	9.935	10.822	10.014
	R^2	0.926	0.912	0.925

Table 3. Statistics results of the training performances (FAO-56 PM ET_o)

6.4 Testing performance

The neural networks model is tested by determining whether the model meets the objectives of modeling within some preestablished criteria or not. Of course, the optimal parameters, which are calculated during the training performance, are applied for the testing performance of the neural networks model (Kim, 2004). For the testing performance, the monthly climatic data (01/1985-12/1990) in Gwangju and Haenam stations were used. The total amount of data used for the testing performance consisted of 72 data for the monthly time series. The testing performance applied the cross-validation method in order to overcome the over-fitting problem of SVM-NNM. The cross-validation method is not to train all the training data until SVM-NNM reaches the minimum RMSE, but is to cross-validate with the testing data at the end of each training performance. If the over-fitting problem occurs, the convergence process over the mean square error of the testing data will not decrease but will increase as the training data are still trained (Bishop, 1994; Haykin, 2009).

Table 4 shows the statistics results of the testing performances for 3 training patterns of PE modeling. In PE of Gwangju station, from the Table 4, 100/PARMA(1,1)/SVM-NNM training pattern produced the statistics results with CC value of 0.955, RMSE value of 12.239 mm and R^2 value of 0.908. 500/PARMA(1,1)/SVM-NNM training pattern produced the statistics results with CC value of 0.956, RMSE value of 13.501 mm and R^2 value of 0.888. 1000/PARMA(1,1)/SVM-NNM training pattern produced the statistics results with CC value of 0.953, RMSE value of 15.103 mm and R^2 value of 0.860, respectively. In PE of Haenam station, from the Table 4, 100/PARMA(1,1)/SVM-NNM training pattern produced the statistics results with CC value of 0.966, RMSE value of 9.581 mm and R^2 value of 0.932. 500/PARMA(1,1)/SVM-NNM training pattern produced the statistics results with CC value of 0.968, RMSE value of 9.370 mm and R^2 value of 0.935. 1000/PARMA(1,1)/SVM-NNM training pattern produced the statistics results with CC value of 0.953, RMSE value of 11.313 mm and R^2 value of 0.905, respectively. From the above results, the statistics results of the testing performance for 100/PARMA(1,1)/SVM-NNM training pattern were better than those of the testing performances for 500/PARMA(1,1)/SVM-NNM and 1000/PARMA(1,1)/SVM-NNM training patterns for PE of Gwangju station. And, the statistics results of the testing performance for 500/PARMA(1,1)/SVM-NNM training pattern were better than those of the testing performances for 100/PARMA(1,1)/SVM-NNM and 1000/PARMA(1,1)/SVM-NNM training patterns for PE of Haenam station, respectively.

Station	Statistics Indexes	100/PARMA(1,1)/ SVM-NNM	500/PARMA(1,1)/ SVM-NNM	1000/PARMA(1,1) SVM-NNM
Gwangju	CC	0.955	0.956	0.953
	RMSE (mm)	12.239	13.501	15.103
	R^2	0.908	0.888	0.860
Haenam	CC	0.966	0.968	0.953
	RMSE (mm)	9.581	9.370	11.313
	R^2	0.932	0.935	0.905

Table 4. Statistics results of the testing performances (PE)

Table 5 shows the statistics results of the testing performances for 3 training patterns of FAO-56 PM ET_0 modeling. In FAO-56 PM ET_0 of Gwangju station, from the Table 5, 100/PARMA(1,1)/SVM-NNM training pattern produced the statistics results with CC value of 0.981, RMSE value of 7.300 mm and R^2 value of 0.962. 500/PARMA(1,1)/SVM-NNM training pattern produced the statistics results with CC value of 0.974, RMSE value of 8.803 mm and R^2 value of 0.944. 1000/PARMA(1,1)/SVM-NNM training pattern produced the statistics results with CC value of 0.976, RMSE value of 9.448 mm and R^2 value of 0.936, respectively. In FAO-56 PM ET_0 of Haenam station, from the Table 5, 100/PARMA(1,1)/SVM-NNM training pattern produced the statistics results with CC value of 0.971, RMSE value of 9.007 mm and R^2 value of 0.939. 500/PARMA(1,1)/SVM-NNM training pattern produced the statistics results with CC value of 0.970, RMSE value of 8.882 mm and R^2 value of 0.941. 1000/PARMA(1,1)/SVM-NNM training pattern produced the statistics results with CC value of 0.972, RMSE value of 8.581 mm and R^2 value of 0.945, respectively. From the above results, the statistics results of the testing performance for 100/PARMA(1,1)/SVM-NNM training pattern were better than those of the testing performances for 500/PARMA(1,1)/SVM-NNM and 1000/PARMA(1,1)/SVM-NNM training patterns for FAO-56 PM ET_0 of Gwangju station. And, the statistics results of the testing performance for 1000/PARMA(1,1)/SVM-NNM training pattern were better than those of the testing performances for 100/PARMA(1,1)/SVM-NNM and 500/PARMA(1,1)/SVM-NNM training patterns for FAO-56 PM ET_0 of Haenam station, respectively. Kim (2011) suggested, however, that the statistics results of testing performance for 1000/ PARMA(1,1)/GRNNM-GA training pattern were better than those of testing performances for 100/PARMA(1,1)/GRNNM-GA and 500/PARMA(1,1)/GRNNM-GA training patterns for the modeling of PE and ET_r, South Korea. The continuous research will be needed to establish the neural networks models available for the optimal training patterns and modeling of PE and FAO-56 PM ET_0.

From the statistics results of the testing performances for PE and FAO-56 PM ET_0, the statistics results of FAO-56 PM ET_0 were better than those of PE. PE is the observed data as a reason and represents the natural phenomenon including strong nonlinear patterns and various uncertainties, whereas ET_o is calculated by FAO-56 PM equation with the constant operation processes. In FAO-56 PM ET_o, furthermore, the strong nonlinear patterns of the natural phenomena are transformed into linear patterns including the constant uncertainty. The author can consider that the modeling of FAO-56 PM ET_o has significantly less uncertainty compared to that of PE. Kim (2011) suggested the similar results for the modeling of PE and ET_r using the neural networks models. He suggested that the statistics results of ET_r were better than those of PE for the modeling of PE and ET_r using GRNNM-BP and GRNNM-GA, South Korea. Fig. 5(a)-(f) show comparison plots of observed and calculated PE/FAO-56 PM ET_o for the testing performances of 3 training patterns for Gwangju station, respectively. Fig. 6(a)-(f) show comparison plots of observed and calculated PE/FAO-56 PM ET_o for the testing performances of 3 training patterns for Haenam station, respectively.

Station	Statistics Indexes	100/PARMA(1,1)/ SVM-NNM	500/PARMA(1,1)/ SVM-NNM	1000/PARMA(1,1) SVM-NNM
Gwangju	CC	0.981	0.974	0.976
	RMSE (mm)	7.300	8.803	9.448
	R²	0.962	0.944	0.936
Haenam	CC	0.971	0.970	0.972
	RMSE (mm)	9.007	8.882	8.581
	R²	0.939	0.941	0.945

Table 5. Statistics results of the testing performances (FAO-56 PM ET_o)

6.5 Homogeneity evaluation

FAO-56 PM ET_o equation, which has been unanimously accepted by the FAO consultation members for ET_o calculation (Allen et al., 1998), was used to calculate ET_o since there are no observed data for ET_o using a lysimeter, South Korea. Even if FAO-56 PM ET_o is not observed data, the reliability for the calculated FAO-56 PM ET_o is adequate and proper. Homogeneity evaluation was performed to compare the observed PE/FAO-56 PM ET_o with the calculated PE/FAO-56 PM ET_o for the results of the test performances, respectively. In this study, homogeneity evaluation consisted of the One-way ANOVA and the Mann-Whitney U test, respectively.

6.5.1 The One-way ANOVA

The One-way ANOVA is a class of statistical analysis that is widely used because it encourages systematic decision making for the underlying problems that involve considerable uncertainty. It enables inferences to be made in such a way that sample data can be combined with statistical theory. It supposedly removes the effects of the biases of the individual, which leads to more rational and accurate decision making. The One-way ANOVA is the formal procedure for using statistical concepts and measures in performing decision making. The following six steps can be used to make statistical analysis of the One-way ANOVA on the means and variances: 1) Formulation of hypotheses 2) Define the test statistic and its distribution 3) Specify the level of significance 4) Collect data and compute test statistic 5) Determine the critical value of the test statistic 6) Make a decision (McCuen, 1993; Salas et al., 2001; Ayyub and McCuen, 2003).

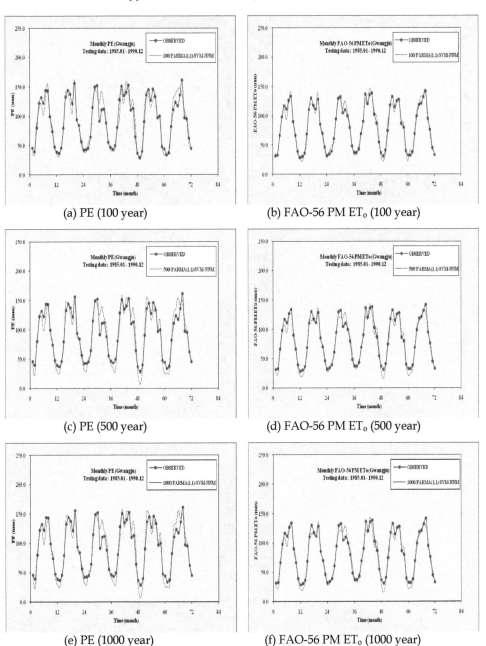

(a) PE (100 year)   (b) FAO-56 PM ET_o (100 year)

(c) PE (500 year)   (d) FAO-56 PM ET_o (500 year)

(e) PE (1000 year)   (f) FAO-56 PM ET_o (1000 year)

Fig. 5. Comparison plots of observed and calculated PE and FAO-56 PM ET_o (Gwangju)

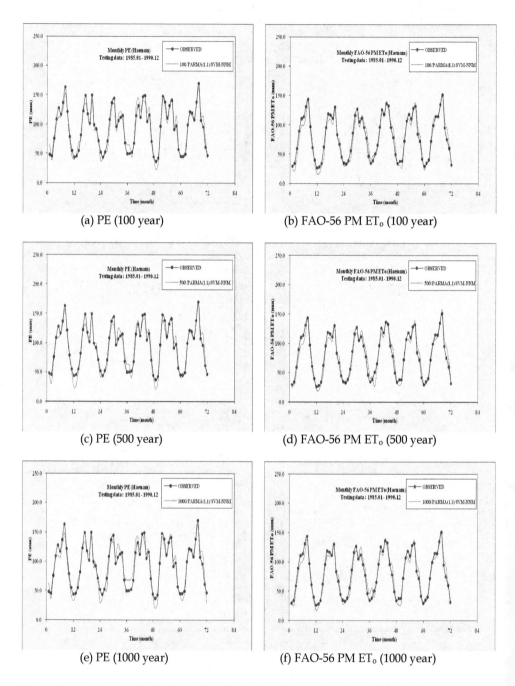

Fig. 6. Comparison plots of observed and calculated PE and FAO-56 PM ET$_o$ (Haenam)

The One-way ANOVA on the means was performed and computed t statistic using two-sample t test between the observed PE/FAO-56 PM ET_0 and the calculated PE/FAO-56 PM ET_0, respectively. The critical value of t statistic was computed for the level of significance 5 percent (5%) and 1 percent (1%). If the computed value of t statistic is greater than the critical value of t statistic, the null hypothesis, which is the means are equal, should be rejected and the alternative hypothesis should be accepted. The One-way ANOVA on the variances was performed and computed F statistic using two-sample F test between the observed PE/FAO-56 PM ET_0 and the calculated PE/FAO-56 PM ET_0, respectively. The critical value of F statistic was computed for the level of significance 5 percent (5%) and 1 percent (1%). If the computed value of F statistic is greater than the critical value of F statistic, the null hypothesis, which is the population variances are equal, should be rejected and the alternative hypothesis should be accepted.

Table 6 shows the results of the One-way ANOVA on the means of PE. The critical value of t statistic was computed as $t_{0.05}$ =1.981 and $t_{0.01}$=2.620 for the level of significance 5 percent (5%) and 1 percent (1%), respectively. The computed values of t statistic with 0.045 for 100/PARMA/SVM-NNM training pattern, 0.111 for 500/PARMA/SVM-NNM training pattern, 0.390 for 1000/PARMA/SVM-NNM training pattern were not significant for PE of Gwangju station. So, the null hypothesis, which is the means are equal, was accepted for PE of Gwangju station. Furthermore, the computed values of t statistic with 0.145 for 100/PARMA/SVM-NNM training pattern, 0.169 for 500/PARMA/SVM-NNM training pattern, 0.103 for 1000/PARMA/SVM-NNM training pattern were not significant for PE of Haenam station. So, the null hypothesis, which is the means are equal, was accepted for PE of Haenam station.

Station	Training Pattern	Level of Significance	Two-sample t test		
			Critical t Statistic	Computed t Statistic	Null Hypothesis
Gwangju	100/PARMA(1,1)/SVM-NNM	0.05/0.01	1.981/2.620	0.045	Accept/Accept
	500/PARMA(1,1)/SVM-NNM	0.05/0.01	1.981/2.620	0.111	Accept/Accept
	1000/PARMA(1,1)/SVM-NNM	0.05/0.01	1.981/2.620	0.390	Accept/Accept
Haenam	100/PARMA(1,1)/SVM-NNM	0.05/0.01	1.981/2.620	0.145	Accept/Accept
	500/PARMA(1,1)/SVM-NNM	0.05/0.01	1.981/2.620	0.169	Accept/Accept
	1000/PARMA(1,1)/SVM-NNM	0.05/0.01	1.981/2.620	0.103	Accept/Accept

Table 6. Results of the One-way ANOVA on the means of PE

Table 7 shows the results of the One-way ANOVA on the means of FAO-56 PM ET_0. The critical value of t statistic was computed as $t_{0.05}$ =1.981 and $t_{0.01}$=2.620 for the level of significance 5 percent (5%) and 1 percent (1%), respectively. The computed values of t statistic with 0.040 for 100/PARMA/SVM-NNM training pattern, 0.283 for 500/PARMA /SVM-NNM training pattern, 0.483 for 1000/PARMA/SVM-NNM training pattern were not significant for FAO-56 PM ET_0 of Gwangju station. So, the null hypothesis, which is the means are equal, was accepted for FAO-56 PM ET_0 of Gwangju station. Furthermore, the computed values of t statistic with 0.231 for 100/PARMA/SVM-NNM training pattern, 0.071 for 500/PARMA/SVM-NNM training pattern, 0.018 for 1000/PARMA/SVM-NNM training pattern were not significant for FAO-56 PM ET_0 of Haenam station. So, the null hypothesis, which is the means are equal, was accepted for FAO-56 PM ET_0 of Haenam station.

Station	Training Pattern	Level of Significance	Two-sample t test		
			Critical t Statistic	Computed t Statistic	Null Hypothesis
Gwangju	100/PARMA(1,1)/SVM-NNM	0.05/0.01	1.981/2.620	0.040	Accept/Accept
	500/PARMA(1,1)/SVM-NNM	0.05/0.01	1.981/2.620	0.283	Accept/Accept
	1000/PARMA(1,1)/SVM-NNM	0.05/0.01	1.981/2.620	0.483	Accept/Accept
Haenam	100/PARMA(1,1)/SVM-NNM	0.05/0.01	1.981/2.620	0.231	Accept/Accept
	500/PARMA(1,1)/SVM-NNM	0.05/0.01	1.981/2.620	0.071	Accept/Accept
	1000/PARMA(1,1)/SVM-NNM	0.05/0.01	1.981/2.620	0.018	Accept/Accept

Table 7. Results of the One-way ANOVA on the means of FAO-56 PM ET_0

Table 8 shows the results of the One-way ANOVA on the variances of PE. The critical value of F statistic was computed as $F_{0.05}$ =1.981 and $F_{0.01}$=2.620 for the level of significance 5 percent (5%) and 1 percent (1%), respectively. The computed values of F statistic with 1.040 for 100/PARMA/SVM-NNM training pattern, 1.249 for 500/PARMA/SVM-NNM training pattern, 1.343 for 1000/PARMA/SVM-NNM training pattern were not significant for PE of Gwangju station. So, the null hypothesis, which is the variances are equal, was accepted for PE of Gwangju station. Furthermore, the computed values of F statistic with 1.033 for 100/PARMA/SVM-NNM training pattern, 1.030 for 500/PARMA/SVM-NNM training pattern, 1.036 for 1000/PARMA/SVM-NNM training pattern were not significant for PE of Haenam station. So, the null hypothesis, which is the variances are equal, was accepted for PE of Haenam station.

Station	Training Pattern	Level of Significance	Two-sample F test		
			Critical F Statistic	Computed F Statistic	Null Hypothesis
Gwangju	100/PARMA(1,1)/SVM-NNM	0.05/0.01	1.981/2.620	1.040	Accept/Accept
	500/PARMA(1,1)/SVM-NNM	0.05/0.01	1.981/2.620	1.249	Accept/Accept
	1000/PARMA(1,1)/SVM-NNM	0.05/0.01	1.981/2.620	1.343	Accept/Accept
Haenam	100/PARMA(1,1)/SVM-NNM	0.05/0.01	1.981/2.620	1.033	Accept/Accept
	500/PARMA(1,1)/SVM-NNM	0.05/0.01	1.981/2.620	1.030	Accept/Accept
	1000/PARMA(1,1)/SVM-NNM	0.05/0.01	1.981/2.620	1.036	Accept/Accept

Table 8. Results of the One-way ANOVA on the variances of PE

Table 9 shows the results of the One-way ANOVA on the variances of FAO-56 PM ET_0. The critical value of F statistic was computed as $F_{0.05}$ =1.981 and $F_{0.01}$=2.620 for the level of significance 5 percent (5%) and 1 percent (1%), respectively. The computed values of F statistic with 1.055 for 100/PARMA/SVM-NNM training pattern, 1.045 for 500/PARMA/SVM-NNM training pattern, 1.154 for 1000/PARMA/SVM-NNM training pattern were not significant for FAO-56 PM ET_0 of Gwangju station. So, the null hypothesis, which is the variances are equal, was accepted for FAO-56 PM ET_0 of Gwangju station. Furthermore, the computed values of F statistic with 1.033 for 100/PARMA/SVM-NNM training pattern, 1.021 for 500/PARMA/SVM-NNM training pattern, 1.031 for 1000/PARMA/SVM-NNM training pattern were not significant for FAO-56 PM ET_0 of Haenam station. So, the null hypothesis, which is the variances are equal, was accepted for FAO-56 PM ET_0 of Haenam station.

Station	Training Pattern	Level of Significance	Two-sample F test		
			Critical F Statistic	Computed F Statistic	Null Hypothesis
Gwangju	100/PARMA(1,1)/SVM-NNM	0.05/0.01	1.981/2.620	1.055	Accept/Accept
	500/PARMA(1,1)/SVM-NNM	0.05/0.01	1.981/2.620	1.045	Accept/Accept
	1000/PARMA(1,1)/SVM-NNM	0.05/0.01	1.981/2.620	1.154	Accept/Accept
Haenam	100/PARMA(1,1)/SVM-NNM	0.05/0.01	1.981/2.620	1.033	Accept/Accept
	500/PARMA(1,1)/SVM-NNM	0.05/0.01	1.981/2.620	1.021	Accept/Accept
	1000/PARMA(1,1)/SVM-NNM	0.05/0.01	1.981/2.620	1.031	Accept/Accept

Table 9. Results of the One-way ANOVA on the variances of FAO-56 PM ET_0

6.5.2 The Mann-Whitney U test

The Mann-Whitney U test is a nonparametric alternative to the two-sample t test for two independent samples and can be used to test whether two independent samples have been taken from the same population. It is the most powerful alternative to the two-sample t test. Therefore, when the assumptions of the two-sample t test are violated or are difficult to evaluate such as with small samples, the Mann-Whitney U test should be applied. The Mann-Whitney U test is to be used in the case of two independent samples, and the Kruskal-Wallis test is an extension of the Mann-Whitney U test for the case of more than two independent samples (McCuen, 1993; Salas et al., 2001; Ayyub and McCuen, 2003).

The Mann-Whitney U test was performed and computed z statistic between the observed PE/FAO-56 PM ET_0 and the calculated PE/FAO-56 PM ET_0, respectively. The critical value of z statistic was computed for the level of significance 5 percent (5%) and 1 percent (1%). If the computed value of z statistic is greater than the critical value of z statistic, the null hypothesis, which is the two independent samples are from the same population, should be rejected and the alternative hypothesis should be accepted in this study.

Table 10 shows the results of the Mann-Whitney U test of PE. The critical value of z statistic was computed as $z_{0.05}$ =1.645 and $z_{0.01}$=2.327 for the level of significance 5 percent (5%) and 1 percent (1%), respectively. The computed values of z statistic with -0.196 for 100/PARMA/ SVM-NNM training pattern, -0.136 for 500/PARMA/SVM-NNM training pattern, -0.288 for 1000/PARMA/SVM-NNM training pattern were not significant for PE of Gwangju station. So, the null hypothesis, which is the two independent samples are from the same population, was accepted for PE of Gwangju station. Furthermore, the computed values of z statistic with -0.172 for 100/PARMA/SVM-NNM training pattern, -0.124 for 500/PARMA/SVM-NNM training pattern, -0.076 for 1000/PARMA/SVM-NNM training pattern were not significant for PE of Haenam station. So, the null hypothesis, which is the two independent samples are from the same population, was accepted for PE of Haenam station.

Table 11 shows the results of the Mann-Whitney U test of FAO-56 PM ET_0. The critical value of z statistic was computed as $z_{0.05}$ =1.645 and $z_{0.01}$=2.327 for the level of significance 5 percent (5%) and 1 percent (1%), respectively. The computed values of z statistic with -0.056 for 100/PARMA/SVM-NNM training pattern, -0.515 for 500/PARMA/SVM-NNM training pattern, -0.711 for 1000/PARMA/SVM-NNM training pattern were not significant for FAO-56 PM ET_0 of Gwangju station. So, the null hypothesis, which is the two independent samples are from the same population, was accepted for FAO-56 PM ET_0 of Gwangju station. Furthermore, the computed values of z statistic with -0.380 for 100/PARMA/SVM-NNM training pattern, -0.212 for 500/PARMA/SVM-NNM training pattern, -0.176 for 1000/PARMA/SVM-NNM training pattern were not significant for FAO-56 PM ET_0 of Haenam station. So, the null hypothesis, which is the two independent samples are from the same population, was accepted for FAO-56 PM ET_0 of Haenam station.

Station	Training Pattern	Level of Significance	Mann-Whitney U test		
			Critical z Statistic	Computed z Statistic	Null Hypothesis
Gwangju	100/PARMA(1,1)/SVM-NNM	0.05/0.01	1.645/2.327	-0.196	Accept/Accept
	500/PARMA(1,1)/SVM-NNM	0.05/0.01	1.645/2.327	-0.136	Accept/Accept
	1000/PARMA(1,1)/SVM-NNM	0.05/0.01	1.645/2.327	-0.288	Accept/Accept
Haenam	100/PARMA(1,1)/SVM-NNM	0.05/0.01	1.645/2.327	-0.172	Accept/Accept
	500/PARMA(1,1)/SVM-NNM	0.05/0.01	1.645/2.327	-0.124	Accept/Accept
	1000/PARMA(1,1)/SVM-NNM	0.05/0.01	1.645/2.327	-0.076	Accept/Accept

Table 10. Results of the Mann-Whitney U test of PE

Station	Training Pattern	Level of Significance	Mann-Whitney U test		
			Critical z Statistic	Computed z Statistic	Null Hypothesis
Gwangju	100/PARMA(1,1)/SVM-NNM	0.05/0.01	1.645/2.327	-0.056	Accept/Accept
	500/PARMA(1,1)/SVM-NNM	0.05/0.01	1.645/2.327	-0.515	Accept/Accept
	1000/PARMA(1,1)/SVM-NNM	0.05/0.01	1.645/2.327	-0.711	Accept/Accept
Haenam	100/PARMA(1,1)/SVM-NNM	0.05/0.01	1.645/2.327	-0.380	Accept/Accept
	500/PARMA(1,1)/SVM-NNM	0.05/0.01	1.645/2.327	-0.212	Accept/Accept
	1000/PARMA(1,1)/SVM-NNM	0.05/0.01	1.645/2.327	-0.176	Accept/Accept

Table 11. Results of the Mann-Whitney U test of FAO-56 PM ET_0

7. Construction of the Bivariate Linear Regression Analysis Model

The bivariate linear regression analysis model (BLRAM) was adopted to calculate FAO-56 PM ET_0 simply using the observed PE and compare the observed PE and the calculated FAO-56 PM ET_0. The BLRAM is a conventional and universal model, which can calculate FAO-56 PM ET_0 simply using the observed PE for Gwangju and Haenam stations, respectively. The BLRAM consists of two variables; PE as the independent variable X_t, and FAO-56 PM ET_0 as the dependent variable Y_t. The mathematical expression can be written as the following equation (12) (McCuen, 1993; Salas et al., 2001; Ayyub and McCuen, 2003).

$$Y_t = b_0 + b_1 \cdot X_t \tag{12}$$

where b_1 = the slope coefficient, which is also known as the regression coefficient because it is calculated by the result of a regression analysis. Using the BLRAM, the correlation relationship was investigated between the observed PE and FAO-56 PM ET_0 for 3 training patterns. A very good relationship was found with the BLRAM, which could calculate FAO-56 PM ET_0 in this study. The results of the BLRAM were the same for 3 training patterns. Therefore, it can be considered that the observed PE and FAO-56 PM ET_0 for 3 training patterns are homogeneous groups. It can be inferred from the homogeneity evaluation of the previous chapter.

Table 12 shows the BLRAM, goodness-of-fit test, and regression coefficient between the observed PE and FAO-56 PM ET_0 for 3 training patterns of Gwangju and Haenam stations, respectively. From the Table 12, for Gwangju station, the regression coefficient of the BLRAM indicates that FAO-56 PM ET_0 increases 0.9060 mm as each 1 mm increase in PE. R^2 = 0.966 indicates that the total variance of FAO-56 PM ET_0 corresponds to 96.6%. The ratio of $S(e)/S(y)$ = 0.187 suggests a very good level of accuracy. In addition, the standard error of estimate (SEE) decreases from 37.471 mm ($S(y)$) to 7.007 mm ($S(e)$). where $S(y)$ = the standard deviation of FAO-56 PM ET_0; and $S(e)$ = the standard error of FAO-56 PM ET_0 using the equation (12). The overall deviations are nearly zero, and this tendency always occurs for the BLRAM. The standard error ratios of the regression coefficient (b_1) and the intercept (b_0) are 0.023 and 0.911, which indicates that the regression coefficient is relatively more accurate than the intercept. For Haenam station, the regression coefficient of the BLRAM indicates that FAO-56 PM ET_0 increases 0.9574 mm as each 1 mm increase in PE. R^2 = 0.919 indicates that the total variance of FAO-56 PM ET_0 corresponds to 91.9%. The ratio of $S(e)/S(y)$ = 0.287 suggests a very good level of accuracy. In addition, the standard error of estimate (SEE) decreases from 36.794 mm ($S(y)$) to 10.560 mm ($S(e)$). where $S(y)$ = the standard deviation of FAO-56 PM ET_0; and $S(e)$ = the standard error of FAO-56 PM ET_0 using the equation (12). The overall deviations are nearly zero, and this tendency always occurs for the BLRAM. The standard error ratios of the regression coefficient (b_1) and the intercept (b_0) are 0.035 and 0.345, which indicates that the regression coefficient is relatively more accurate than the intercept. If a large and negative intercept exists, it can create some problems for forecasting or modeling (McCuen, 1993). Fig. 7(a)-(b) show comparison plots of the observed PE and FAO-56 PM ET_0 for 1000/PARMA(1,1)/SVM-NNM training pattern of Gwangju and Haenam station, respectively.

	BLRAM (3 Training Patterns)	Goodness-of-fit test			Regression coefficient analysis	
		$S(e)$ (mm)	$S(e)/S(y)$	R^2	$Se(b_1)/b_1$	$Se(b_0)/b_0$
Gwangju	$ET_0 = 0.9060\ PE - 2.2901$	7.007	0.187	0.966	0.023	0.911
Haenam	$ET_0 = 0.9574\ PE - 9.9968$	10.560	0.287	0.919	0.035	0.345

Table 12. Regression analysis between the observed PE and FAO-56 PM ET_0

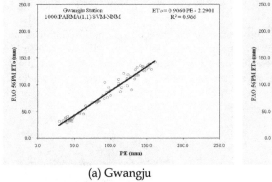
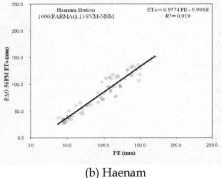

(a) Gwangju   (b) Haenam

Fig. 7. Comparison plots of the observed PE and FAO-56 PM ET_0

8. Conclusions

The hybrid method was developed for the modeling of the monthly PE and FAO-56 PM ET_o simultaneously. The author determined in advance 4 kinds of Univariate Seasonal PARMA(p,q) models including PARMA(1,1), PARMA(1,2), PARMA (2,1), and PARMA(2,2), which are the low-order models and contain the seasonal properties. As a result, the author selected Univariate Seasonal PARMA(1,1) model, which show the best statistical properties and is simple in parameter estimation. The data which were generated by Univariate Seasonal PARMA(1,1) model consisted of 100 years (Short-term), 500 years (Mid-term), and 1000 years (Long-term), respectively. The following conclusions can be drawn from this study.

[1] The statistics results of the testing performance for 100/PARMA(1,1)/SVM-NNM training pattern were better than those of the testing performances for 500/PARMA(1,1) /SVM-NNM and 1000/PARMA(1,1)/SVM-NNM training patterns for PE of Gwangju station. And, the statistics results of the testing performance for 500/PARMA(1,1)/SVM-NNM training pattern were better than those of the testing performances for 100 /PARMA(1,1)/SVM-NNM and 1000/ PARMA(1,1)/SVM-NNM training patterns for PE of Haenam station, respectively

[2] The statistics results of the testing performance for 100/PARMA(1,1)/SVM-NNM training pattern were better than those of the testing performances for 500/PARMA(1,1) /SVM-NNM and 1000/PARMA(1,1)/SVM-NNM training patterns for FAO-56 PM ET_o of Gwangju station. And, the statistics results of the testing performance for 1000 /PARMA(1,1)/SVM-NNM training pattern were better than those of the testing performances for 100/PARMA(1,1)/SVM-NNM and 500/PARMA(1,1)/SVM-NNM training patterns for FAO-56 PM ET_o of Haenam station, respectively

[3] Homogeneity evaluation consisted of the One-way ANOVA and the Mann-Whitney U test. The null hypothesis, which is the means are equal, was accepted using the One-way ANOVA on the means for PE and FAO-56 PM ET_o of Gwangju and Haenam stations, respectively. And, the null hypothesis, which is the variances are equal, was accepted using the One-way ANOVA on the variances for PE and FAO-56 PM ET_o of Gwangju and Haenam stations, respectively. The null hypothesis, which is the two independent samples are from the same population, was accepted using the Mann-Whitney U test for PE and FAO-56 PM ET_o of Gwangju and Haenam stations, respectively.

[4] The BLRAM was adopted to calculate FAO-56 PM ET_o simply using the observed PE and compare the observed PE and the calculated FAO-56 PM ET_o of Gwangju and Haenam stations, respectively. A very good relationship was found with the BLRAM, which could calculate FAO-56 PM ET_o.

As PE and FAO-56 PM ET_o are relatively important for the design of irrigation facilities and agricultural reservoirs, the spread of an automatic measuring system for PE and FAO-56 PM ET_o is important and urgent to ensure the reliable and accurate data from the measurements of PE and FAO-56 PM ET_o. Furthermore, the continuous research will be needed to establish the neural networks models available for the optimal training patterns and modeling of PE and FAO-56 PM ET_o.

9. References

Allen, R.G.; Jensen, M.E.; Wright, J.L. & Burman, R.D. (1989). Operational estimates of reference evapotranspiration. *Agronomy Journal*, Vol. 81, No. 4, pp. 650-662, ISSN: 0002-1962.

Allen, R.G.; Pereira, L.S.; Raes, D. & Smith, M. (1998). *Crop evapotranspiration. Guidelines for computing crop water requirement,* Irrigation and Drainage Paper No. 56, FAO, Rome, Italy.

ASCE Task Committee on Definition of Criteria for Evaluation of Watershed Models. (1993). Criteria for evaluation of watershed models. *Journal of Irrigation and Drainage Engineering, ASCE,* Vol. 119, No. 3, pp. 429-442, ISSN: 0733-9437.

Ayyub, B.M. & McCuen, R.H. (2003). *Probability, Statistics, and Reliability for Engineers and Scientists 2nd Edition,* Taylor & Francis, ISBN: 1584882867, Boca Raton, FL, USA.

Bishop, C.M. (1994). Neural networks and their applications. *Review of Scientific Instruments,* Vol. 65, pp. 1803-1832, ISSN: 0034-6748.

Bruton, J.M.; McClendon, R.W. & Hoogenboom, G. (2000). Estimating daily pan evaporation with artificial neural networks. *Transactions of the American Society Agricultural Engineers, ASAE,* Vol. 43, No. 2, pp. 491-496, ISSN:0001-2351.

Chang, F.J.; Chang, L.C.; Kao, H.S. & Wu, G.R. (2010). Assessing the effort of meteorological variables for evaporation estimation by self-organizing map neural network. *Journal of Hydrology,* Vol. 384, pp. 118-129, ISSN: 0022-1694.

Choudhury, B.J. (1999). Evaluation of empirical equation for annual evaporation using field observations and results from a biophysical model. *Journal of Hydrology,* Vol. 216, pp. 99-110, ISSN: 0022-1694.

Cover, T.M. (1965). Geometrical and statistical properties of systems of linear inequalities with applications in pattern recognition. *IEEE Transactions on Electronic Computers EC-14,* pp. 326-334.

Dibike, Y.B.; Velickov, S.; Solomatine, D. & Abbott, M.B. (2001). Model induction with support vector machines: introductions and applications. *Journal of Computing in Civil Engineering, ASCE,* Vol. 15, No. 3, pp. 208-216, ISSN: 0887-3801.

Doorenbos, J. & Pruitt, W.O. (1977). *Guidelines for predicting crop water requirement,* Irrigation and Drainage Paper No. 24 2nd Edition, FAO, Rome, Italy.

Finch, J.W. (2001). A comparison between measured and modeled open water evaporation from a reservoir in south-east England. *Hydrological Processes,* Vol. 15, pp. 2771-2778, ISSN: 0885-6087.

Guven, A. & Kisi, O. (2011). Daily pan evaporation modeling using linear genetic programming technique. *Irrigation Science,* Vol. 29, No. 2, pp. 135-145, ISSN: 0342-7188.

Haykin, S. (2009). *Neural networks and learning machines 3rd Edition,* Prentice Hall, ISBN: 0131471392, NJ, USA.

Hirsch R.M. (1979). Synthetic hydrology and water supply reliability. *Water Resources Research,* Vol. 15, pp.1603-1615, ISSN: 0043-1397.

Hush, D.R. & Horne, B.G. (1993). Progress in supervised neural network : What's new since Lippmann ?. *IEEE Signal Processing Magazine,* Vol. 10, pp. 8-39, ISSN: 1053-5888.

Jain, S.K.; Nayak, P.C. & Sudheer, K.P. (2008). Models for estimating evapotranspiration using artificial neural networks, and their physical interpretation. *Hydrological Processes,* Vol. 22, pp. 2225-2234, ISSN: 0885-6087.

Jensen, M.E.; Burman, R.D. & Allen, R.G. (1990). *Evapotranspiration and irrigation water requirements,* ASCE Manual and Report on Engineering Practice No. 70, ASCE, NY, USA.

Keskin, M.E. & Terzi, O. (2006). Artificial neural networks models of daily pan evaporation. *Journal of Hydrologic Engineering, ASCE,* Vol. 11, No. 1, pp. 65-70, ISSN: 1084-0699.

Khadam, I.M. & Kaluarachchi, J.J. (2004). Use of soft information to describe the relative uncertainty of calibration data in hydrologic models. *Water Resources Research,* Vol. 40, No. 11, W11505, ISSN: 0043-1397.

Kim, S. (2004). Neural Networks Model and Embedded Stochastic Processes for Hydrological Analysis in South Korea. *KSCE Journal of Civil Engineering,* Vol. 8, No. 1, pp. 141-148, ISSN: 1226-7988.

Kim, S. (2011). Nonlinear hydrologic modeling using the stochastic and neural networks approach. *Disaster Advances,* Vol. 4, No. 1, pp. 53-63, ISSN: 0974-262X.

Kim, S.; Kim, J.H. & Park, K.B. (2009). Neural networks models for the flood forecasting and disaster prevention system in the small catchment. *Disaster Advances,* Vol. 2, No. 3, pp. 51-63, ISSN: 0974-262X.

Kim, S. & Kim, H.S. (2008). Neural networks and genetic algorithm approach for nonlinear evaporation and evapotranspiration modeling. *Journal of Hydrology,* Vol. 351, pp. 299-317, ISSN: 0022-1694.

Kisi, O. (2006). Generalized regression neural networks for evapotranspiration modeling. *Hydrological Sciences Journal,* Vol. 51, No. 6, pp. 1092-1105, ISSN: 0262-6667.

Kisi, O. (2007). Evapotranspiration modeling from climatic data using a neural computing technique. *Hydrological Processes,* Vol. 21, pp. 1925-1934, ISSN: 0885-6087.

Kisi, O. (2009). Modeling monthly evaporation using two different neural computing technique. *Irrigation Science,* Vol. 27, No. 5, pp. 417-430, ISSN: 0342-7188.

Kisi, O. & Ozturk, O. (2007). Adaptive neurofuzzy computing technique for evapotranspiration estimation. *Journal of Irrigation and Drainage Engineering, ASCE,* Vol. 133, No. 4, pp. 368-379, ISSN: 0733-9437.

Kumar, M.; Bandyopadhyay, A.; Raghuwanshi, N.S. & Singh, R. (2008). Comparative study of conventional and artificial neural network-based ET_0 estimation models. *Irrigation Science,* Vol. 26, No. 6, pp. 531-545, ISSN: 0342-7188.

Kumar, M.; Raghuwanshi, N.S. & Singh, R. (2009). Development and validation of GANN model for evapotranspiration estimation. *Journal of Hydrologic Engineering, ASCE,* Vol. 14, No. 2, pp. 131-140, ISSN: 1084-0699.

Kumar, M.; Raghuwanshi, N.S.; Singh, R.; Wallender, W.W. & Pruitt, W.O. (2002). Estimating evapotranspiration using artificial neural network. *Journal of Irrigation and Drainage Engineering, ASCE,* Vol. 128, No. 4, pp. 224-233, ISSN: 0733-9437.

Landeras, G.; Ortiz-Barredo, A. & Lopez, J.J. (2008). Comparison of artificial neural network models and empirical and semi-empirical equations for daily reference evapotranspiration estimation in the Basque country (Northern Spain). *Agricultural Water Management,* Vol. 95, No. 5, pp. 553-565, ISSN: 0378-3774.

McCuen, R.H. (1993). *Microcomputer applications in statistical hydrology,* Prentice Hall, ISBN: 0135852900, Englewood Cliffs, NJ, USA.

McKenzie, R.S. & Craig, A.R. (2001). Evaluation of river losses from the Orange River using hydraulic modeling. *Journal of Hydrology,* Vol. 241, pp. 62-69, ISSN: 0022-1694.

Mishra, A.K.; Desai, V.R. & Singh, V.P. (2007). Drought forecasting using a hybrid stochastic and neural network model. *Journal of Hydrologic Engineering, ASCE,* Vol. 12, No. 6, pp. 626-638, ISSN: 1084-0699.

Monteith, J.L. (1965). The state and movement of water in living organism, *Proceedings of Evaporation and Environment,* pp. 205-234, Swansea, Cambridge University Press, NY, USA.

Nash, J.E. & Sutcliffe, J.V. (1970). River flow forecasting through conceptual models, Part 1 - A discussion of principles. *Journal of Hydrology*, Vol. 10, No. 3, pp. 282-290, ISSN: 0022-1694.

Penman, H.L. (1948). Natural evaporation from open water, bare soil and grass, *Proceedings of the Royal Society of London*, Vol. 193, pp. 120-146, London, England.

Principe, J.C.; Euliano, N.R. & Lefebvre, W.C. (2000). *Neural and adaptive systems: fundamentals through simulation*, Wiley, John & Sons, ISBN: 0471351679, NY, USA.

Rosenberry, D.O.; Winter, T.C.; Buso, D.C. & Likens, G.E. (2007). Comparison of 15 evaporation methods applied to a small mountain lake in the northeastern USA. *Journal of Hydrology*, Vol. 340, pp. 149-166, ISSN: 0022-1694.

Salas, J.D. (1993). Analysis and modeling of hydrologic time series. In: *Handbook of Hydrology, Chapter 19*, Maidment, D.R., (Ed.), pp. 19.1-19.72, McGraw-Hill, ISBN: 0070397325, NY, USA.

Salas, J.D. & Abdelmohsen, M. (1993). Initialization for generating single site and multisite low order PARMA processes. *Water Resources Research*, Vol.29, pp.1771-1776, ISSN: 0043-1397.

Salas J.D.; Boes D.C.; & Smith, R.A. (1982). Estimation of ARMA models with seasonal parameters. *Water Resources Research*, Vol. 18, pp.1006-1010, ISSN: 0043-1397.

Salas J.D.; Delleur J.R.; Yevjevich, V & Lane, W.L. (1980). *Applied modeling of hydrologic time series*, Water Resources Publication, ISBN: 0918334373, CO, USA.

Salas, J.D.; Smith, R.A., Tabios III, G.O. & Heo, J.H. (2001). *Statistical computing techniques in water resources and environmental engineering*, Unpublished book in CE622, Colorado State University, Fort Collins, CO, USA.

Shiri, J. & Kisi, O. (2011). Application of artificial intelligence to estimate daily pan evaporation using available and estimated climatic data in the Khozestan Province (Southwestern Iran). *Journal of Irrigation and Drainage Engineering, ASCE*, Vol. 137, No. 7, pp. 412-425, ISSN: 0733-9437.

Singh, V.P. (1988). *Hydrologic system rainfall-runoff modeling. Vol. 1*, Prentice Hall, ISBN: 0134480511, NJ, USA.

Sivakumar B.; Jayawardena, A.W. & Fernando, T.M.K.G. (2002). River flow forecasting : use of phase-space reconstruction and artificial neural networks approaches. *Journal of Hydrology*, Vol. 265, pp. 225-245, ISSN: 0022-1694

Sudheer, K.P.; Gosain, A.K.; Rangan, D.M. & Saheb, S.M. (2002). Modeling evaporation using an artificial neural network algorithm. *Hydrological Processes*, Vol. 16, pp. 3189-3202, ISSN: 0885-6087.

Sudheer, K.P.; Gosain, A.K. & Ramasastri, K.S. (2003). Estimating actual evapotranspiration from limited climatic data using neural computing technique. *Journal of Irrigation and Drainage Engineering, ASCE*, Vol. 129, No. 3, pp. 214-218, ISSN: 0733-9437.

Tabari, H.; Marofi, S. & Sabziparvar, A.A. (2009). Estimation of daily pan evaporation using artificial neural network and multivariate non-linear regression. *Irrigation Science*, Vol. 28, No. 5, pp. 399-406, ISSN: 0342-7188.

Tao, P.C. & Delleur, J.W. (1976). Seasonal and nonseasonal ARMA models in hydrology. *Journal of Hydraulic Division, ASCE*, Vol. 102, pp.1591-1599, ISSN: 0733-9429.

Tokar, A.S. & Johnson, P.A. (1999). Rainfall-runoff modeling using artificial neural networks. *Journal of Hydrologic Engineering, ASCE*, Vol. 4, No. 3, pp. 232-239, ISSN: 1084-0699.

Trajkovic, S. (2005). Temperature-based approaches for estimating reference evapotranspiration. *Journal of Irrigation and Drainage Engineering, ASCE,* Vol. 131, No. 4, pp. 316-323, ISSN: 0733-9437.

Trajkovic, S; Todorovic, B. & Stankovic, M. (2003). Forecasting reference evapotranspiration by artificial neural networks. *Journal of Irrigation and Drainage Engineering, ASCE,* Vol. 129, No. 6, pp. 454-457, ISSN: 0733-9437.

Tripathi, S.; Srinivas, V.V. & Nanjundish, R.S. (2006). Downscaling of precipitation for climate change scenarios: A support vector machine approach. *Journal of Hydrology,* Vol. 330, pp. 621-640, ISSN: 0022-1694.

Vallet-Coulomb, C.; Legesse, D.; Gasse, F.; Travi, Y. & Chernet, T. (2001). Lake evaporation estimates in tropical Africa (Lake Ziway, Ethiopia). *Journal of Hydrology,* Vol. 245, pp. 1-18, ISSN: 0022-1694.

Vapnik, V.N. (1992). Principles of risk minimization for learning theory. In: *Advances in Neural Information Processing Systems Vol. 4,* Moody, Hanson & Lippmann, (Ed.), pp. 831-838, Elsevier, ISBN: 1558602224, NY, USA.

Vapnik, V.N. (2010). *The nature of statistical learning theory 2nd Edition,* Springer-Verlag, ISBN: 0387987800, NY, USA.

Zanetti, S.S.; Sousa, E.F.; Oliveira, V.P.S.; Almeida, F.T. & Bernardo, S. (2007). Estimating evapotranspiration using artificial neural network and minimum climatological data. *Journal of Irrigation and Drainage Engineering, ASCE,* Vol. 133, No. 2, pp. 83-89, ISSN: 0733-9437.

6

Modelling Evapotranspiration and the Surface Energy Budget in Alpine Catchments

Giacomo Bertoldi[1], Riccardo Rigon[2] and Ulrike Tappeiner[1,3]

[1] *Institute for Alpine Environment, EURAC Research, Bolzano.*
[2] *University of Trento.*
[3] *Institute of Ecology, University of Innsbruck.*
[1,2] *Italy*
[3] *Austria.*

1. Introduction

Accurate modelling of evapotranspiration (ET) is required to predict the effects of climate and land use changes on water resources, agriculture and ecosystems. Significant progress has been made in estimating ET at the global and regional scale. However, further efforts are needed to improve spatial accuracy and modeling capabilities in alpine regions (Brooks & Vivoni, 2008b). This chapter will point out the components of the energy budget needed to model ET, to discuss the fundamental equations and to provide an extended review of the parametrizations available in the hydrological and land surface models literature. The second part of the chapter will explore the complexity of the energy budget with special reference to mountain environments.

2. The energy budget components

Evapotranspiration is controlled by the surface water and energy budget. In this section the single components of the energy budget will be discussed: radiation, soil heat flux, sensible and latent heat fluxes.

The surface energy budget inside a control volume can be written as:

$$\Delta t \left(R_n - H - LE - G \right) = \Delta E \tag{1}$$

where the energy fluxes concerning the soil-atmosphere interface in the time interval Δt are:

R_n net radiative flux;
H sensible heat flux;
LE latent heat flux;
G heat flux in the soil;
ΔE internal energy variation in the control volume;

The control volume is assumed with a thickness of some meters, so as to include the soil layer close to the surface and the first meters of the atmosphere, including the possible vegetation cover.

Besides the surface energy budget also the mass budget must be considered in order to quantify ET, namely the water conservation inside the control volume.

$$\frac{\Delta S}{\Delta t} = P - ET - R - R_G - R_S \tag{2}$$

where ΔS is the storing of the various supplies (underground and surface storage, soil moisture, vegetation interception, storing in channels); P is precipitation; $ET = \lambda LE$ is evapotranspiration; λ is the latent heat of vaporization; R is the surface runoff; R_G is the runoff towards the deep water table; R_S is the sub-surface runoff.

In the next sections it will be explained how the different components of the energy budget are usually described in hydrological and Land Surface Models (LSMs): radiation, sensible and latent heat flux, soil heat flux.

2.1 Radiation

The radiation is usually divided in short-wave components - indicated here as SW (including visible light, part of the ultraviolet radiation and the close infrared) and long-wave components - indicated here as LW (infrared radiation), with wavelength λ ranging respectively from 0.1 to 3 μm (98% of extraterrestrial radiation) and from 3 to 100 μm (2% of extraterrestrial radiation). In photosynthetic processes the short-wave radiation is further divided in photosynthetically active radiation (0.4 $< \lambda <$ 0.7μm) and near infrared (Bonan, 1996).

Moreover, there is a further distinction between diffuse radiation D (coming by diffusion in the atmosphere from any direction) and direct radiation P (coming only from the sun), and also between radiation from the sky downwards ("down" $\downarrow$) and radiation from the soil upwards ("up" $\uparrow$).

The net radiation at the soil level can be factorized as follows:

$$R_n = sw \cdot (R \downarrow_{SW P} + R \downarrow_{LW P}) + V \cdot (R \downarrow_{SW D} + R \downarrow_{LW D}) - \tag{3}$$
$$R \uparrow_{SW} - R \uparrow_{LW R} - R \uparrow_{LW} + R \downarrow_{SW O} + R \downarrow_{LW O}$$

with:

$R \downarrow_{SW P}$	short-wave direct radiation;
$R \downarrow_{LW P}$	long-wave direct radiation;
$R \downarrow_{SW D}$	short-wave diffuse radiation;
$R \downarrow_{LW D}$	long-wave diffuse radiation;
$R \uparrow_{SW}$	short-wave reflected radiation;
$R \uparrow_{LW R}$	long-wave reflected radiation;
$R \uparrow_{LW}$	radiation emitted by surface;
$R \downarrow_{SW O} + R \downarrow_{LW O}$	reflected radiation emitted by the surrounding surfaces;

The effects of topography on the diffused radiation can be expressed through the sky view factor V, indicating the sky fraction visible in one point.

In the presence of reliefs only part of the horizon is visible and this has consequences on the radiative exchanges.

V is defined as:

$$V = \omega / 2\pi \tag{4}$$

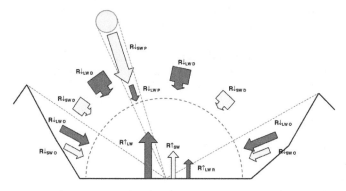

Fig. 1. Scheme of the solar radiation components.

where ω is the solid angle seen from the point considered.

The presence of shadows due to surrounding mountains can be expressed through a factor sw, a function of topography and sun position, defined as:

$$sw = \begin{cases} 1 \text{ if the point is in the sun} \\ 0 \text{ if the point is in the shadow} \end{cases} \qquad (5)$$

All direct radiation terms have to be multiplied by this factor.

In the next paragraphs we analyze in detail the parametrization of the single terms composing the radiation flux.

2.1.1 Direct radiation $\quad R\downarrow_{SW\ P}$ and $R\downarrow_{LW\ P}$

The direct long-wave radiation $R\downarrow_{LW\ P}$ is emitted directly by the sun and therefore it is negligible at the soil level (differently from the long-wave diffuse radiation).

Usually the short-wave radiation $R\downarrow_{SW\ P}$ is assumed as an input variable, measured or calculated by an atmospheric model. The direct radiation can be written as the product of the extraterrestrial radiation R_{Extr} by an attenuation factor varying in time and space.

$$R\downarrow_{SW\ P} = F_{att}R_{Extr} \qquad (6)$$

The extraterrestrial radiation can be easily calculated on the basis of geometric formulas (Iqbal, 1983). The atmospheric attenuation is due to Rayleigh diffusion, to the absorption on behalf of ozone and water vapor, to the extinction (both diffusion and absorption) due to atmospheric dust and shielding caused by the possible cloud cover. Moreover the absorption entity depends on the ray path length through the atmosphere, a function of the incidence angle and of the measurement point elevation. The effect of the latter can be very important in a mountain environment, where it is necessary to consider the shading effects.

Part of the dispersed radiation is then returned as short-wave diffuse radiation ($R\downarrow_{SW\ D}$) and part of the energy absorbed by atmosphere is then re-emitted as long-wave diffuse radiation ($R\downarrow_{LW\ D}$).

From a practical point of view, according to the application type and depending on the measured data possessed, the attenuation coefficient can be calculated with different degrees of complexity. The radiation transfer through the atmosphere is a well studied phenomenon and there exist many models providing the soil incident radiation spectrum in a detailed way, considering the various attenuation effects separately (Kondratyev, 1969).

2.1.2 Diffuse downward short-wave radiation $\quad R \downarrow_{SW\,D}$

This term is a function of the atmospheric radiation due to Rayleigh dispersion and to the aereosols dispersion, as well as to the presence of cloud cover. The $R \downarrow_{SW\,D}$ actually is not isotropic and it depends on the sun position above the horizon. For its parametrization, see, for example, Paltrige & Platt (1976).

2.1.3 Diffuse downward long-wave radiation $\quad R \downarrow_{LW\,D}$

Often this term in not provided by standard meteorological measurements, and many LSMs provide expressions to calculate it. This term indicates the long-wave radiation emitted by atmosphere towards the earth. It can be calculated starting from the knowledge of the distribution of temperature, humidity and carbon dioxide of the air column above. If this information is not available, various formulas, based only on ground measurements, can be found in literature with expressions as follows:

$$R \downarrow_{LW\,D} = \epsilon_a \sigma T_a^4 \tag{7}$$

with:

T_a air temperature [K];

ϵ_a atmosphere emissivity $f(e_a, T_a,$ cloud cover);

e_a vapor pressure [mb];

Usually for ϵ_a empirical formulas have been used, but it is also possible to provide a derivation based on physical topics like in Prata (1996). The cloud cover effect on this term is significant and not easy to consider in a simple way. Cloud cover data can be provided during the day by ground or satellite observations but, especially on night, is difficult to collect.

2.1.4 Reflected short-wave radiation $\quad R \uparrow_{SW}$

This term indicates the short-wave energy reflection.

$$R \uparrow_{SW} = a(R \downarrow_{SW\,P} + R \downarrow_{SW\,D}) \tag{8}$$

where a is the albedo.

The albedo depends strongly on the wave length, but generally a mean value is used for the whole visible spectrum. Besides its dependance on the surface type, it is important to consider its dependence on soil water content, vegetation state and surface roughness. The albedo depends moreover on the sun rays inclination, in particular for smooth surfaces: for small angles it increases. There is very rich literature about albedo description, it being a key parameter in the radiative exchange models, see for example Kondratyev (1969). Albedo is often divided in visible, near infrared, direct and diffuse albedo, as in Bonan (1996).

2.1.5 Long-wave radiation emitted by the surface $\quad R \uparrow_{LW}$

This term indicates the long-wave radiation emitted by the earth surface, considered as a grey body with emissivity ε_s (values from 0.95 to 0.98). The surface temperature $T_s\ [K]$ is unknown and must be calculated by a LSM. $\sigma = 5.6704 \cdot 10^{-8}\ W/(m^2K^4)$ is Stefan-Boltzman constant.

$$R\uparrow_{LW} = \varepsilon_s \sigma T_s^4 \tag{9}$$

2.1.6 Reflected long-wave radiation $R\uparrow_{LW\ R}$

This term is small and can be subtracted by the incoming long-wave radiation, assuming surface emissivity ε_s equal to surface absorptivity:

$$R\downarrow_{LW\ D} = \varepsilon_s \cdot \epsilon_a \sigma T_a^4 \tag{10}$$

2.1.7 Radiation emitted and reflected by surrounding surfaces $R\downarrow_{SW\ O} + R\downarrow_{LW\ O}$

It indicates the radiation reflected ($R\uparrow_{SW} + R\uparrow_{LW\ R}$) and emitted ($R\uparrow_{LW}$) by the surfaces adjacent to the point considered. This term is important at small scale, in the presence of artificial obstructions or in the case of a very uneven orography. To calculate it with precision it is necessary to consider reciprocal orientation, illumination, emissivity and the albedo of every element, through a recurring procedure (Helbig et al., 2009). A simple solution is proposed for example in Bertoldi et al. (2005).

If the intervisible surfaces are hypothesized to be in radiative equilibrium, i.e. they absorb as much as they emit, these terms can be quantified in a simplified way:

$$\begin{aligned} R\downarrow_{SW\ O} &= (1 - V)R\uparrow_{SW} \\ R\downarrow_{LW\ O} &= (1 - V)(R\uparrow_{LW} + R\uparrow_{LW\ R}) \end{aligned} \tag{11}$$

2.1.8 Net radiation

Inserting expressions (7) and (9) in the (3), the net radiation is:

$$R_n = [sw \cdot R\downarrow_{SW\ P} + V \cdot R\downarrow_{SW\ D}](1 - V \cdot a) + V \cdot \varepsilon_s \cdot \varepsilon_a \cdot \sigma \cdot T_a^4 - V \cdot \varepsilon_s \cdot \sigma \cdot T_s^4 \tag{12}$$

with $\varepsilon_a = f(e_a, T_a, \text{cloud cover})$ as for example in Brutsaert (1975).

Equation (12) is not invariant with respect to the spatial scale of integration: indeed it contains non-linear terms in T_a, T_s, e_a, consequently the same results are not obtained if the local values of these quantities are substituted by the mean values of a certain surface. Therefore, the shift from a treatment valid at local level to a distributed model valid over a certain spatial scale must be done with a certain caution.

2.1.9 Radiation adsorption and backscattering by vegetation

Expression (12) needs to be modified to take into account the radiation adsorption and backscattering by vegetation, as shown in Figure 2. This effect is very important to obtain a correct soil surface skin temperature (Deardorff, 1978). From Best (1998) it is possible to derive the following relationship:

$$\begin{aligned} R_n &= [sw \cdot R\downarrow_{SW\ P} + V \cdot R\downarrow_{SW\ D}](1 - V \cdot a) * (f_{trasm} + a_v) \\ &\quad + (1 - \varepsilon_v) \cdot V \cdot \varepsilon_s \cdot \varepsilon_a \cdot \sigma \cdot T_a^4 + \varepsilon_v \cdot \varepsilon_s \cdot \sigma \cdot T_v^4 \end{aligned} \tag{13}$$

where T_v is vegetation temperature, ε_v vegetation emissivity (supposed equal to absorption), a_v vegetation albedo (downward albedo supposed equal to upward albedo) and f_{trasm}

vegetation transmissivity, depending on plant type, leaf area index and photosynthetic activity.

Models oriented versus ecological applications have a very detailed parametrization of this term (Dickinson et al., 1986). Bonan (1996) uses a two-layers canopy model. Law et al. (1999) explicit the relationship between leaf area distribution and radiative transfer. A first energy budget is made at the canopy cover layer, and the energy fluxes are solved to find the canopy temperature, then a second energy budget is made at the soil surface. Usually a fraction of the grid cell is supposed covered by canopy and another fraction by bare ground.

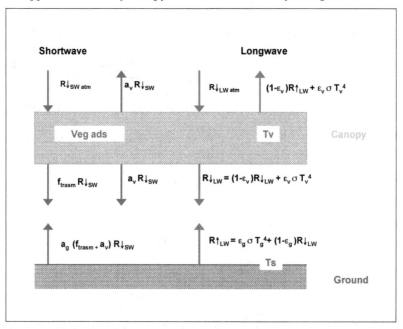

Fig. 2. Schematic diagram of short-wave radiation (left) and long-wave radiation (right) absorbed, transmitted and reflected by vegetation and ground , as in equation 13 (from Bonan (1996), modified).

2.2 Soil heat flux

The soil heat flux G at a certain depth z depends on the temperature gradient as follows:

$$G = -\lambda_s \frac{\partial T_s}{\partial z} \tag{14}$$

where λ_s is the soil thermal conductivity ($\lambda_s = \rho_s c_s \kappa_s$ with ρ_s density, c_s specific heat and κ_s soil thermal diffusivity) depending strongly on the soil saturation degree. The heat transfer inside the soil can be described in first approximation with Fourier conduction law:

$$\frac{\partial T_s}{\partial t} = \kappa_s \frac{\partial^2 T_s}{\partial z^2} \tag{15}$$

Equation (14) neglects the heat associated to the vapor transportation due to a vertical gradient of the soil humidity content as well as the horizontal heat conduction in the soil. The vapor transportation can be important in the case of dry climates (Saravanapavan & Salvucci, 2000). The soil heat flux can be calculated with different degrees of complexity. The most simple assumption (common in weather forecast models) is to calculate G as a fraction of net radiation (Stull (1988) suggests $G = 0.1R_n$). Another simple approach is to use the analytical solution for a sinusoidal temperature wave. A compromise between precision and computational work is the force restore method (Deardorff, 1978; Montaldo & Albertson, 2001), still used in many hydrological models (Mengelkamp et al., 1999). The main advantage is that only two soil layers have to be defined: a surface thin layer, and a layer getting down to a depth where the daily flux is almost zero. The method uses some results of the analytical solution for a sinusoidal forcing and therefore, in the case of days with irregular temperature trend, it provides less precise results.

The most general solution is the finite difference integration of the soil heat equation in a multilayered soil model (Daamen & Simmonds, 1997). However, this method is computationally demanding and it requires short time steps to assure numerical stability, given the non-linearity and stationarity of the surface energy budget, which is the upper boundary condition of the equation.

2.2.1 Snowmelt and freezing soil

In mountain environments snow-melt and freezing soil should be solved at the same time as soil heat flux. A simple snow melt model is presented in Zanotti et al. (2004), which has a lumped approach, using as state variable the internal energy of the snow-pack and of the first layer of soil. Other models consider a multi-layer parametrization of the snowpack (e.g. Bartelt & Lehning, 2002; Endrizzi et al., 2006). Snow interception by canopy is described for example in Bonan (1996). A state of the art freezing soil modeling approach can be found in Dall'Amico (2010) and Dall'Amico et al. (2011).

2.3 Turbulent fluxes

A modeling of the ground heat and vapor fluxes cannot leave out of consideration the schematization of the atmospheric boundary layer (ABL), meant as the lower part of atmosphere where the earth surface properties influence directly the characteristics of the motion, which is turbulent. For a review see Brutsaert (1982); Garratt (1992); Stull (1988).

A flux of a passive tracer x in a turbulent field (as for example heat and vapor close to the ground), averaged on a suitable time interval, is composed of three terms: the first indicates the transportation due to the mean motion v, the second the turbulent transportation $\overline{x' \, v'}$, the third the molecular diffusion k.

$$\overline{\mathbf{F}} = \overline{x}\,\overline{\mathbf{v}} + \overline{x' \, \mathbf{v}'} - k\nabla x \tag{16}$$

The fluxes parametrization used in LSMs usually only considers as significant the turbulent term only. The molecular flux is not negligible only in the few centimeters close the surface, and the horizontal homogeneity hypothesis makes negligible the convective term.

2.3.1 The conservation equations

The first approximation done by all hydrological and LSMs in dealing with turbulent fluxes is considering the Atmospheric Boundary Layer (ABL) as subject to a stationary, uniform motion, parallel to a plane surface.

This assumption can become limitative if the grid size becomes comparable to the vertical heterogeneity scale (for example for a grid of 10 m and a canopy height of 10 m). In this situation horizontal turbulent fluxes become relevant. A possible approach is the Large Eddy Simulation (Albertson et al., 2001).

If previous assumptions are made, then the conservation equations assume the form:

- Specific humidity conservation, failing moisture sources and phase transitions:

$$k_v \frac{\partial^2 \bar{q}}{\partial z^2} - \frac{\partial}{\partial z}(\overline{w'q'}) = 0 \tag{17}$$

where:

k_v is the vapor molecular diffusion coefficient $[m^2/s]$

$q = \frac{m_v}{m_v + m_d}$ is the specific humidity [vapor mass out of humid air mass].

- Energy conservation:

$$k_h \frac{\partial^2 \bar{\theta}}{\partial z^2} - \frac{\partial}{\partial z}(\overline{w'\theta'}) - \frac{1}{\rho c_p}\frac{\partial H_R}{\partial z} = 0 \tag{18}$$

where:

k_h is the thermal diffusivity $[m^2/s]$
H_R is the radiative flux $[W/m^2]$
θ is the potential temperature [K]
ρ is the air density $[kg/m^3]$
w is the vertical velocity $[m/s]$.

- The horizontal mean motion equations are obtained from the momentum conservation by simplifying Reynolds equations (Stull, 1988; Brutsaert, 1982 cap.3):

$$-\frac{1}{\rho}\frac{\partial \bar{p}}{\partial x} + 2\omega \sin\phi\, \bar{v} + v\frac{\partial^2 \bar{u}}{\partial z^2} - \frac{\partial}{\partial z}(\overline{w'u'}) = 0 \tag{19}$$

$$-\frac{1}{\rho}\frac{\partial \bar{p}}{\partial y} - 2\omega \sin\phi\, \bar{u} + v\frac{\partial^2 \bar{v}}{\partial z^2} - \frac{\partial}{\partial z}(\overline{w'v'}) = 0 \tag{20}$$

where:

v is the kinematic viscosity $[m^2/s]$
ω is the earth angular rotation velocity [rad/s]
ϕ is the latitude [rad] .

The vertical motion equation can be reduced to the hydrostatic equation:

$$\frac{\partial p}{\partial z} = -\rho g. \tag{21}$$

In a turbulent motion the molecular transportation terms of the momentum, heat and vapor quantity, respectively v, k_h and k_v, are several orders of magnitude smaller than Reynolds fluxes and can be neglected.

2.3.2 Wind, heat and vapor profile at the surface

Inside the ABL we can consider, with a good approximation, that the decrease in the fluxes intensity is linear with elevation. This means that in the first meters of the air column the fluxes and the friction velocity u^* can be considered constant. Considering the momentum flux constant with elevation implies that also the wind direction does not change with elevation (in the layer closest to the soil, where the geostrofic forcing is negligible). In this way the alignment with the mean motion allows the use of only one component for the velocity vector, and the problem of mean quantities on uniform terrain becomes essentially one-dimensional, as these become functions of the only elevation z.

In the first centimeters of air the energy transportation is dominated by the molecular diffusion. Close to the soil there can be very strong temperature gradients, for example during a hot summer day. Soil can warm up much more quickly than air. The air temperature diminishes very rapidly through a very thin layer called *micro layer*, where the molecular processes are dominant. The strong ground gradients support the molecular conduction, while the gradients in the remaining part of the surface layer drive the turbulent diffusion. In the remaining part of the surface layer the potential temperature diminishes slowly with elevation.

The effective turbulent flux in the interface sublayer is the sum of molecular and turbulent fluxes. At the surface, where there is no perceptible turbulent flux, the effective flux is equal to the molecular one, and above the first cm the molecular contribution is neglegible. According to Stull (1988), the turbulent flux measured at a standard height of 2 m provides a good approximation of the effective ground turbulent flux.

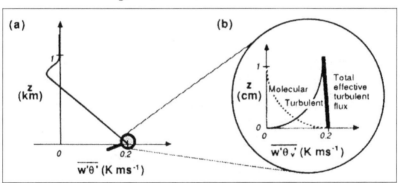

Fig. 3. (a) The effective turbulent flux in diurnal convective conditions can be different from zero on the surface. (b) The effective flux is the sum of the turbulent flux and the molecular flux (from Stull, 1988).

Applying the concept of effective turbulent flux, the molecular diffusion term can be neglected, while the hypothesis of uniform and stationary limit layer leads to neglect the convective terms due to the mean vertical motion and the horizontal flux. The vertical flux at the surface can then be reduced to the turbulent term only:

$$\overline{\mathbf{F_z}} = \overline{x' \, w'} \tag{22}$$

In the case of the water vapor, equation (17) shows that, if there is no condensation, the flux is:

$$ET = \lambda \rho \overline{w'q'} \tag{23}$$

where ET is the evaporation quantity at the surface, ρ the air density and λ is the latent heat of vaporization.

Similarly, as to sensible heat, equation (18) shows that the heat flux at the surface H is:

$$H = \rho c_p \overline{w'\theta'} \tag{24}$$

where c_p is the air specific heat at constant pressure.

The entity of the fluctuating terms $\overline{w'u'}$, $\overline{w'\theta'}$ and $\overline{w'q'}$ remains unknown if further hypotheses (called closing hypotheses) about the nature of the turbulent motion are not introduced. The closing model adopted by the LSMs is Bousinnesq model: it assumes that the fluctuating terms can be expressed as a function of the vertical gradients of the quantities considered (diffusive closure).

$$\tau_x = -\rho \overline{u'w'} = \rho K_M \partial \overline{u}/\partial z \tag{25}$$

$$H = -\rho c_p \overline{w'\theta'} = -\rho c_p K_H \partial \overline{\theta}/\partial z \tag{26}$$

$$ET = -\lambda \rho \overline{w'q'} = -\rho K_W \partial \overline{q}/\partial z \tag{27}$$

where K_M is the turbulent viscosity, K_H and K_W $[m^2/s]$ are turbulent diffusivity. Moreover a logarithmic velocity profile in atmospheric neutrality conditions is assumed:

$$\frac{k\,u}{u_{*0}} = \ln(\frac{z}{z_0}) \tag{28}$$

where k is the Von Karman constant, z_0 is the aerodynamic roughness, evaluated in first approximation as a function of the height of the obstacles as $z_0/h_c \simeq 0.1$ (for more precise estimates see Stull (1988) p.379; Brutsaert (1982) ch.5; Garratt (1992) p.87). In the case of compact obstacles (e.g. thick forests), the profile can be thought of as starting at a height d_0, and the height z can be substituted with a fictitious height $z - d_0$.

Surface type	z_0 [cm]
Large water surfaces	0.01-0.06
Grass, height 1 cm	0.1
Grass, height 10 cm	2.3
Grass, height 50 cm	5
Vegetation, height 1-2 m	20
Trees, height 10-15 m	40-70
Big towns	165

Table 1. Values of aerodynamic roughness length z_0 for various natural surfaces (from Brutsaert, 1982).

Also the other quantities θ and q have an analogous distribution. Using as scale quantities $\theta_{*0} = -\overline{w'\theta'}_0/u_{*0}$ e $q_{*0} = -\overline{w'q'}_0/u_{*0}$ and substituting them in the (25), the following integration is obtained:

$$\frac{k(\theta - \theta_0)}{\theta_{*0}} = \ln\left(\frac{z}{z_T}\right) \tag{29}$$

$$\frac{k(q - q_0)}{q_{*0}} = \ln\left(\frac{z}{z_q}\right). \tag{30}$$

The boundary condition chosen is $\theta = \theta_0$ in $z = z_T$ and $q = q_0$ in $z = z_q$. The temperature θ_0 then is not the ground temperature, but that at the elevation z_T. The roughness height z_T is the height where temperature assumes the value necessary to extrapolate a logarithmic profile. Analogously, z_q is the elevation where the vapor concentration assumes the value necessary to extrapolate a logarithmic profile.

Indeed, close to the soil (interface sublayer) the logarithmic profile is not valid and then, to estimate z_T and z_q, it would be necessary to study in a detailed way the dynamics of the heat and mass transfer from the soil to the first meters of air.

If we consider a real surface instead of a single point, the detail requested to reconstruct accurately the air motion in the upper soil meters is impossible to obtain. Then there is a practical problem of difficult solution: on the one hand, the energy transfer mechanisms from the soil to the atmosphere operate on spatial scales of few meters and even of few cm, on the other hand models generally work with a spatial resolution ranging from tens of m (as in the case of our approach) to tens of km (in the case of mesoscale models). Models often apply to local scale the same parametrizations used for mesoscale. Therefore a careful validation test, even for established theories, is always important.

Observations and theory (Brutsaert, 1982, p.121) show that z_T and z_q generally have the same order of magnitude, while the ratio $\frac{z_T}{z_0}$ is roughly included between $\frac{1}{5} - \frac{1}{10}$.

2.3.3 The atmospheric stability

In conditions different from neutrality, when thermal stratification allows the development of buoyancies, Monin & Obukhov (1954) similarity theory is used in LSMs. The similarity theory wants to include the effects of thermal stratification in the description of turbulent transportation. The stability degree is expressed as a function of Monin-Obukhov length, defined as:

$$L_{MO} = -\frac{u_{0*}^3 \theta_0}{kg \overline{w'\theta'}} \tag{31}$$

where θ_0 is the potential temperature at the surface.

Expressions of the stability functions can be found in many texts of Physics of the Atmosphere, for example Katul & Parlange (1992); Parlange et al. (1995). The most known formulation is to be found in Businger et al. (1971). Yet stability is often expressed as a function of bulk Richardson number Ri_B between two reference heights, expresses as:

$$Ri_B = \frac{g \, z \, \Delta\theta}{\overline{\theta} u^2} \tag{32}$$

where $\Delta\theta$ is the potential temperature difference between two reference heights, and $\overline{\theta}$ is the mean potential temperature.

If $Ri_B > 0$ atmosphere is steady, if $Ri_B < 0$ atmosphere is unsteady. Differently from L_{MO}, Ri_B is also a function of the dimensionless variables z/z_0 e z/z_T. The use of Ri_B has the advantage that it does not require an iterative scheme.

Expressions of the stability functions as a function of Ri_B are provided by Louis (1979) and more recently by Kot & Song (1998). Many LSMs use empirical functions to modify the wind profile inside the canopy cover.

From the soil up to an elevation $h_d = f(z_0)$, limit of the interface sublayer, the logarithmic universal profile and Reynolds analogy are no more valid. For smooth surfaces the interface sublayer coincides with the viscous sublayer and the molecular transport becomes important. For rough surfaces the profile depends on the distribution of the elements present, in a way which is not easy to parametrize. Particular experimental relations can be used up to elevation h_d, to connect them up with the logarithmic profile (Garratt, 1992, p. 90 and Brutsaert, 1982, p. 88). These are expressions of non-easy practical application and they are still little tested.

2.3.4 Latent and sensible heat fluxes

As consequence of the theory explained in the previous paragraph, the turbulent latent and sensible fluxes H and LE can be expressed as:

$$H = \rho c_p \overline{w'\theta'} = \rho c_p C_H u(\theta_0 - \theta) \tag{33}$$

$$ET = \lambda \rho \overline{w'q'} = \lambda \rho C_E u(q_0 - q), \tag{34}$$

where $\theta_0 - \theta$ and $q_0 - q$ are the difference between surface and measurement height of potential temperature and specific humidity respectively. C_H and C_E are usually assumed to be equal and depending on the bulk Richardson number (or on Monin-Obukhov lenght):

$$C_H = C_{Hn} F_H(Ri_B), \tag{35}$$

where C_{Hn} is the heat bulk coefficient for neutral conditions:

$$C_{Hn} = C_{En} = \frac{k^2}{[\ln(z/z_0)][\ln(z_a/z_T)]} \tag{36}$$

derived on Eq. 29 and depending on the wind speed u, the measurement height z, the temperature (or moisture) measurement height z_a, the momentum roughness length z_0 and the heat roughness length z_T.

A common approach is the 'electrical resistance analogy' (Bonan, 1996), where the atmospheric resistance is expressed as:

$$r_{aH} = r_{aE} = (C_H u)^{-1} \tag{37}$$

3. Evapotranspiration processes

In order to convert latent heat flux in evapotranspiration the energy conservation must be solved at the same time as water mass budget. In fact, there must be a sufficient water quantity available for evaporation. Moreover, vegetation plays a key role.

3.1 Unsaturated soil evaporation

If the availability of water supply permits to reach the surface saturation level, then evaporation is potential $ET = EP$ and then we have air saturation at the surface $q(T_s) = q^*(T_s)$ (the superscript * stands here for saturation). If the soil is unsaturated, $q(T_s) \neq q^*(T_s)$ and different approaches are possible to quantify the water content at the surface, in dependance of the water budget scheme adopted.

1. A first possibility is to introduce then the concept of surface resistance r_g to consider the moisture reduction with respect to the saturation value. As it follows from equation (34):

$$ET = \lambda \rho C_E u(q_0 - q) = \lambda \rho \frac{1}{r_a}(q_0 - q) = \lambda \rho \frac{1}{r_a + r_g}(q_0^* - q) \tag{38}$$

2. As an alternative, we can define a soil-surface relative moisture

$$r_h = q_0/q_0^* \tag{39}$$

and then the expression for evaporation becomes:

$$ET = \lambda \rho \frac{1}{r_a}(r_h q_0^* - q) \tag{40}$$

An expression of r_h as a function of the potential ψ_s [m] (work required to extract water from the soil against the capillarity forces) and of the ratio of the soil water content η to the saturation water content η_s is given in Philip & Vries (1957):

$$r_h = exp(-(g/R_v T_s)\psi_s(\eta/\eta_s)^{-b}) \tag{41}$$

where $R_v = 461.53 [J/(kg\,K)]$ is the gas constant for water vapor, T_s is the soil temperature, b an empirical constant. Tables of these parameters for different soil types can be found in Clapp & Hornberger (1978).

Another more simple expression frequently applied in models to link the value r_h with the soil water content η is provided by Noilhan & Planton (1989):

$$r_h = \begin{cases} 0.5(1 - \cos(\frac{\eta}{\eta_k}\pi)) & se\ \eta < \eta_k \\ 1 & if\ \eta \geq \eta_k \end{cases} \tag{42}$$

where η is the moisture content of a soil layer with thickness d_1, and η_k is a critical value depending on the saturation water content: $\eta_k \simeq 0.75\eta_s$.

3. A third possibility, very used in large-scale models, is that of expressing the potential/real evaporation ration through a simple coefficient:

$$ET = x\ EP = x\ \lambda \rho \frac{1}{r_a}(q_0^* - q) \tag{43}$$

The value of x can be connected to the soil water content η through the expression (Deardorff, 1978) (see Figure 4):

$$x = min(1, \frac{\eta}{\eta_k}) \tag{44}$$

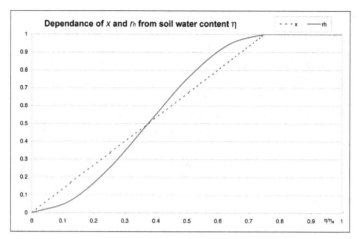

Fig. 4. Dependence of x and r_h on the soil water content η (Eq. 44-42)

3.2 Transpiration

Usually transpiration takes into account the canopy resistance r_c to add to the atmospheric resistance r_a:

$$ET = x\,EP = x\,\lambda\rho\frac{1}{r_a + r_c}(q_0^* - q) \tag{45}$$

The canopy resistance depends on plant type, leaf area index, solar radiation, vapor pressure deficit, temperature and water content in the root layer. There is a wide literature regarding such dependence, see for example Feddes et al. (1978); Wigmosta et al. (1994).

Canopy interception and evaporation from wet leaves are important processes modeled that should be modelled, according to Deardorff (1978). It is possible to distinguish two fundamental approaches: single-layer canopy models and multi-layer canopy models.

Single-layer canopy models (or "big leaf" models)

The vegetation resistance is entirely determined by stomal resistance and only one temperature value, representative of both vegetation and soil, is considered. Moreover a vegetation interception function can be defined so as to define when the foliage is wet or when the evaporation is controlled by stomal resistance.

Multi-layer canopy models

These are more complex models in which a soil temperature T_g, different from the foliage temperature T_f, is considered. Therefore, two pairs of equations of latent and sensible heat flux transfer, from the soil level to the foliage level, and from the latter to the free atmosphere, must be considered (Best, 1998). Moreover the equation for the net radiation calculation must consider the energy absorption and the radiation reflection by the vegetation layer.

Deardorff (1978) is the first author who presents a two-layer model with a linear interpolation between zones covered with vegetation and bare soil, to be inserted into atmosphere general circulation models. Over the last years many detailed models have been developed, above all with the purpose of evaluating the CO_2 fluxes between vegetation and atmosphere. Particularly complex is the case of scattered vegetation, where evaporation is due to a combination of soil/vegetation effects, which cannot be schematized as a single layer (Scanlon & Albertson, 2003).

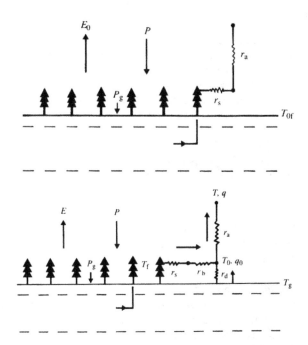

Fig. 5. **Above**: scheme representing a single-layer vegetation model. Linked both with atmosphere (with resistance r_a) and with the deep soil (through evapotranspiration with resistance r_s), vegetation and soil surface layer are assumed to have the same temperature T_{0f}. **Below**: scheme representing a multilayer vegetation model. Linked both with atmosphere (with resistances r_b and r_a), and with the deep soil (through evapotranspiration with resistance r_s), as well as with the soil under the vegetation (r_d), vegetation and soil surface layer assumed to have different temperature T_f and T_g. P_g is the rainfall reaching the soil surface (from Garratt, 1992).

Given the many uncertainties regarding the forcing data and the components involved (soil, atmosphere), and the numerous simplifying hypotheses, the detail requested in a vegetation cover scheme is not yet clear.

A single-layer description of vegetation cover (big-leaf) and a two-level description of soil represent probably the minimum level of detail requested. In general, if the horizontal scale is far larger than the vegetation scale, a single-layer model is sufficient (Garratt, 1992, p. 242), as in the case of the general circulation atmospheric models or of mesoscale hydrologic models for large basins. These models determine evaporation as if the vegetation cover were but a partially humid plane at the atmosphere basis. In an approach of this kind surface resistance, friction length, albedo and vegetation interception must be specified. The surface resistance must include the dependence on solar radiation or on soil moisture, as transpiration decreases when humidity becomes smaller than the withering point (Jarvis & Morrison, 1981). For the soil, different coefficients depending on moisture are requested, together with a functional relation of evaporation to the soil moisture.

4. Water in soils

Real evaporation is coupled to the infiltration process occurring in the soil, and its physically-based estimate cannot leave the estimation of soil water content consideration.
The most simple schemes to account water in soils used in LSMs single-layer and two-layer methods. The most general approach, which allows water transport for unsaturated stratified soil, is based on the integration of Richards (1931) equation, under different degrees of approximations.

4.1 Single layer or bucket method
In this method the whole soil layer is considered as a bucket and real evaporation E_0 is a fraction x of potential evaporation E_p, with x proportional to the saturation of the whole soil.

$$E_0 = xE_p \tag{46}$$

with x expressed by Eq. (44). The main problem of this method is that evaporation does not respond to short precipitation, leading to surface saturation but not to a saturation of the whole soil layer (Manabe, 1969).

4.2 Two-layer or force restore method
This method is analogous to the one developed to calculate the soil heat flux, but it requires calibration parameters which are unlikely to be known. With this method it is possible to consider the water quantity used by plants for transpiration, considering a water extraction by roots in the deepest soil layer (Deardorff, 1978).

4.3 Multilayer methods and Richards equation
Richards (1931) equation and Darcy-Buckingham law govern the unsaturated water transport in isobar and isothermal conditions:

$$\vec{q} = -K\nabla\left(z + \psi\right) \tag{47}$$

$$\frac{\partial\psi}{\partial\eta}\nabla \cdot (K\nabla\psi) - \frac{\partial K}{\partial z} = \frac{\partial\psi}{\partial t} \tag{48}$$

where $\vec{q} = (q_x, q_y, q_z)$ is the specific discharge, K is the hydraulic conductivity tensor, z is the upward vertical coordinate and ψ is the suction potential or matrix potential.
The determination of the suction potential allows also a more correct schematization of the plant transpiration and it lets us describe properly flow phenomena from the water table to the surface, necessary to the maintenance of evaporation from the soils.
Richards equation is, rightfully, an energy balance equation, even if this is not evident in the modes from which it has been derived. Then the solutions of the equation (48) must be searched by assigning the water retention curve which relates ψ with the soil water content η and an explicit relation of the hydraulic conductivity as a function of ψ (or η). Both relationships depend on the type of terrain and are variable in every point. K augments with η, until it reaches the maximum value K_s which is reached at saturation.

Although the integration of the Richards equation is the only physically based approach, it requires remarkable computational effort because of the non linearity of the water retention curve. It is difficult to find a representative water retention curve because of the high degree of spatial variability in soil properties (Cordano & Rigon, 2008).

4.4 Spatial variability in soil moisture and evapotranspiration

Topography controls the catchment-scale soil moisture distribution (Beven & Freer, 2001) and therefore water availability for ET. Two methods most frequently used to incorporate sub-grid variability in soil moisture and runoff production SVATs models are the variable infiltration capacity approach (Wood, 1991) and the topographic index approach (Beven & Kirkby, 1979). They represent computationally efficient ways to represent hydrologic processes within the context of regional and global modeling. A review and a comparison of the two methods can be found in Warrach et al. (2002).

More detailed approaches need to track surface or subsurface flow within a catchment explicitly. Such approaches, which require to couple the ET model with a distributed hydrological model, are particularly useful in mountain regions, as presented in the next section.

5. Evapotranspiration in Alpine Regions

In alpine areas, evapotranspiration (ET) spatial distribution is controlled by the complex interplay of topography, incoming radiation and atmospheric processes, as well as soil moisture distribution, different land covers and vegetation types.

1. Elevation, slope and aspect exert a direct control on the incoming solar radiation (Dubayah et al., 1990). Moreover, elevation and the atmospheric boundary layer of the valley affect the air temperature, moisture and wind distribution (e.g., Bertoldi et al., 2008; Chow et al., 2006; Garen & Marks, 2005).

2. Vegetation is organized along altitudinal gradients, and canopy structural properties influence turbulent heat transfer processes, radiation divergence (Wohlfahrt et al., 2003), surface temperature (Bertoldi et al., 2010), therefore transpiration, and, consequently, ET.

3. Soil moisture influences sensible and latent heat partitioning, therefore ET. Topography controls the catchment-scale soil moisture distribution (Beven & Kirkby, 1979) in combination with soil properties (Romano & Palladino, 2002), soil thickness (Heimsath et al., 1997) and vegetation (Brooks & Vivoni, 2008a).

Spatially distributed hydrological and land surface models (e.g., Ivanov et al., 2004; Kunstmann & Stadler, 2005; Wigmosta et al., 1994) are able to describe land surface interactions in complex terrain, both in the temporal and spatial domains. In the next section we show an example of the simulation of the ET spatial distribution in an Alpine catchment simulated with the hydrological model GEOtop (Endrizzi & Marsh, 2010; Rigon et al., 2006).

6. Evapotranspiration in the GEOtop model

The GEOtop model describes the energy and mass exchanges between soil, vegetation and atmosphere. It takes account of land cover, soil moisture and the implications of topography on solar radiation. The model is open-source, and the code can be freely obtained from the web site: http://www.geotop.org/. There, we provide a brief description of the 0.875 version of the model (Bertoldi et al., 2005), used in this example. For details of the most recent numerical implementation, see Endrizzi & Marsh (2010).

The model has been proved to simulate realistic values for the spatial and temporal dynamics of soil moisture, evapotranspiration, snow cover (Zanotti et al., 2004) and runoff production, depending on soil properties, land cover, land use intensity and catchment morphology (Bertoldi et al., 2010; 2006).

The model is able to simulate the following processes: (i) coupled soil vertical water and energy budgets, through the resolution of the heat and Richard's equations, with temperature and water pressure as prognostic variables (ii) surface energy balance in complex topography, including shadows, shortwave and longwave radiation, turbulent fluxes of sensible and latent heat, as well as considering the effects of vegetation as a boundary condition of the heat equation (iii) ponding, infiltration, exfiltration, root water extraction as a boundary condition of Richard's equation (iv) subsurface lateral flow, solved explicitly and considered as a source/sink term of the vertical Richard's equation (v) surface runoff by kinematic wave, and (vi) multi-layer glacier and snow cover, with a solution of snow water and energy balance fully integrated with soil.

The incoming direct shortwave radiation is computed for each grid cell according to the local solar incidence angle, including shadowing (Iqbal, 1983). It is also split into a direct and diffuse component according to atmospheric and cloud transmissivity (Erbs et al., 1982). The diffuse incoming shortwave and longwave radiation is adjusted according to the theory described in Par. 2.1. The soil column is discretized in several layers of different thicknesses. The heat and Richards' equations are written respectively as:

$$C_t(P)\frac{\partial T}{\partial t} - \frac{\partial}{\partial z}\left[K_t(P)\frac{\partial T}{\partial z}\right] = 0 \tag{49}$$

$$C_h(P)\frac{\partial P}{\partial t} - \frac{\partial}{\partial z}\left[K_h(P)\left(\frac{\partial P}{\partial z}+1\right)\right] - q_s = 0 \tag{50}$$

Where T is soil temperature, P the water pressure, C_t the thermal capacity, K_t the thermal conductivity, C_h the specific volumetric storativity, K_h the hydraulic conductivity, and q_s the source term associated with lateral flow. The variables C_t, K_t, C_h, and K_h depend on water content, and, in turn, on water pressure, and are therefore a source of non-linearity. At the bottom of the soil column a boundary condition of zero fluxes has been imposed.

The boundary conditions at the surface are consistent with the infiltration and surface energy balance, and are given in terms of surface fluxes of water (Q_h) and heat (Q_t) at the surface, namely:

$$Q_h = min\left[p_{net}, K_{h1}\frac{(h-P_1)}{dz/2} + K_{h1}\right] - E(T_1, P_1) \tag{51}$$

$$Q_t = SW_{in} - SW_{out} + LW_{in} - LW_{out}(T_1) - H(T_1) - LE(T_1) \tag{52}$$

Where p_{net} is the net precipitation, K_{h1} and P_1 are the hydraulic conductivity and water pressure of the first layer, h is the pressure of ponding water, dz the thickness of the first layer, T_1 the temperature of the first layer. E is evapotranspiration (as water flux), SW_{in} and SW_{out} are the incoming and outgoing shortwave radiation, LW_{in} and LW_{out} the incoming and outgoing longwave radiation, H the sensible heat flux and LE the latent heat flux. H and LE are calculated taking into consideration the effects of atmospheric stability (Monin & Obukhov, 1954).

E is partitioned by evaporation or sublimation from the soil or snow surface E_G, transpiration from the vegetation E_{TC}, evaporation of the precipitation intercepted by the vegetation E_{VC}. Every cell has a fraction covered by vegetation and a fraction covered by bare soil. In the 0.875 version of the model, a one-level model of vegetation is employed, as in Garratt (1992) and in Mengelkamp et al. (1999): only one temperature is assumed to be representative of both soil and vegetation. In the most recent version, a two-layer canopy model has been introduced. Bare soil evaporation E_g is related to the water content of the first layer through the soil resistance analogy (Bonan, 1996):

$$E_G = (1 - cop)\, E_P\, \frac{r_a}{r_a + r_s} \tag{53}$$

where cop is fraction of soil covered by the vegetation EP is the potential ET calculated with equation 34 and r_a the aerodynamic resistance:

$$r_a = 1/\left(\rho\, C_E\, \hat{u}\right) \tag{54}$$

The soil resistance r_s is function of the water content of the first layer.

$$r_s = r_a\, \frac{1.0 - (\eta_1 - \eta_r)/(\eta_s - \eta_r)}{(\eta_1 - \eta_r)/(\eta_s - \eta_r)} \tag{55}$$

where η_1 is the water content of the first soil layer close the surface, η_r is the residual water content (defined following Van Genuchten, 1980) and η_s is the saturated water content, both in the first soil layer.

The evaporation from wet vegetation is calculated following Deardorff (1978):

$$E_{VC} = cop\, E_P\, \delta_W \tag{56}$$

where δ_W is the wet vegetation fraction.

The transpiration from dry vegetation is calculated as:

$$E_{TC} = cop\, E_P (1 - \delta_W) \sum_i^n \frac{f_{root}^i\, r_a}{r_a + r_c^i} \tag{57}$$

The root fraction f_{root}^i of each soil layer i is calculated decreasing linearly from the surface to a maximum root depth, depending from the cover type. The canopy resistance r_c depends on solar radiation, vapor pressure deficit, temperature as in Best (1998) and on water content in the root zone as in Wigmosta et al. (1994).

6.1 The energy balance at small basin scale: application to the Serraia Lake.

An application of the model to a small basin is shown here, in order to bring out the problems arising when passing from local one-dimensional scale to basin-scale. The Serraia Lake basin is a mountain basin of 9 km^2, with an elevation ranging from 900 to 1900 m, located in Trentino, Italy. Within the basin there is a lake of about 0.5 km^2. During the year 2000 a study to calculate the yearly water balance was performed (Bertola & al., 2002).

The model was forced with meteorological measurement of a station located in the lower part of the basin at about 1000 m, and the stream-flow was calibrated for the sub-catchment of Foss Grand, of about 4 km^2. Then the model was applied to the whole basin. Further details on the calibration can be found in Salvaterra (2001). Meteorological data are assumed to be constant across the basin, except for temperature, which varies linearly with elevation (0.6 ^oC / 100 m) and solar radiation, which slightly increases with elevation and is affected by shadow and aspect.

With the GEO_{TOP} model it is possible to simulate the water and energy balance, aggregated for the whole basin (see figure 6 and 7) and its distribution across the basin. Figure 7 shows the map of the seasonal latent heat flux (ET) in the basin. During winter and fall ET is low (less than 40 W/m^2), with the lowest values in drier convex areas. During summer and spring ET increases (up to 120 W/m^2), with highest values in the bottom of the main valley (where indeed there are a lake and a wetland) and lowest values in north-facing, high-elevation areas.

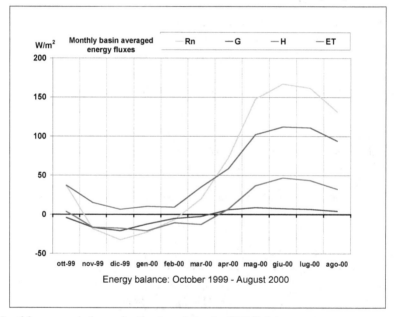

Fig. 6. Monthly energy balance for the Serraia basin (TN, Italy).

The main factors controlling the ET pattern in a mountain environment (see Figure 8) are also: elevation, which controls temperature, aspect, which influences radiation, soil thickness, which determines storage capacity, topographic convergence, which controls the moisture availability. In particular, aspect has a primary effect on net radiation and a secondary effect more on sensible rather than on latent heat flux, as in Figure 9, where south aspect locations have larger R_n and H, but similar behavior for the other energy budget components). Water content changes essentially the rate between latent and heat flux, as in Figure 10 where wet locations have larger ET and lower H.

Therefore, the surface fluxes distribution seems to agree with experience and current hydrology theory, but the high degree of variability poses some relevant issues because the hypothesis of homogenous turbulence at the basis of the fluxes calculation is no more valid

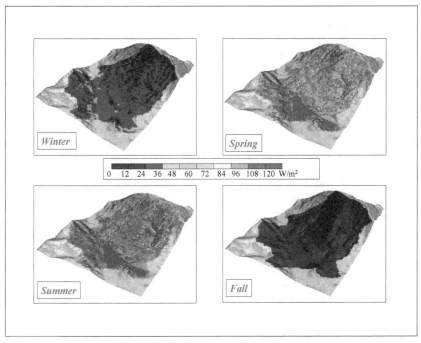

Fig. 7. Seasonal latent heat maps ET $[W/m^2]$ for the Serraia basin (TN, Italy).

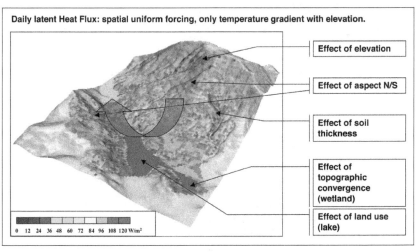

Fig. 8. Example of evapotranspiration ET for the Serraia basin, Italy. Notice the elevation effect (areas more elevated have less evaporation); the aspect effect (more evaporation in southern slopes, left part of the image); the topographic convergence effect on water availability (at the bottom of the valley).

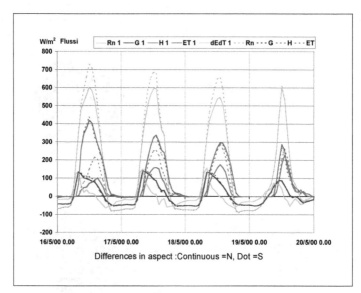

Fig. 9. Difference in energy balance between locations with the same properties but different aspect. Dotted lines are for a south aspect location, while continuos lines are for a north aspect location. It can be noticed how south aspect locations have larger R_n and H, but similar behavior for the other fluxes.

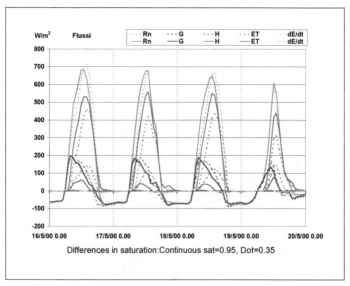

Fig. 10. Difference in energy balance between locations with the same properties but different soil saturation. Dotted lines are for a dry location, while continuos lines are for wet location. It can be noticed how wet locations have larger ET and lower H, but similar behavior for the other fluxes. The time lag in R_n is due to differences in aspect.

(Albertson & Parlange, 1999). Moreover, horizontal differences in surface fluxes can start local air circulations, which can affect temperature and wind surface values with a feedback effect. How much such processes may affect the energy and water balance of the whole basin is easy to quantify, but GEO_{TOP} can be a powerful tool to explore these issues.

7. Conclusion

This chapter illustrates the components of the energy budget needed to model evapotranspiration (ET) and provides an extended review of the fundamental equations and parametrizations available in the hydrological and land surface models literature. In alpine areas, ET spatial distribution is controlled by the complex interplay of topography, incoming radiation and atmospheric processes, as well as soil moisture distribution, different land covers and vegetation types. An application of the distributed hydrological model GEOtop to a small basin is shown here, in order to bring out the problems arising when passing from local one-dimensional scale to basin-scale ET models.

8. References

Albertson, J., Kustas, W. P. & Scanlon, T. M. (2001). Large-eddy simulation over heterogeneous terrain with remotely sensed land surface conditions, *Water Resour. Res.* 37: 1939–1953.

Albertson, J. & Parlange, M. B. (1999). Natural integration of scalar fluxes from complex terrain, *Adv. Water Resour.* 23: 239–252.

Bartelt, P. & Lehning, M. (2002). A physical snowpack model for the swiss avalanche warning: Part i: numerical model, *Cold Regions Science and Technology* 35(3): 123–145.

Bertola, P. & al. (2002). "studio integrato dell'eutrofizzazione del lago della serraia", *Atti del XXVIII Convegno di Idraulica e Costruzioni Idrauliche* 3: 403–413.

Bertoldi, G., Kustas, W. P. & Albertson, J. D. (2008). Estimating spatial variability in atmospheric properties over remotely sensed land-surface conditions, *J. Appl. Met. and Clim.* 47(doi: 10.1175/2007JAMC1828.1): 2147–2165.

Bertoldi, G., Notarnicola, C., Leitinger, G., Endrizzi, S., ad S. Della Chiesa, M. Z. & Tappeiner, U. (2010). Topographical and ecohydrological controls on land surface temperature in an alpine catchment, *Ecohydrology* 3(doi:10.1002/eco.129): 189 – 204.

Bertoldi, G., Rigon, R. & Over, T. (2006). Impact of watershed geomorphic characteristics on the energy and water budgets., *Journal of Hydrometeorology*, 7: 389–403.

Bertoldi, G., Tamanini, D., Endrizzi, S., Zanotti, F. & Rigon, R. (2005). GEOtop 0.875: the programmer's manual, *Technical Report DICA-05-002*, University of Trento E-Prints. In preparation.

Best, M. J. (1998). A model to predict surface temperatures, *Bound. Layer Meteorol.* 88(2): 279–306.

Beven, K. J. & Freer, J. (2001). A dynamic TOPMODEL, *Hydrol. Proc.* 15: 1993–2011.

Beven, K. J. & Kirkby, M. J. (1979). A physically-based variable contributing area model of basin hydrology, *Hydrol. Sci. Bull.* 24(1): 43–49.

Bonan, G. (1996). A land surface model for ecological, hydrological, and atmospheric studies: technical description and user's guide., *Technical Note NCAR/TN-417+STR*, NCAR, Boulder, CO.

Brooks, P. D. & Vivoni, E. R. (2008a). Mountain ecohydrology: quantifying the role of vegetation in the water balance of montane catchments, *Ecohydrol.* 1(DOI: 10.1002/eco.27): 187 – 192.

Brooks, P. & Vivoni, E. R. (2008b). Mountain ecohydrology: quantifying the role of vegetation in the water balance of montane catchments., *Ecohydrology* 1: 187–192.

Brutsaert, W. (1975). On a derivable formula for long-wave radiation from clear skies, *Water Resour. Res.* 11(5): 742–744.

Brutsaert, W. (1982). *Evaporation into the Atmosphere: Theory, Hystory and Applications*, Kluver Academic Publisher.

Businger, J. A., Wyngaard, J. C., Izumi, Y. & Bradley, E. F. (1971). Flux profile relationships in the atmospheric surface layer, *J. Atmospheric Sciences* 28: 181–189.

Chow, F. K., Weigel, A. P., Street, R. L., Rotach, M. W. & Xue, M. (2006). High-resolution large-eddy simulations of flow in a steep alpine valley. Part I: methodology, verification, and sensitivity experiments, *J. Appl. Met. and Clim.* pp. 63–86.

Clapp, R. B. & Hornberger, G. M. (1978). Empirical equations for some hydraulic properties, *Water Resour. Res.* 14: 601–605.

Cordano, E. & Rigon, R. (2008). A perturbative view on the subsurface water pressure response at hillslope scale, *Water Resour. Res.* 44(W05407): doi:10.1029/2006WR005740.

Daamen, C. C. & Simmonds, L. P. (1997). Soil, water, energy and transpiration, a numerical model of water and energy fluxes in soil profiles and sparse canopies, *Technichal report*, University of Reading.

Dall'Amico, M. (2010). *Coupled water and heat transfer in permafrost modeling*, PhD thesis, Institute of Civil and Environmental Engineering, Universita' degli Studi di Trento, Trento. Available from http://eprints-phd.biblio.unitn.it/335/.

Dall'Amico, M., Endrizzi, S., Gruber, S. & Rigon, R. (2011). A robust and energy-conserving model of freezing variably-saturated soil, *The Cryosphere* 5(2): 469–484.
URL: *http://www.the-cryosphere.net/5/469/2011/*

Deardorff, J. W. (1978). Efficient prediction of ground surface temperature and moisture with inclusion of a layer of vegetation, *J. Geophys. Res.* 83(C4): 1889–1903.

Dickinson, R. E., Heanderson-Sellers, A., Kennedy, P. J. & Wilson, M. (1986). Biosphere Atmosphere Transfer Scheme (BATS) for the NCAR Community Climate Model, *Technical Note NCAR/TN-275+STR*, NCAR.

Dubayah, A., Dozier, J. & Davis, F. W. (1990). Topographic distribution of clear-sky radiation over the Konza Prairie, Kansas, *Water Resour. Res.* 26(4): 679–690.

Endrizzi, S., Bertoldi, G., Neteler, M. & Rigon, R. (2006). Snow cover patterns and evolution at basin scale: GEOtop model simulations and remote sensing observations, *Proceedings of the 63rd Eastern Snow Conference*, Newark, Delaware USA, pp. 195–209.

Endrizzi, S. & Marsh, P. (2010). Observations and modeling of turbulent fluxes during melt at the shrub-tundra transition zone 1: point scale variations, *Hydrology Research* 41(6): 471–490.

Erbs, D. G., Klein, S. A. & Duffie, J. A. (1982). Estimation of the diffuse radiation fraction for hourly, daily and monthly average global radiation., *Sol. Energy* 28(4): 293–304.

Feddes, R., Kowalik, P. & Zaradny, H. (1978). Simulation of field water use and crop yield, *Simulation Monographs*, PUDOC, Wageningen, p. 188pp.

Garen, D. C. & Marks, D. (2005). Spatially distributed energy balance snowmelt modeling in a mountainous river basin: estimation of meteorological inputs and verification of model results, *J. Hydrol.* 315: 126–153.

Garratt, J. R. (1992). *The Atmospheric Boundary Layer*, Cambridge University Press.

Heimsath, M. A., Dietrich, W. E., Nishiizumi, K. & Finkel, R. (1997). The soil production function and landscape equilibrium, *Nature* 388: 358–361.

Helbig, N., Lowe, H.&Lehning, M. (2009). Radiosity approach for the short wave surface radiation balance in complex terrain., *J.Atmos.Sci.* 66(doi:10.1175/2009JAS2940.1): 2900–2912.

Iqbal, M. (1983). *An Introduction to Solar Radiation*, Academic Press.

Ivanov, V. Y., Vivoni, E. R., Bras, R. L. & Entekhabi, D. (2004). Catchment hydrologic response with a fully distributed triangulated irregular network model, *Water Resour. Res.* 40: doi:10.1029/2004WR003218.

Jarvis, P. & Morrison, J. (1981). The control of transpiration and photosynthesis by the stomata., *in* P. Jarvis & T. Mansfield (eds), *Stomatal Physiology*, Cambridge Univ. Press, UK, pp. 247–279.

Katul, G. G. & Parlange, M. B. (1992). A penman-brutsaert model for wet surface evaporation, *Water Resour. Res.* 28(1): 121–126.

Kondratyev, K. Y. (1969). *Radiation in the atmosphere*, Academic Press, New York.

Kot, S. C. & Song, Y. (1998). An improvement of the Louis scheme for the surface layer in an atmospheric modelling system, *Bound. Layer Meteorol.* 88(2): 239–254.

Kunstmann, H. & Stadler, C. (2005). High resolution distributed atmospheric-hydrological modelling for alpine catchments, *J. Hydrol.* 314: 105–124.

Law, B. E., Cescatti, A. & Baldocchi, D. D. (1999). Leaf area distribution and radiative transfer in open-canopy forests: Implications to mass and energy exchange., *Tree Physiol.* 21: 287–298.

Louis, J. F. (1979). A parametric model of vertical eddy fluxes in the atmosphere, *Bound. Layer Meteorol.* 17: 187–202.

Manabe, S. (1969). Climate and ocean circulation. i. the atmospheric circulation and the hydrology of the earth's surface., *Monthly Weather Review* 97: 739–774.

Mengelkamp, H.,Warrach, K. & Raschke, E. (1999). SEWAB - a parametrization of the surface energy and water balance for atmospheric and hydrologic models, *Adv.Water Resour.* 23: 165–175.

Monin, A. S. & Obukhov, A. M. (1954). Basic laws of turbulent mixing in the ground layer of the atmosphere, *Trans. Geophys. Inst. Akad.* 151: 163–187.

Montaldo, N. & Albertson, J. (2001). On the use of the force-restore svat model formulation for stratified soils, *J. Hydromet.* 2(6): 571–578.

Noilhan, J. & Planton, S. (1989). A simple parametriztion of land surface processes for meteorological models, *Mon. Wea. Rev* 117: 536–585.

Paltrige, G. W. & Platt, C. M. R. (1976). *Radiative Processes in Meteorology and Climatology*, Elsevier.

Parlange, M. B., Eichinger,W. E. & Albertson, J. D. (1995). Regional scale evaporation and the atmospheric boundary layer, *Reviews of Geophysic* 33(1): 99–124.

Philip, J. R. & Vries, D. A. D. (1957). Moisture movement in porous materials under temperature gradients, *Trans. Am. Geophys. Union* 38(2): 222–232.

Prata, A. J. (1996).   A new long-wave formula for estimating downward clear-sky radiation at the surface., *Quarterly Journal of the Royal Meteorological Society* 122(doi: 10.1002/qj.49712253306): 1127–1151.

Richards, L. A. (1931). Capillary conduction of liquids in porous mediums, *Physics* 1: 318–333.

Rigon, R., Bertoldi, G. & Over, T. M. (2006). GEOtop: a distributed hydrological model with coupled water and energy budgets, *Journal of Hydrometeorology* 7: 371–388.

Romano, N. & Palladino, M. (2002). Prediction of soil water retention using soil physical data and terrain attributes, *J. Hydrol.* 265: 56–75.

Salvaterra, M. (2001). *Applicazione di un modello di bilancio idrologico al bacino del Lago di Serraia (TN)*, Tesi di diploma, Corso di diploma in Ingegneria per l'Ambiente e le Risorse.

Saravanapavan, T. & Salvucci, G. D. (2000).   Analysis of rate-limiting processes in soil evaporation with implications for soil resistance models, *Adv. Water Resour.* 23: 493–502.

Scanlon, T. M. & Albertson, J. D. (2003).   Water availability and the spatial complexity of co2, water, and energy fluxes over a heterogeneous sparse canopy, *J. Hydromet.* 4(5): 798–809.

Stull, R. B. (1988). *An Introduction to Boundary Layer Meteorology*, Kluwer Academic Publisher.

Van Genuchten, M. T. (1980). A closed-form equation for predicting the hydraulic conductivity of unsaturated soils., *Soil Sci. Soc. Am. J.* 44: 892–898.

Warrach, K., Stieglitz, M., Mengelkamp, H. & Raschke, E. (2002).   Advantages of a topographically controlled runoff simulation in a SVAT model, *J. Hydromet.* 3: 131–148.

Wigmosta, M. S., Vail, L. & Lettenmaier, D. (1994). A Distributed Hydrology-Vegetation Model for complex terrain, *Water Resour. Res.* 30(6): 1665–1679.

Wohlfahrt, G., Bahn, M., Newesely, C. H., Sapinsky, S., Tappeiner, U. & Cernusca, A. (2003). Canopy structure versus physiology effects on net photosynthesis of mountain grasslands differing in land use., *Ecological modelling.* 170: 407–426.

Wood, E. F. (1991). Land-surface-atmosphere interactions for climate modelling. observations, models and analysis, *Surv. in Geophys.* 12: 1–3.

Zanotti, F., Endrizzi, S., Bertoldi, G. & Rigon, R. (2004). The geotop snow module, *Hydrol. Proc.* 18: 3667–3679. DOI:10.1002/hyp.5794.

7

Critical Review of Methods for the Estimation of Actual Evapotranspiration in Hydrological Models

Nebo Jovanovic and Sumaya Israel
Council for Scientific and Industrial Research, Stellenbosch
South Africa

1. Introduction

The quantification of a catchment water balance is a fundamental requirement in the assessment and management of water resources, in particular under the impacts of human-induced land use and climate changes. The description and quantification of the water cycle is often very complex, particularly because of the spatial and temporal dimensions, variabilities and uncertainties inherent to the system. The advent of powerful computers, numerical modelling, and GIS is making it possible to describe the complexities of hydrological systems with statistically acceptable accuracy (Duan et al., 2004). Both local (e.g. on-farm) and catchment scale models, physically-based numerical models and simple conceptual balance models are now available to support water resource assessment, management, allocation as well as adaptation to climate change. In particular, the coupling of dedicated atmospheric, hydrological, unsaturated zone and groundwater models is becoming a powerful means of evaluating and managing water resources.

Evapotranspiration (ET) is a key process of the hydrological balance and arguably the most difficult component to determine, especially in arid and semi-arid areas where a large proportion of low and sporadic precipitation is returned to the atmosphere via ET. In these areas, vegetation is often subject to water stress and plant species adapt in different ways to prolonged drought conditions. This makes the process of ET very dynamic over time and variable in space. The focus of this chapter is on the methodologies used in hydrological models for the estimation of actual ET, which may be limited (adjusted) by water or other stresses. The chapter includes: i) a theoretical overview of ET processes, including the principle of atmospheric evaporative demand-soil water supply; ii) a schematic review of methods and techniques to measure and estimate ET; and iii) a review of methods for the estimation of ET in hydrological models.

2. Theoretical overview of evapotranspiration processes

ET is the combination of two separate processes, where liquid water is converted to water vapour (vaporization) from the soil, wet vegetation, open water or other surfaces, as well as from plants by transpiration through stomata (Allen et al., 1998). Evaporation and transpiration occur simultaneously and they are difficult to separate out. ET rate depends on weather conditions, water availability, vegetation characteristics, management and environmental constraints. The main weather variables affecting ET are temperature, solar radiation, wind speed and vapour pressure. The nature of the soil, its hydraulic properties and water retention capacity determine plant available water. Under natural conditions, water stored in the soil is replenished through precipitation, surface and groundwater. The type and developmental stage of vegetation, its adaptation to drought, structure and roughness, albedo, ground cover, root density and depth also affect ET rates. ET rates can be managed through different tillage practices, the establishment of windbreaks, different planting densities and thinning of vegetation, by reducing soil evaporation using, for example, localized irrigation targeting the root zone or mulching, and by reducing transpiration with herbicides or anti-transpirants (substances that induce closing of stomata, envelop vegetation with a surface film or change its albedo). Besides water stress, vegetation may be subject to other types of environmental stresses that are likely to result in a reduction of ET rates and plant growth, like for example pests, diseases, nutrient shortages, exposure to toxic substances and salinization (Allen et al., 1998).

Reference ET is the evaporation from a reference surface of the Earth and it depends on weather conditions. The reference surface can be an open water surface (open pan) or it can be related to weather variables (temperature, radiation, sunshine hours, wind speed, air humidity etc.). Many semi-empirical equations exist that relate reference ET to weather variables. Some of the most commonly adopted are Blaney-Criddle (Blaney and Criddle, 1950), Jensen-Haise (Jensen and Haise, 1963), Hargreaves (1983) and Thornthwaite (1948). Lu et al. (2005) compared the performance of three temperature-based methods, namely Thornthwaite (1948), Hamon (1963) and Hargreaves-Samani (1985), and three radiation-based methods, namely Turc (1961), Makkink (1957) and Priestley-Taylor (1972) for application in large scale hydrological studies in the south-eastern United States. Similarly, Oudin et al. (2005) tested the performance of 27 reference ET models in rainfall-runoff modelling of catchments located in France, Australia and the United States. Both Lu et al. (2005) and Oudin et al. (2005) proposed simple temperature-based methods for calculation of reference ET at catchment scale, in particular when availability of weather data sets is limited.

Theoretical equations that describe the mechanisms of the evaporation process are also available. For example, reference evaporation from an open water surface was first described by Penman (1948) and consisted of a radiation and a vapour pressure deficit term, representing the available energy for the endothermal evaporation process. Priestley and Taylor (1972) proposed the Priestley-Taylor equation, where the radiation term dominates over the advection term by a factor of 1.26, suitable for large forest catchments and humid environments. Based on decades of data and knowledge gained, the FAO (United Nations Food and Agricultural Organization) proposed a grass reference evapotranspiration (ETo) calculated with the Penman-Monteith equation (Monteith, 1965). The FAO Penman-Monteith ETo is defined as the evapotranspiration rate from a reference surface not short of water. The reference surface is a hypothetical grass reference crop with an assumed height of 0.12 m, a fixed surface resistance of 70 s m^{-1} and an albedo of 0.23 (Allen et al., 1998). The Penman-Monteith ETo is a function of the four main factors affecting evaporation, namely temperature, solar radiation, wind speed, and vapour pressure:

$$ETo = \frac{\dfrac{\Delta}{\lambda}(R_n - G) + \dfrac{\rho_a C_p}{\lambda} \dfrac{e_s - e_a}{r_a}}{\Delta + \gamma\left(1 + \dfrac{r_s}{r_a}\right)} \tag{1}$$

where λ is the latent heat of vaporization of water (MJ kg^{-1}); Δ is the gradient of the saturation vapour pressure-temperature function (kPa °C^{-1}); R_n is the net radiation (MJ m^{-2} d^{-1}); G is the soil heat flux (MJ m^{-2} d^{-1}); ρ_a is the air density (kg m^{-3}); C_p is the specific heat of the air at constant pressure = 1.013 kJ kg^{-1} K^{-1}; e_s is the saturated vapour pressure of the air (kPa), a function of air temperature measured at height z; e_a is the mean actual vapour pressure of the air measured at height z (kPa); r_a is the aerodynamic resistance to water vapour diffusion into the atmospheric boundary layer (s m^{-1}); γ is the psychrometric constant (kPa °C^{-1}); and r_s is the vegetation canopy resistance to water vapor transfer (s m^{-1}). Equation (1) uses standard climatic data that can be easily measured or derived from commonly collected weather data. Allen et al. (1998) also recommended procedures for the calculation of missing variables in equation (1).

In equation (1), the type of vegetation is accounted for through canopy resistance to gas exchange fluxes (r_s), vegetation height determining surface roughness (implicitly in r_a) and albedo (implicitly in R_n). Theoretically, the Penman-Monteith equation allows for direct calculation of actual ET through the introduction of canopy and air resistances to water vapour diffusion. However, this one-step approach is difficult to apply because canopy and air resistances are not known for many plant species and they are complex to measure. A two-step approach is then commonly used to determine actual ET, where the potential evapotranspiration (PET) is first calculated using a minimum value of canopy resistance for a specific crop/vegetation and the actual air resistance from weather data and vegetation height. In a second step, actual ET is calculated taking into account reduction in root water uptake due to water (and/or other) stress and reduction in soil evaporation due to drying of the top soil.

ET of crops or other vegetation differs distinctly from ETo because the ground cover, canopy properties, physiological adaptation and aerodynamic resistance of vegetation may be different from grass. These differences can be integrated into a factor Kc, commonly known as the crop coefficient because it is used to calculate crop water requirements (Allen et al., 1998). The FAO-56 model (Allen et al., 1998) provides a means of calculating reference and crop ET from meteorological data and crop coefficients. The effect of climate on crop water requirements is given by the reference evapotranspiration (ETo), and the effect of the crop by the crop coefficient Kc. Crop evapotranspiration under standard conditions (ETc) is the evapotranspiration from disease-free, well-fertilized crops, grown in large fields, under optimum soil water conditions, and achieving full production under the given climatic conditions. ETc can be calculated as:

$$ETc = Kc\,ETo \tag{2}$$

The Kc factor approach is applicable to uniform conditions, e.g. uniform crop fields with adequate fetch distance to minimize micrometeorological effects of field edges. Caution should therefore be exercised in the application of Kc under conditions where spatial variability of soil properties and crop management occur, in natural vegetation etc. The Kc factor can be split into two separate coefficients Kcb + Ke, where Kcb is the basal crop coefficient referred to crop transpiration and Ke is referred to direct evaporation from the soil (Allen et al., 1998).

The term ETc in equation (2) corresponds to evapotranspiration of vegetation at potential rates (PET) under given climatic conditions. In nature, PET seldom occurs, especially in semi-arid areas. When water is a limiting factor, physiological adaptation of plants occurs, stomata close and ET rates are below potential rates. This mechanism of stomatal control is described schematically in Figure 1.

In the soil-plant-atmosphere continuum (SPAC), water fluxes are driven by atmospheric evaporative demand and limited by soil water supply. Under wet soil conditions, the ratio of actual transpiration (T) and potential transpiration (PT), or relative transpiration (T/PT) is close to 1, showing that the root system is able to supply the canopy with water fast enough to keep up with the atmospheric evaporative demand and thereby preventing wilting. Under these conditions, transpiration is atmospheric demand-limited. As the soil dries beyond field capacity (FC) and beyond a threshold value of water content, T/PT drops below 1. Under these conditions, transpiration is soil water supply-limited as the root system can no longer supply water fast enough to keep up with demand and the soil water can be seen to be less available. Beyond soil water content at permanent wilting point (PWP), transpiration does not occur and T/PT = 0. The same mechanism can be represented for ratios of actual to potential evapotranspiration (ET/PET) as well as actual to maximum yield or productivity (Y/Ym). Plant available water depends on rooting depth, soil texture and structure. A similar mechanism occurs for direct evaporation from the soil surface. Canopy cover is generally used to split evaporation and transpiration, and approximates the available solar energy intercepted by the canopy compared to that reaching the soil surface (Ritchie, 1972).

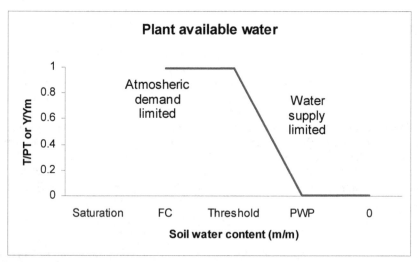

Fig. 1. Schematic representation of the plant available water graph. T – Actual transpiration; PT – Potential transpiration; Y – Actual yield or productivity; Ym – Maximum yield or productivity; FC – Soil water content at field capacity; PWP – Soil water content at permanent wilting point.

The original publication of Denmead and Shaw (1962) included the first scientific evidence of the concept of atmospheric evaporative demand-soil water supply (Figure 2) and this was followed in the last few decades by a large number of research studies on crop productivity-water functions (Doorenbos and Kassam, 1977; Hsiao et al., 2009; Raes et al., 2009; Steduto et al., 2009). This concept is applicable both to wet climates where the limiting factor for ET is generally atmospheric evaporative demand, and to dry climates where the dominant limiting factor is soil water supply.

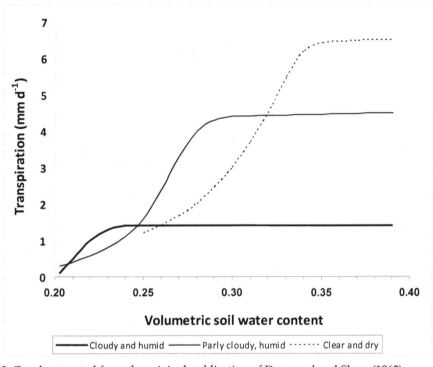

Fig. 2. Graph extracted from the original publication of Denmead and Shaw (1962), supplying scientific evidence of the dependence of transpiration on soil water supply and atmospheric demand.

3. Brief review of methods and techniques to measure and estimate actual evapotranspiration

A large number of methods and techniques for measurement and estimation of ET are available. These can be categorized into the following:
- Lysimeters (Allen et al., 1991): This is the only direct method to measure actual ET.
- Atmospheric measurements
  - Energy balance and micrometeorological methods: These methods are based on the computation of water fluxes based on measurements of atmospheric variables and they are therefore often referred to as direct measurements. Methods and techniques (e.g. Bowen ratio (Bowen, 1926), eddy correlation, scintillometry etc.) were widely discussed by Jarmain et al. (2008).

- • Weather data: These methods are based on the calculation of ET from weather data (e.g. Penman-Monteith equation for reference grass ETo).
- • Plant measurements
  - • Remote sensing from aircraft and satellite: Reflected electromagnetic energy is measured using sensors to generate multi- or hyper-spectral digital images. These data can then be translated into spatial variables such as surface temperature, surface reflectance, and vegetation indices (e.g. the Normalized Difference Vegetation Index NDVI) that describe the vegetation activity and its energy status. These methods were not feasible in the past at large scale and high frequency; however, with the latest technological advances, these techniques show promise (e.g. SEBAL) (Bastiaanssen et al., 1998a and b).
- • Soil measurements
  - • Soil water balance:

$$ET = P - R - D + \Delta S \tag{3}$$

where P is precipitation; R is runoff or run-on (a component of lateral subsurface inflow/outflow can also be included); D is drainage (or capillary rise), it approximates vertical recharge; ΔS is the change in soil water content, usually measured continuously or manually with a variety of techniques like gravimetric method, soil water sensors, neutron probe, time domain reflectometry etc. (Hillel, 1982). All units are usually expressed in mm per time.

4. Estimation of actual evapotranspiration in hydrological models

Although methodologies for the estimation of ETo and PET are widely adopted, actual (below-potential) ET is difficult to quantify and it usually requires the reduction of PET through a factor that describes the level of stress experienced by plants (two-step approach). The level of stress can be mathematically expressed linearly (slope of line in Figure 1) or through more complex functions. Currently, many models developed for different purposes and operating at different scales apply different functions to reduce PET based on the concept of atmospheric evaporative demand-soil water supply limited ET.

4.1 Field scale models

One-dimensional, field (point) scale hydrological models generally use more detailed functions to predict ET compared to large scale catchment models. The Soil Water Balance (SWB) is an example of a one-dimensional crop model for uniform canopies (Annandale et al., 1999). It is a daily time step model that includes a multi-layer soil water reservoir, where infiltrating water cascades from the top soil layer towards the bottom of the soil profile. Actual transpiration is limited by the evaporative demand (T_{max}) and root water uptake determined by soil wetness (Figure 3). Soil water potential translates into leaf water potential taking into account resistances to water flow in the SPAC (parallel line intersecting the curve in Figure 3) (Annandale et al., 2000).

WATCROS is another example of one-dimensional, cascading water balance and dry matter production simulation model based on climate, soil and plant variables and parameters (Aslyng and Hansen, 1982). It calculates reference ET from grass using a modified formula of Makkink (1957), and it assumes that this grass reference represents any dense, green, growing agricultural crop under Nordic conditions. Such potential evapotranspiration is partitioned into potential evaporation from the soil and crop transpiration using Beer's law.

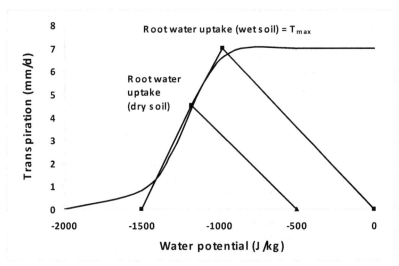

Fig. 3. Schematic representation of the root water uptake function adopted in SWB (adapted from Annandale et al., 2000). T_{max} – Maximum transpiration loss rate (mm d^{-1}).

In order to calculate actual transpiration, water is extracted at a potential rate when the actual soil water content is bigger than half the capacity of the root zone reservoir. Beyond this threshold, actual transpiration is decreased linearly as a function of the remaining water in the reservoir. If no water is left in the root zone reservoir, the transpiration rate equals 0. The size of the root zone reservoir depends on the soil and effective root depth (Hansen, 1984).

GLEAMS (Groundwater Loading Effects of Agricultural Management System) (Knisel, 1993) is also a one-dimensional, piston-flow water balance model used to simulate processes affecting water quality events in agricultural fields. It is the modified version of the well-validated CREAMS model (Knisel, 1980). PET is calculated with the Priestley and Taylor (1972) or with the Penman-Monteith equation (Allen et al., 1998). The model calculates actual soil evaporation and crop transpiration as a function of soil water content and leaf area index.

Cascading soil water balance models based on soil water reservoirs are often employed because of their conceptual simplicity and they are not data intensive. However, soil water movement in porous media can be best described physically with Richards' mass balance continuity equation for unsaturated water flow (Richards, 1931). Richards' equation equilibrates water between specified points (nodes) based on gradients in water energy and hydraulic conductivity:

$$\frac{\partial \theta}{\partial z} = \frac{d}{dz}\left[K(h,z)\frac{dh}{dz} - K(h,z)\right] - S(z,t) \qquad (4)$$

where θ is the volumetric soil water content (m^3 m^{-3}); t is time (h); z is soil depth (m, assumed positive downward); h is the soil water pressure head (m); K is the unsaturated hydraulic conductivity (m h^{-1}), a function of h and z; S (z, t) is the sink term (h^{-1}). The conversion of soil water pressure heads into soil water contents and vice versa can be done using different forms of the soil water retention curve (van Genuchten, 1980). The unsaturated hydraulic conductivity-soil water pressure head functions were also described by van Genuchten (1980).

The sink term $S(z, t)$ in equation (4) may include various sinks (or gains with a negative sign) like for example root water uptake. Root water uptake can be calculated with the approach of Nimah and Hanks (1973):

$$-S(z,t)=\frac{\left[H_r+(RRES\cdot z)-h(z,t)-s(z,t)\right]R(z)\cdot K(h)}{\Delta x\cdot\Delta z} \tag{5}$$

where H_r is the effective root water pressure head (m); RRES is a root resistance term; s is the osmotic pressure head (m); Δz is the soil depth increment (m); Δx is the horizontal distance increment; R is the proportion of the total root activity in the depth increment Δz. S cannot exceed potential transpiration.

Richards' equation (4) is non-linear and it can be solved iteratively through a finite-difference solution. It is adopted in several hydrological models to simulate water redistribution in the root zone and for accurate estimates of root water uptake and ET. For example, the RZWQM (Root Zone Water Quality Model) is a physically-based contaminant transport model that includes sub-models to simulate infiltration, runoff, water distribution and chemical movement in the soil (Ahuja et al., 2000). RZWQM simulates PET with a modified Penman-Monteith model and actual ET is constrained by water availability as estimated from Richards' equation.

Soil-Water-Atmosphere-Plant (SWAP) is a 2-D, transient model for water flow and solute transport in the unsaturated and saturated zones (Kroes and van Dam, 2003). It is applied to agrohydrological problems at field scale and it makes use of Richards' equation for soil water redistribution. The relative plant water uptake (T/PT) calculated with this model as a function of soil water potential is shown in Figure 4 (Feddes et al., 1978). The soil water potential values h_1, h_2, and h_4 are inputs. Threshold soil water potentials for reduction in T/PT vary in the range between h_{3h} and h_{3l} and they are applied depending on high (T_{high}) or low (T_{low}) transpiration demand. The h_4 input is wilting point. Reduction in T/PT occurs also in the wet soil range (close to saturation between h_2 and h_1) to simulate the effects of water-logging. The plant water uptake solution in SWAP (Feddes et al., 1978) is also adopted in the HYDRUS unsaturated flow and solute transport model (Simunek et al., 2007) as well as in the SIMGRO (SIMulation of GROundwater and surface water levels) catchment model (van Walsum et al., 2004).

MACRO (Jarvis, 1994) is a deterministic, one-dimensional, transient model for water and solute transport in field soils. It also uses the water uptake function proposed by Feddes et al. (1978). It accounts for conditions that are too wet (close to saturation h_1 in Figure 4) and too dry (close to wilting point h_4 in Figure 4). A dimensionless water stress index ω is used to calculate the ratio of actual to potential root water uptake. This stress index combines two functions describing the distribution of roots and water content in the multi-layered soil profile:

$$\omega = \sum_{i=1}^{i=k} r_i\,\omega_i \tag{6}$$

where k is the number of soil layers in the profile containing roots, and r_i and ω_i are the proportion of the total root length and a water stress reduction factor in layer i. Root length is distributed logarithmically with depth, whilst the stress factor ω_i depends on the soil water content in the particular layer. The root system is usually represented as an inverted cone and its distribution with depth is often non-linear (Yang et al., 2009). The shape of root distribution can therefore be represented with two inputs, namely root depth and an extractable water parameter (Gardner, 1991).

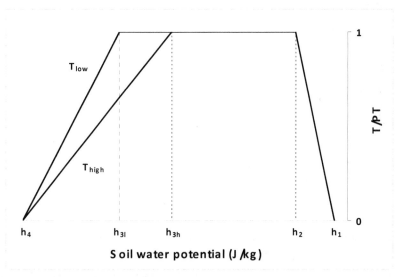

Fig. 4. Schematic representation of the plant available water graph adopted in SWAP (adapted from Feddes et al., 1978). T/PT – Relative plant water uptake; T_{low} – Low transpiration; T_{high} –High transpiration; h_n – Inputs of soil water potential.

The importance of knowing the root depth of vegetation in order to define the size of the soil reservoir and plant available water was underlined by Ritchie (1998) and illustrated in Figure 5. Ritchie (1998) proposed a linear relation between root water uptake and soil water content. Maximum, minimum and usual range of root water uptake are indicated in Figure 5. These depend on root length density Lv and the ability of plants to explore a certain volume of soil.

Another example of a model with a fairly detailed description of root distribution is WAVES (Dawes and Short, 1993; Zhang et al., 1996). WAVES is a water balance model that simulates surface runoff, soil infiltration, ET, soil water redistribution, drainage and water table interactions. Daily transpiration is calculated with the Penman-Monteith equation and reduced using weighting factors determined by the modelled root density and a normalized weighted sum of the matric and osmotic soil water potentials of each layer. The model has been parameterised and used to simulate the water use of various vegetation types in South Africa (Dye et al., 2008).

Feddes et al. (2001) discussed that deep-rooted vegetation and increased water availability may have an effect even on global climate. Deep rooting systems result in large volumes of soil being explored by the roots, large amounts of soil profile available water and large transpiration rates. This is even more prominent in the presence of shallow groundwater. Jovanovic et al. (2004) proved that the contribution of shallow water tables to root water uptake through capillary rise can be a substantial component of the water balance.

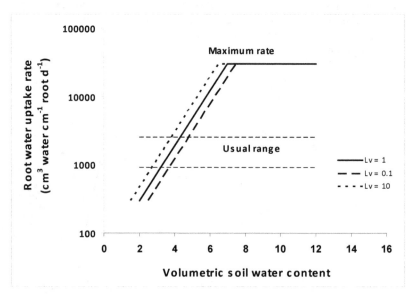

Fig. 5. Relationship between root water uptake rate, volumetric soil water content and root length density (Lv in cm cm^{-3}) (adapted from Ritchie, 1998).

The DRAINMOD computer model was primarily developed to simulate the effects of drainage and associated water management practices on water table depths, the soil water regime and crop yields (Skaggs, 1978). ET is calculated according to the relationship of Norero (1969):

$$ET=\frac{PET}{1+\left(\dfrac{h}{h*}\right)^{k}} \tag{7}$$

where k is a constant that can be defined using methods given in Taylor and Ashcroft (1972) and Norero (1969), h is the soil water potential in the root zone which could be obtained from the soil water characteristics using the average root zone water content, and h* is the value of h when ET = 0.5 PET. Equation (7) is graphically illustrated in Figure 6.

Given the purpose of the DRAINMOD model, direct evaporation from the soil can be estimated using the simplified Gardner (1958) equation relating maximum evaporation rate in terms of water table depth and unsaturated soil hydraulic conductivity:

$$\frac{d}{dz}\left[K(h,z)\frac{dh}{dz}-K(h,z)\right]=0 \tag{8}$$

The symbols and units are the same as those defined for equation (4). Maximum soil evaporation rate for a given water table depth can be approximated by solving equation (8), using a large negative h value (for example h = -1000 cm) at the surface (z = 0) and h = 0 at the water table depth. An example of solution of equation (8) is shown in Figure 7 for a loamy sand.

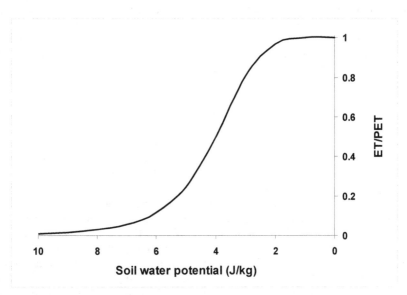

Fig. 6. Schematic of relative evapotranspiration (ET/PET), as affected by soil water potential in the root zone (adapted from Skaggs, 1978)

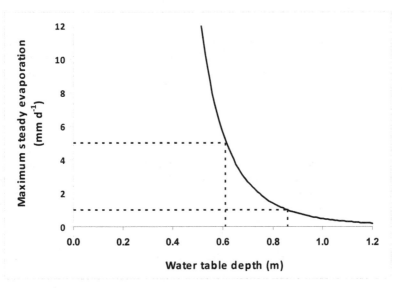

Fig. 7. Relationship between maximum upward movement of water versus water table depth for a loamy sand (adapted from Skaggs, 1978).

4.2 Catchment scale models

Many catchment scale models account for soil moisture in the estimate of ET (Viviroli et al., 2009) using more or less sophisticated approaches. For example, Zhang et al. (2001) developed a semi-empirical water balance model for forested and non-forested catchments in the Murray-Darling basin of Australia. This was based on the assumption that actual ET is equal to precipitation under very dry conditions, and that it equals PET under very wet conditions. On the other hand, Gurtz et al. (1999) applied the PREVAH (PRecipitation-Runoff-EVApotranspiration HRU Model) hydrological model in an alpine basin. They calculated ET using the Penman-Monteith equation by changing the canopy stomatal resistance (equation (1)) below a given threshold of soil moisture.

Barr et al. (1997) reviewed a number of studies where the dependence of ET on soil moisture was evidenced. In their study, they evaluated three methods for estimating ET in the SLURP mesoscale hydrological model (Kite, 1995), namely: i) the complementary relationship areal ET model (Morton, 1983), ii) the Granger (1991) modification of Penman's method and iii) the Spittlehouse (1989) energy-limited versus soil moisture-limited method. The method of Morton (1983) makes use of ET estimated with the Penman (1948) equation and reduced by an amount proportional to vapour pressure deficit, without taking into account the effects of soil moisture on ET. The method of Granger (1991) is a modification to the Penman (1948) equation that includes a relative evaporation variable in the vapour pressure deficit term. The Spittlehouse (1989) method takes into account soil moisture and it calculates actual ET withdrawal from the soil store as the lesser of the soil store and energy-limited rates. The energy-limited rate is calculated with the Priestley-Taylor equation (Priestley and Taylor, 1972) and the soil store-limited rate is calculated as a function of the fraction of extractable soil moisture (Spittlehouse, 1989). The formulation of all three methods was based on forests and grasslands in large catchments. Amongst the three methods tested over a 5-year period in the Kootenay Basin of eastern British Columbia, the Spittlehouse (1989) method including the soil moisture feedback to ET estimates gave the best agreement between simulated and recorded streamflow.

Zhou et al. (2006) used the Shuttleworth and Wallace (1985) model and NDVI to estimate ET from sparse canopies to feed in the BTOPMC distributed hydrological model (Takeuchi et al., 1999). The methodology adopted the Penman-Monteith ETo with an increase of stomatal resistance based on the generic equation:

$$r_s = \frac{r_{s\,min}}{LAI\,F_i(X_i)} \qquad (9)$$

where $r_{s\,min}$ represents the minimal stomatal resistance of individual leaves under optimal conditions (s m^{-1}), LAI is the effective leaf area index and $F_i(X_i)$ is the stress function for a factor X_i (nutrients, pests, water etc.). The water stress function was expressed as a function of volumetric soil water content θ, in the range between field capacity θ_{fc} and residual soil water content θ_r:

$$f(\theta) = \frac{\theta - \theta_r}{\theta_{fc} - \theta_r} \qquad (10)$$

The Agricultural Catchments Research Unit (ACRU) (Schulze, 1994) is a catchment scale agrohydrological modeling system. It calculates relative evapotranspiration (ET/PET) as a function of plant available water (Figure 8). The reduction of ET/PET on the left side of the graph in Figure 8 describes the effect of water-logging. The threshold f_s is user-specified, or it is calculated as a function of a critical leaf water potential ψ^{cr} and ET/PET:

$$f_s = F\left(user\,specified\,PAW\right) \tag{11}$$

$$f_s = 0.94 + \frac{0.0026\,\psi^{cr}}{ET\,/\,PET} \tag{12}$$

Precipitation-Runoff Modular System (PRMS) is a hydrological modular modeling system for large scale basins (Leavesley et al., 1983, 1996). It calculates actual ET for four types of vegetation/land use and three types of soil texture (Figure 9).

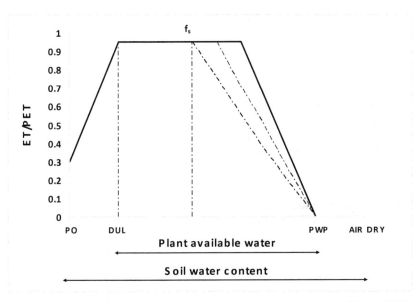

Fig. 8. Schematic representation of the plant available water graph adopted in ACRU (adapted from Schulze, 1994). ET – Actual evapotranspiration; PET – Potential evapotranspiration; PO – Soil water content at saturation; DUL – Drainage upper limit; PWP – Permanent wilting point; f_s – Threshold of reduction of relative evapotranspiration (ET/PET).

MIKE SHE is a physically-based, distributed, integrated hydrological and water quality modeling system (Abbott et al., 1986). ET is calculated based on PET, leaf area index and root depth, soil water content and physical characteristics as well as a set of empirical parameters (Kristensen and Jansen, 1975). Specifically, the ratio of ET to PET is calculated with two functions, the one describing the leaf area index and the other describing the soil water status.

More empirical approaches aimed at describing the hydrological cycle also take into consideration ET. A semi-empirical model called EARTH (Extended model for Aquifer Recharge and moisture Transport through unsaturated Hardrock; Department of Water

Affairs and Forestry, 2006) was developed in South Africa to estimate large scale groundwater recharge by accounting for the variables of the hydrological cycle. EARTH uses modules for vegetation, soil, linear reservoir and saturated flow. The soil module calculates ET as a linear function of soil moisture (Figure 10).

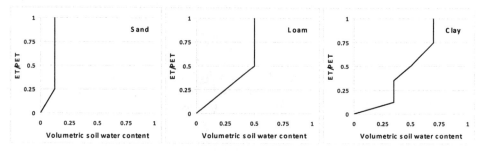

Fig. 9. Schematic representation of the plant available water graph adopted in PRMS for three types of soil texture (adapted from Leavesley et al., 1983). ET – Actual evapotranspiration; PET – Potential evapotranspiration.

The chloride mass balance (CMB) is another method commonly used to estimate groundwater recharge in semi-arid areas (Xu and Beekman, 2003). The estimates of groundwater recharge with CMB refer to long term annual averages, usually over hundreds of years. Implicitly, this technique accounts for the concentrating effects of water by ET in semi-arid regions. Groundwater recharge can be calculated with the following formula:

$$P\,Cl_p = R_T\,Cl_{gw} \qquad (13)$$

where P is precipitation (mm a^{-1}); Cl_p is the chloride concentration in precipitation (mg L^{-1}); R_T is total groundwater recharge (mm a^{-1}), approximated with the term D in equation (3);

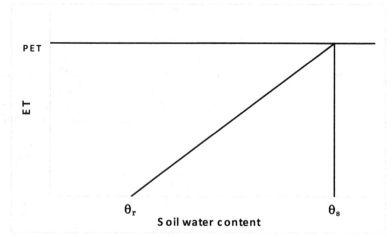

Fig. 10. Schematic representation of the plant available water graph adopted in EARTH (adapted from DWAF, 2006). PET – Potential evapotranspiration; ET – Actual evapotranspiration; θ_r –Soil moisture retained by the soil matrix; θ_s – Maximum soil moisture.

Cl_{gw} is the chloride concentration in groundwater (mg L^{-1}). The source of Cl has to be precipitation solely as other sources may intefere with the interpretation of Cl measurements. Other conservative tracers can also be used. As groundwater recharge can be approximated with term D and ΔS is negligible in the long term, equation (3) can be applied to calculate ET if mean annual runoff data are available.

4.3 Remote sensing applications in the estimate of actual evapotranspiration

The methods discussed above generate point estimates of ET. These values are usually applicable to uniform crop fields, hillslope transects or hydrologically homogeneous areas, and they often need to be upscaled (Oudin et al., 2005). Upscaling can be done through repetitive measurements in all representative areas of interest or through regionalization (Krause, 2002). Due to spatial variations in climate, vegetation, land use and physiographic characteristics, point methods for estimating ET are often too intensive to be applied at large catchment scales. A promising application that may overcome these shortcomings involves areal estimates of ET with remote sensing techniques.

The theory described in the canopy temperature-ET models of Hatfield et al. (1984) was the foundation for surface energy balance approaches based on remote sensing. In these approaches, each pixel of aircraft or satellite images is processed to determine the components of the energy balance equation:

$$\lambda E = R_n - H - G \tag{14}$$

where R_n is net radiation, λE is the latent heat of vaporization, H is the sensible heat flux, G is the soil heat flux and all terms are usually expressed in W m^{-2}. Algorithms such as the Surface Energy Balance Algorithm over Land (SEBAL) use remote sensing imagery, empirical relationships and physical modules to calculate the terms of the energy balance equation and estimate ET (converted from λE in equation (14)) (Bastiaanssen et al., 1998a, 1998b; Tasumi et al., 2005). In particular, SEBAL requires visible, near-infrared and thermal infrared input data obtained from satellite images. Instantaneous net radiation can be calculated from incoming solar radiation measured at ground stations and outgoing thermal radiation estimated from surface albedo, surface emissivity and temperature. Soil heat flux can be computed from surface temperature, albedo and NDVI. The sensible heat flux is calculated with an algorithm of standard heat and momentum transport equations including pixel-based Monin–Obukhov stability corrections. Both wet and dry surface pixels are required because these represent extreme limits in the studied domain at the specific time when the satellite images are taken. The sensible heat flux is constrained by a dry limit (surface with latent heat flux $\lambda E = 0$; sensible heat flux $H = R_n - G$) and wet limit (surface with sensible heat flux $H = 0$; vertical difference in air temperature $dT_a = 0$). A value of dT_a is assigned to all other pixels assuming it varies linearly between the dry and wet ranges. H is then calculated as a function of dT_a and λE computed as the residual of the energy balance. Instantaneous λE values are extrapolated over time assuming that the instantaneous evaporative fraction in equation (14) is stable for the given time period.

Other remote sensing based methods to estimate ET are also available. The Surface Energy Balance System (SEBS) is an energy balance algorithm for the estimation of ET (Su, 2002) that works on similar principles as SEBAL. The MODIS evapotranspiration (ET – MOD 16) algorithm is based on the Penman-Monteith equation (Allen et al., 1998). Land cover, fraction of absorbed photosynthetically active radiation, leaf area index and global surface meteorology information derived from MODIS are used to estimate daily ET and PET, which is then composited over an 8-day interval. ET is expressed in mm d^{-1} and calculated globally every day at 1 km resolution. METRIC (Mapping EvapoTranspiration at high Resolution with Internalized Calibration) is a computer model that uses LandSat data to compute and map ET. These ET maps (i.e. images) provide the means to quantify ET on a field by field basis in terms of both rates and spatial distribution (Allen et al., 2007). Sinclair and Pegram (2010) implemented a real time platform for supplying satellite-based information on ETo and soil moisture in South Africa. Wang et al. (2003) found a significant correlation between deseasonalized time series of NDVI and soil moisture, from where root zone depth can be indirectly estimated. This procedure, however, requires calibration for specific vegetation and climatic conditions.

Although some studies have been carried out in order to test and compare remote sensing methodologies to conventional methods for estimation of ET (Gibson et al., 2011; Kite & Droogers, 2000), more research is required in order to assess the feasibility of application of remote sensing techniques to improve water use efficiency, irrigation management on farms and catchment management, particularly in arid and semi-arid areas. Given the temporal dynamics of ET and its dependance on soil water supply conditions, the interpolation of instantaneous satellite information to estimate ET over a given time period may require verification (Olioso et al., 2005). Processed information from satellite images needs to be supplied at a required frequency for applications in water management on farms and in large catchments. In addition, cloud-free satellite images are required and these are not always available.

5. Conclusion

This chapter discussed the theoretical principles of some hydrological models as examples. It was not meant to provide a review of all models available. The models described here were extensively evaluated in specific studies. Wagener (2003) proposed models should be evaluated for performance (e.g. by minimizing the objective function which can be the difference between simulated and observed data), uncertainty (e.g. by analyzing reasonable ranges of model inputs, parameters and structure) and realism (e.g. by analyzing how consistent the model output is with our understanding of reality). No unique approach for model evaluation exists and, therefore, there is no easy answer to the question on which model is the most accurate. Models should be used for the purpose that they were developed and evaluated with different techniques and for different conditions.

The quantification of actual ET is of utmost importance for various applications in hydrology and water management, such as resource allocation, water footprinting, quantification of water use efficiency etc. This review has highlighted that a large number of both field (point) scale, one-dimensional models and catchment scale spatial GIS-based models adopt conceptually similar approaches to the estimation of actual ET. These approaches are based on the concept of atmospheric evaporative demand-soil water supply limited ET. Such a concept is applicable both to wet climates (limiting factor is atmospheric evaporative demand) and to dry climates (limiting factor is soil water supply). Some models make use of a one-step approach to increase canopy stomatal resistance directly in the Penman-Monteith equation, which represents a mechanistic and physically sound solution to the estimation of actual ET (e.g. BTOPMC). This methodology is, however, hampered by the difficulty in estimating the canopy resistance term. Other models adopt a more conventional two-step approach to calculate PET and reduce it using a water stress index generally based on soil water content (e.g. WATCROS). Some models make use of the data intensive and physically sound principles embedded in Richards' equation to redistribute water in the root zone (e.g. SWAP). Other models make use of a simplistic soil reservoir-based cascading water balance as finite differences are difficult to apply to complex and large scale systems (e.g. ACRU). In addition, abrupt and large changes in soil water content in space and time may lead to numerical instabilities in the finite difference solution of Richards' equation, or in longer simulation times compared to cascading soil water balance models because equilibrium conditions, usually solved through an iterative process, may not be reached easily.

When applying specific models, it is essential to be aware of the specific assumptions around which they were built, their advantages and limitations. Field scale models are generally more data intensive than catchment scale models. For example, dedicated crop and soil water balance models usually include moving thresholds in the atmospheric demand-soil water supply function (e.g. SWB). Models that estimate leaf area provide the opportunity to partition the energy available for soil evaporation and plant transpiration, and those that calculate root growth and depth facilitate the estimation of plant available water in the soil. If properly calibrated, such models are more accurate in predicting field (point) scale ET, but they are also more data intensive compared to large scale models. Large scale catchment models require ET-related inputs in the spatial domain and make use of less detailed ET calculation sub-routines as trade-off (e.g. PRMS).

Given the principles governing soil water redistribution, the soil water dynamics and ET, it is recommended that a daily time step be used in the calculation of water balance variables. Root depth is a very important variable that determines the volume of soil explored by plant roots. This is not often easily measured resulting in uncertainties in the estimation of ET and the water balance. Promising technologies for large scale spatial estimation of ET, soil moisture, and indirectly root depth include remote sensing. These techniques, however, need to be tested and validated for applicability to a wide range of water management conditions in arid and semi-arid areas. The purpose and applicability of remote sensing methods depend on the spatial resolution of the images and their temporal resolution (frequency).

6. Acknowledgment

The authors acknowledge the Water Research Commission (Pretoria, South Africa) for funding this study emanating from project No. K5/1909 on "Reducing Uncertainties of Evapotranspiration and Preferential Flow in the Estimation of Groundwater Recharge".

7. References

Abbott, M. B.; Bathurst, J. C.; Cunge, J. A.; O'onnell, P. E. & Rasmussen, J. (1986). An introduction to the European Hydrological System - Systeme Hydrologique Europeen (SHE). 1: History and Philosophy of a Physically-Based Distributed Modelling System. *Journal of Hydrology*, Vol. 87, pp. 45-59.

Ahuja, L.R.; Rojas, K.W.; Hanson, J.D.; Shaffer, M.J. & Ma, L. (2000). *Root Zone Water Quality Model. Modeling Management Effects on Water Quality and Crop Production*, Water Resources Publications, Colorado, USA, 356 pp.

Allen, R.G.; Howell, T.A.; Pruitt, W.O.; Walter, I.A. & Jensen, M.E. (Eds.) (1991). *Lysimeters for Evapotranspiration and Environmental Measurements*, American Society of Civil Engineers, New York, USA.

Allen, R.G.; Pereira, L.S.; Raes, D. & Smith, M. (1998). *Crop Evapotranspiration: Guidelines for Computing Crop Water Requirements*, United Nations Food and Agriculture Organization, Irrigation and Drainage Paper 56. Rome, Italy, 300 pp.

Allen, R.G.; Tasumi, M.; Morse, A.; Trezza, R.; Wright, J.L.; Bastiaanssen, W.; Kramber, W.; Lorite, I. & Robinson, C.W. (2007). Satellite-Based Energy Balance for Mapping Evapotranspiration with Internalized Calibration (METRIC) – Applications. *J. Irrig. and Drain. Eng.*, Vol. 133(4), pp. 395-406.

Annandale, J.G. ; Benade, N. ; Jovanovic, N.Z. ; Steyn, J.M.; Du Sautoy, N. & Marais, D. (1999). *Facilitating Irrigation Scheduling by Means of the Soil Water Balance Model*, Water Research Commission Report No. 753/1/99, Pretoria, South Africa.

Annandale, J.G.; Campbell, G.S.; Olivier, F.C. & Jovanovic, N.Z. (2000). Predicting crop water uptake under full and deficit irrigation: An example using pea (*Pisum sativum* cv. Puget). *Irrigation Science*, Vol. 19, pp. 65-72.

Aslyng, H.C. & Hansen, S. (1982). *Water Balance and Crop Production Simulation. Model WATCROS for Local and Regional Application*. Hydrotechnical Laboratory, The Royal Vet. and Agric. Univ., Copenhagen, 200 pp.

Barr, A. G.; Kite, G.W.; Granger, R. & Smith, C. (1997). Evaluating Three Evapotranspiration Methods in the SLURP Macroscale Hydrological Model. *Hydrological processes*, Vol. 11, pp. 1685-1705.

Bastiaanssen, W.G.M.; Menenti, M.; Feddes, R.A. & Holtslag, A.A.M. (1998a). A Remote Sensing Surface Energy Balance Algorithm for Land (SEBAL) 1. Formulation. *Journal of Hydrology*, Vol. 212-213, pp. 198-212.

Bastiaanssen, W.G.M.; Pelgrum, H.; Wang, J.; Ma, Y.; Moreno, J.F.; Roerink, G.J. & van der Wal, T. (1998b). A Remote Sensing Surface Energy Balance Algorithm for Land (SEBAL) 2. Validation. *Journal of Hydrology*, Vol. 212-213, pp. 213-229.

Blaney, H.F. & Criddle, W.D. (1950). *Determining Water Requirements in Irrigated Areas from Climatological and Irrigation Data*. USDA Soil Conserv. Serv. SCS-TP96, 44 pp.

Bowen, I. S. (1926). The Ratio of Heat Losses by Conduction and by Evaporation from any Water Surface. *Phys. Rev.*, Vol. 27, pp. 779--787.

Dawes, W.R. & Short, D.L. (1993). *The Efficient Numerical Solution of Differential Equations for Coupled Water and Solute Dynamics: the WAVES model*. CSIRO Division of Water Resources Technical Memorandum 93/18, Canberra, ACT, Australia.

Denmead, O.T. & Shaw, R.H. (1962). Availability of soil water to plants as affected by soil moisture content and meteorological conditions. *Agronomy Journal*, Vol. 54, 385-390.

Department of Water Affairs and Forestry. (2006). *Groundwater Resource Assessment II: Recharge Literature Review Report 3aA*, Project No. 2003-150, Department of Water Affairs and Forestry, Pretoria, South Africa.

Doorenbos, J. & Kassam, A.H. (1979) *Yield Response to Water*. Irrigation and Drainage Paper n. 33. FAO, Rome, Italy, 193 pp.

Duan, Q.; Gupta, H.V.; Sorooshian, S.; Rousseau, A. & Turcotte, R. (Eds.) (2004). *Calibration of Watershed Models*, AGU Monograph Series, Water Science and Application 6, American Geophysical union, Washington, D.C., 345 pp.

Dye, P.J.; Jarmain, C.; Le Maitre, D.; Everson, C.S.; Gush, M. & Clulow, A. (2008). *Modelling Vegetation Water Use for General Application in Different Categories of Vegetation.* Water Research Commission Report No.1319/1/08, ISBN 978-1-77005-559-9, Pretoria, South Africa.

Feddes, R.A.; Hoff, H.; Bruen, M.; Dawson, T.; de Rosnay, P.; Dirmeyer, P.; Jackson, R.B.; Kabat, P.; Kleidon, A.; Lilli, A. & Pitman, A.J. (2001). Modeling Root Water Uptake in Hydrological and Climate Models. *Bulletin of the American Meteorological Society,* Vol. 82(12), 2797-2809.

Feddes, R.A.; Kowalik, R.J. & Zaradny, H. (1978). *Simulation of Field Water Use and Crop Yield. Simulation Monographs,* Pudoc, Wageningen, The Netherlands, 189 pp.

Gardner, W.R. (1958). Some Steady-State Solutions of the Unsaturated Moisture Flow Equation with Application to Evaporation from a Water Table. *Soil Science,* Vol. 85, pp. 228-232.

Gardner, W.R. (1991). Modeling Water Uptake by Roots. *Irrigation Science,* Vol. 12, 109-114.

Gibson, L.A., Munch, Z. & Engelbrecht, J. (2011). Particular Uncertainties Encountered in Using a Pre-Packaged SEBS Model to Derive Evapotranspiration in a Heterogeneous Study Area in South Africa. *Hydrology and Earth System Science,* Vol. 15, pp. 295-310.

Granger, R.J. (1991). *Evaporation from Natural Non-Saturated Surfaces,* PhD Thesis, Department of Agricultural Engineering, University of Saskatchewan, Saskatoon, 140 pp.

Gurtz, J.; Baltensweiler, A. & Lang, H. (1999). Spatially Distributed Hydrotope-Based Modelling of Evapotranspiration and Runoff in Mountainous Basins. *Hydrological Processes,* Vol. 13, pp. 2751-2768.

Hamon, W.R. (1963). Computation of Direct Runoff Amounts from Storm Rainfall. *Int. Assoc. Sci. Hydrol. Pub.,* Vol. 63, pp. 52-62.

Hansen, S. (1984). Estimation of Potential and Actual Evapotranspiration. *Nordic Hydrology,* Vol. 15, pp. 205-212.

Hargreaves, G.H. (1983). Discussion of 'Application of Penman Wind Function' by Cuenca, R.H. and Nicholson, M.J. *J. Irrig. and Drain. Eng. ASCE,* Vol. 109(2), pp. 277-278.

Hargreaves, G.H. & Samani, Z.A. (1985). Reference Crop Evapotranspiration from Temperature. *Applied Eng. in Agric.,* Vol. 1(2), pp. 96-99.

Hatfield, J.L.; Reginato, R.J. & Idso, S.B. (1984). Evaluation of Canopy Temperature-Evapotranspiration Models over Various Crops. *Agricultural and Forest Meteorology,* Vol. 32, pp. 41-53.

Hillel, D. (1982). *Introduction to Soil Physics,* Academic Press Inc., New York, USA.

Hsiao, T.C.; Heng, L.; Steduto, P.; Rojas-Lara, B.; Raes, D. & Fereres, E. (2009). AquaCrop The FAO Crop Model to Simulate Yield Response to Water: III. Parameterization and Testing for Maize. *Agronomy Journal,* Vol. 101(3), pp. 448-459.

Jarmain, C.; Everson, C.S.; Savage, M.J.; Mengistu, M.G.; Clulow, A.D.; Walker, S. & Gush, M.B. (2008). *Refining Tools for Evaporation Monitoring in Support of Water Resources Management,* Water Research Commission Report No. K5/1567/1/08, Pretoria, South Africa.

Jarvis, N.J. (1994). *The MACRO Model (Version 3.1). Technical Description and Sample Simulations,* Reports and Dissertations 19, Dept. Soil Sci., Swedish Univ. Agric. Sci., Uppsala, 51 pp.

Jensen, M.E. & Haise, H.R. (1963). Estimating Evapotranspiration from Solar Radiation. *J.Irrig. and Drain. Div. ASCE*, Vol. 89, pp. 15-41.

Jovanovic, N.Z.; Ehlers, L.; Bennie, A.T.P.; Du Preez, C.C. & Annandale, J.G. (2004). Modelling the Contribution of Root Accessible Water Tables towards Crop Water Requirements. *South African Journal of Plant and Soil*, Vol. 21(3), pp. 171-182.

Kite, G.W. (1995). *Manual for the SLURP Hydrological Model*, NHRI, Saskatoon, 111 pp.

Kite, G.W. & Droogers, P. (2000). Comparing Evapotranspiration Estimates from Satellites, Hydrological Models and Field Data. *Journal of Hydrology*, Vol. 229, pp. 3–18.

Knisel, W.G. (Ed.) (1980). *CREAMS: A Field-Scale Model for Chemical, Runoff, and Erosion from Agricultural Management Systems*, Conservation Research Report 26, U.S. Department of Agriculture, Washington, D.C.

Knisel, W.G. (1993). *GLEAMS: Groundwater Loading Effects of Agricultural Management Systems, V.2.10*, University of Georgia, Coastal Plain Experiment Station, Biological and Agricultural Engineering Department (Publication No. 5).

Krause, P. (2002). Quantifying the Impact of Land Use Changes on the Water Balance of Large Catchments using the J2000 Model. *Physics and Chemistry of the Earth*, Vol. 27, pp. 663–673.

Kristensen, K. J. & Jensen, S.E. (1975). A Model for Estimating Actual Evapotranspiration from Potential Evapotranspiration. *Nordic Hydrology*, Vol. 6, pp. 70-88.

Kroes, J.G. & van Dam, J.C. (2003). *Reference Manual SWAP Version 3.0.3.*, Wageningen, Alterra, Green World Research. Alterra Rep. No. 773, pp. 211.

Leavesley, G.H.; Lichty, R.W.; Troutman, B.M. & Saindon, L.G. (1983). *Precipitation-Runoff Modelling System – User's Manual*, US Geological Survey Water Resources Investigation Report 83-4238, Denver, Colorado, USA.

Leavesley, G.J.; Restrepo, P.J.; Markstrom, S.L.; Dixon, M. & Stannard, L.G. (1996). *The Modular Modeling System (MMS): User's Manual*, Open-File Report 96-151, US Geological Survey, Denver, Colorado, USA.

Lu, J.; Sun, G.; McNulty, S.G. & Amatya, D.M. (2005). A Comparison of Six Potential Evapotranspiration Methods for Regional Use in the South-Eastern United States. *Journal of the American Water Resources Association*, Vol. 41(3), pp. 621-633.

Makkink, G.F. (1957). Testing the Penman Formula by Means of Lysimeters. *J. Inst. of Water Eng.*, Vol. 11, pp. 277-288.

Monteith, J.L. (1965). *Evaporation and the Environment, The State and Movement of Water in Living Organisms, XIXth symposium*, Cambridge University Press, Swansea.

Morton, F.I. (1983). Operational Estimate of Aerial Evapoutranspiration and their Significance to the Science and Practice of Hydrology. *Journal of Hydrology*, Vol. 66, pp. 77-100.

Nimah, M. & Hanks, R.J. (1973). Model for Estimating Soil-Water-Plant-Atmospheric Interrelation: I. Description and Sensitivity. *Soil Science Society of America Proceedings*, Vol. 37, pp. 522-527.

Norero, A.L. (1969). *A Formula to Express Evapotranspiration as a Function of Soil Moisture and Evaporative Demand of the Atmosphere*, PhD Thesis., Utah State University, Logan/Utah, USA.

Olioso, A.; Inoue, Y.; Ortega-Farias, S.; Demarty, J.; Wigneron, J.-P.; Braud, I.; Jacob, F.; Lecharpentier, P.; Ottle, C.; Calvet, J.-C. & Brisson, N. (2005). Future Directions for Advanced Evapotranspiration Modeling: Assimilation of Remote Sensing Data into

Crop Simulation Models and SVAT Models. *Irrigation and Drainage Systems*, Vol. 19, 377–412.

Oudin, L.; Hervieu, F.; Michel, C.; Perrin, C.; Andreassian, V.; Anctil, F. & Loumagne, C. (2005). Which Potential Evapotranspiration Input for a Lumped Rainfall–Runoff Model? Part 2 - Towards a Simple and Efficient Potential Evapotranspiration Model for Rainfall–Runoff Modelling. *Journal of Hydrology*, Vol. 303, pp. 290–306.

Penman, H.L. (1948). Natural Evaporation from Open Water, Bare Soil and Grass. *Proc. Roy. Soc. London A(194), S.* pp. 120-145.

Priestley, C.H.B. & Taylor, R.J. (1972). On the Assessment of Surface Heat Flux and Evaporation Using Large Scale Parameters. *Mon. Weath. Rev.*, Vol. 100, pp. 81-92.

Raes, D.; Steduto, P.; Hsiao, T.C. & Fereres, E. (2009) AquaCrop-The FAO Crop Model to Simulate Yield Response to Water: II. Main Algorithms and Software Description. *Agronomy Journal*, Vol. 101(3), pp. 438-447.

Richards, L.A. (1931). Capillary Conduction of Liquids through Porous Mediums. *Physics*, Vol. 1(5), pp. 318-333.

Ritchie, J.T. (1972). Model for Predicting Evaporation from a Row Crop with Incomplete Cover. *Water Resources Research*, Vol. 8, pp. 1204-1213.

Ritchie, J.T. (1998). Soil Water Balance and Plant Water Stress, In: *Understanding Options for Agricultural Production,* G.Y. Tsuji, G. Hoogenboom & P.K. Thornton (Eds.), pp. 41-53, Kluwer Academic Publishers, UK.

Schulze, R.E. (1994). *Hydrology and Agrohydrology: A Text to Accompany the ACRU-300 Agrohydrological Modelling System,* Agricultural Catchments Research Unit, Department of Agricultural Engineering, University of Natal.

Shuttleworth, W.J. & Wallace, J.S. (1985). Evaporation from Sparse Crops - An energy Combination Theory. *Quart. J. Roy. Meteorol. Soc.*, Vol. 111, pp. 839-855.

Simunek, J.; Sejna, M. & van Genuchten, M.Th. (2007). The HYDRUS Software Package for Simulating Two- and Three-Dimensional Movement of Water, Heat, and Multiple Solutes in Variably-Saturated Media, User Manual, Version 1.0, PC Progress, Prague, Czech Republic.

Sinclair, S. & Pegram, G.G.S. (2010). A Comparison of ASCAT and Modelled Soil Moisture over South Africa, Using TOPKAPI in Land Surface Mode. *Hydrology and Earth System Science*, Vol. 14, pp. 613–626.

Skaggs, R.W. (1978). *A Water Management Model for Shallow Water Tables*, Report No. 134, Water Resources Research Institute, The University of North Carolina, Raleigh, NC, USA.

Spittlehouse, D.L. (1989). Estimating Evapotranspiration from Land Surfaces in British Columbia, In: *Estimation of Areal Evapotranspiration*, pp. 245-253, IAHS, Publ., 177.

Steduto, P.; Hsiao, T.C.; Raes, D. & Fereres, E. (2009) AquaCrop-The FAO Crop Model to Simulate Yield Response to Water: I. Concepts and Underlying Principles. *Agronomy Journal*, Vol. 101(3), pp. 426-437.

Su, Z. (2002). The Surface Energy Balance System (SEBS) for Estimation of Turbulent Heat Fluxes. *Hydrology and Earth System Sciences*, Vol. 6, pp. 85-99.

Takeuchi, A.; Ao, T. & Ishidaira, H. (1999). Introduction of Block-Wise Use of TOPMODEL and Muskingum–Cunge Method for the Hydroenvironmental Simulation of a Large Ungauged Basin. *Hydrol. Sci. J.*, Vol. 44(4), pp. 633–646.

Tasumi, M.; Trezza, R.; Allen, R.G. & Wright, J.L. (2005). Operational Aspects of Satellite Based Energy Balance Models for Irrigated Crops in the Semi-Arid US. *Irrigation and Drainage Systems*, Vol. 19, pp. 355-376.

Taylor, S.A. & Ashcroft, G.L. (1972). *Physical Edaphology. The Physics of Irrigated and Non-Irrigated Soils.* Freeman, San Francisco, pp. 533.

Thornthwaite, C.W. (1948). An approach toward a rational classification of climate. *Geograph. Rev.*, Vol. 38, pp. 55.

Turc, L. (1961). Evaluation de Besoins en Eau d'Irrigation, ET Potentielle. *Ann. Agron.*, Vol. 12, pp. 13-49.

Van Genuchten, M.Th. (1980). A Closed-Form Equation for Predicting the Hydraulic Conductivity of Unsaturated Soils. *Soil Science Society of America Journal*, Vol. 44(5), pp. 892-898.

Van Walsum, P.E.V.; Veldhuizen, A.A.; van Bakel, P.J.T.; van der Bolt, F.J.E.; Dik, P.E.; Groenendijk, P.; Querner, E.P. & Smit, M.R.F. (2004). *SIMGRO 5.0.1, Theory and Model Implementation*, Report No. 913.1, Alterra, Wageningen, The Netherlands.

Viviroli, D.; Zappa, M.; Gurtz, J. & Weingartner, R. (2009). An Introduction to the Hydrological Modelling System PREVAH and its Pre- and Post-Processing Tools. *Environmental Modelling & Software*, Vol. 24, pp. 1209-1222.

Wang, X.; Xie, H.; Guan, H. & Zhou, X. (2003). Different Responses of MODIS-Derived NDVI to Root-Zone Soil Moisture in Semi-Arid and Humid Regions. *Journal of Hydrology*, Vol. 340, pp. 12-24.

Wapener, T. (2003) Evaluation of Catchment Models. *Hydrological Processes*, Vol. 17, pp. 3375-3378.

Xu, Y. & Beekman, H.E. (Eds.) (2003). *Groundwater Recharge Estimation in Southern Africa*, ISBN 92-9220-000-3, UNESCO IHP Series no. 64, UNESCO, Paris.

Yang, D.; Zhang, T.; Zhang, K.; Greenwood, D.; Hammond, J.P. & White P.J. (2009). An Easily Implemented Agro-hydrological Procedure with Dynamic Root Simulation for Water Transfer in the Crop–Soil System: Validation and Application. *Journal of Hydrology*, Vol. 370, 177–190

Zhang, L.; Dawes, W.R. & Hatton, T.J. (1996). Modelling Hydrologic Processes Using a Biophysically Based Model – Application of WAVES to FIFE and HAPEX-MOBILHY. *Journal of Hydrology*, Vol. 185, pp. 147-169.

Zhang, L.; Dawes, W.R. & Walker, G.R. (2001). The Response of Mean Annual Evapotranspiration to Vegetation Changes at Catchment Scale. *Water Resources Research*, Vol. 37, pp. 701-708.

Zhou, M.C.; Ishidaira, H.; Hapuarachchi, H.P.; Magome, J.; Kiem, A.S. & Takeuchi, K. (2006). Estimating Potential Evapoutranspiration Using Shuttleworth-Wallace Model and NOAA-AVHRR NDVI Data to Feed a Distributed Hydrological Model over the Mokeng River Basin. *Journal of Hydrology*, Vol. 327, pp. 151-173.

8

Stomatal Conductance Modeling to Estimate the Evapotranspiration of Natural and Agricultural Ecosystems

Giacomo Gerosa[1], Simone Mereu[2], Angelo Finco[3] and Riccardo Marzuoli[1,4]
[1]Dipartimento Matematica e Fisica, Università Cattolica
del Sacro Cuore, via Musei 41, Brescia
[2]Dipartimento di Economia e Sistemi Arborei, Università di Sassari,
[3]Ecometrics s.r.l., Environmental Monitoring and Assessment,
via Musei 41, Brescia
[4]Fondazione Lombardia per l'Ambiente, Piazza Diaz 7, Milano
Italy

1. Introduction

This chapter presents some of the available modelling techniques to predict stomatal conductance at leaf and canopy level, the key driver of the transpiration component in the evapotranspiration process of vegetated surfaces. The process-based models reported, are able to predict fast variations of stomatal conductance and the related transpiration and evapotranspiration rates, e.g. at hourly scale. This high–time resolution is essential for applications which couple the transpiration process with carbon assimilation or air pollutants uptake by plants.

2. Stomata as key drivers of plant's transpiration

Evapotranspiration from vegetated areas, as suggested by the name, has two different components: evaporation and transpiration. Evaporation refers to the exchange of water from the liquid to the gaseous phase over living and non-living surfaces of an ecosystem, while transpiration indicates the process of water vaporisation from leaf tissues, i.e. the mesophyll cells of leaves. Both processes are driven by the available energy and the drying potential of the surrounding air, but transpiration depends also on the capacity of plants to replenish the leaf tissues with water coming from the roots through their hydraulic conduction system, the xylem. This capacity depends directly on soil water availability (i.e. soil water potential), which contributes to the onset of the water potential gradient within the soil-plant-atmosphere continuum.

Moreover, since the cuticle -a waxy coating covering the leaf surface- is nearly impermeable to water, the main part of leaf transpiration (about 95%) results from the diffusion of water vapour through the stomata. Stomata are little pores in the leaf lamina which provide low-resistance pathways to the diffusional movement of gases (CO_2, H_2O, air pollutants) from outside to inside the leaf and vice versa. Following complex signal pathways, environmental, osmotic and hormonal, stomata regulate their opening area and thus the water vapour loss from leaves. When the evaporative demand is bigger than the water replenishing capability from the xylem, stomata closes partially or even totally. High evaporative demands can be due to elevated air temperature, high leaf-to-air vapour pressure deficit (VPD), and intense winds. Stomatal closure can also be caused by high concentration of carbon dioxide in the mesophyll space.

Stomata, thus, directly control plant transpiration preventing plants from excessive drying, and acting as key drivers of water vapour movements from vegetated surfaces to the atmosphere.

This chapter illustrates the modelling techniques to predict the stomatal behaviour of vegetation at high-resolution time scale, and the related water fluxes.

3. Modelling stomatal behaviour: The Jarvis-Stewart model and the Ball-Berry model

Stomata play an essential role in the regulation of both water losses by transpiration and CO_2 uptake for photosynthesis and plant growth. Stomatal aperture is controlled by the turgor pressure difference between the guard cells surrounding the pore and the bulk leaf epidermis. In order to optimize CO_2 uptake and water losses in rapidly changing environmental conditions, plants have evolved the ability to control stomatal aperture in the order of seconds. Stomatal aperture responds to multiple environmental factors such as, solar radiation, temperature, drought, VPD, wind speed, and sub-stomatal CO_2 concentrations.

The availability of modern physiological instrumentation (diffusion porometers, gas-exchange analyzers) has allowed to measure leaf stomatal conductance (g_s) in field conditions and to study how environmental variables influence this parameter.

However, measurements of g_s by porometers and gas-exchange analyzers can be made only when foliage is dry, and long-term enclosure in measuring chamber may lead to changes in the physiological state of the leaves. Consequently measurements in the field are usually made intensively over selected periods of a few hours in selected days.

Furthermore, stomatal conductance values depend also upon the physiological condition of the plant, which relates to the weather of the previous days as well as to the previous season for perennial species.

Therefore it is important to have continuous g_s measurements over the whole vegetative season in order to improve the interpretation of other physiological data such as photosynthesis rate and carbon assimilation.

An alternative to very frequent measurements of g_s in the field is to predict them from models that describe its dependency on environmental factors. These models can be parameterized using the available field measurements conducted on occasional periods.

Furthermore, modeling appears the most effective tool for integration, simulation and prediction purposes concerning the effects of climatic global change on vegetation.

Stomatal conductance is among the processes that have been most extensively modeled during the last decades. In their excellent review, Damour et al. (2010) describe 35 stomatal conductance models classified as:

1.   models based on climatic control only
2.   models mainly based on the gs-photosynthesis relationship
3.   models mainly based on an Abscisic Acid (ABA) control
4.   models mainly based on the turgor regulation of guard cell.

The next paragraphs provides information on two early developed g_s models which are currently among the most widely used: the multiplicative model of Jarvis (1976) based on climatic control and later modified by Stewart (1988), and the Ball Berry model (1988), based on g_s-photosynthesis relationship.

3.1 The Jarvis-Stewart model

The stomatal conductance model developed by Jarvis (1976) can be defined as an empirical multiplicative model based on the observed responses of g_s to environmental factors. The assumption of this model is that the influence of each environmental factor on g_s is independent of the others and can be determined by boundary line analysis (Webb 1972).

The Jarvis model, in its first form, integrates the responses of g_s to light intensity, leaf temperature, vapour pressure deficit, ambient CO_2 concentration and leaf water potential, according to the following equation:

$$g_s = f(Q) \cdot f(T_l) \cdot f(VPD_l) \cdot f(C_a) \cdot f(\Psi) \tag{1}$$

where Q is the quantum flux density (μE m^{-2}s^{-1}), T_l is the leaf temperature (°C), VPD_l is the leaf-to-air vapour pressure deficit calculated at leaf temperature (kPa), C_a is the ambient CO_2 concentration (ppm) and Ψ is leaf water potential (MPa).

Stewart (1988) further implemented this model adopting the assumption that the functions of environmental variables have values between zero and unit and exert their influence reducing the maximum stomatal conductance of the plant (g_{smax}), a species-specific value depending on leaf stomatal density, that can be defined as the largest value of conductance observed in fully developed leaves – but not senescent – of well-watered plants under optimal climatic conditions (Körner et al., 1979). This value can be derived from field measurements conducted under the above mentioned optimal conditions.

Furthermore, in Stewart formulation, quantum flux density is replaced by global solar radiation, leaf temperature by air temperature, leaf-to-air vapour pressure deficit by air vapour pressure deficit and leaf water potential by soil moisture deficit measured in the first meter of soil (i.e. soil water content, SWC). Stewart also omitted $f(C_a)$ because the effect of CO_2 ambient concentrations was considered negligible: this simplification allows for an easier data collection to run the model, but it must be kept in mind that C_a change considerably among seasons and thus the simplification may lead to a considerable error, especially when the model is used for annual g_s behavior of evergreen species.

The model is defined by the following equation:

$$g_s = g_{smax} \cdot f(Q) \cdot f(T_a) \cdot f(VPD_a) \cdot f(SWC) \tag{2}$$

It is important to notice that $f(Q)$ can also be replaced by the more specific $f(PAR)$, based on the photosynthetically active radiation.

Each function has a characteristic shape described by the following equations:

$$f(Q) = 1 - exp^{-aQ} \tag{3}$$

$$f(T_a) = \frac{(T - T_{min})}{(T_{opt} - T_{min})}\left[\frac{(T_{max} - T)}{(T_{max} - T_{opt})}\right]^b \tag{4}$$

where $b = \dfrac{(T_{max} - T_{opt})}{(T_{opt} - T_{min})}$, $f(T_a) = 0.1$ when $T \leq T_{min}$ or $T \geq T_{max}$ and $f(T_a) = 1$ when $T = T_{opt}$

$$f(VPD) = \frac{(1 - 0.1)\cdot(c - VPD)}{(c - d)} \tag{5}$$

where $f(VPD) = 1$ when VPD $\leq d$ and $f(VPD) = 0.1$ when VPD $\geq c$

$$f(SWC) = 1 - exp\left[k(SWC - SWC_{max})\right] \tag{6}$$

where $f(SWC) = 1$ when SWC=SWC$_{max}$

Since g_s depends on four major variables, field measurements do not usually show a clear relationship with any of the considered variables. Often, g_s is reduced below the value expected for a value of a single independent variable, as the result of the influences of the other variables. As a consequence, the coefficients of each function must be derived with boundary-line analysis, plotting all field measurements of relative g_s ($g_{srel} = g_s/g_{smax}$) against each environmental variable considered separately.

Provided that enough measurements have been adequately performed to cover variable space, the upper limit of the scatter diagram indicates the response of g_s to the particular independent variable, when the other variables are not limiting.

An example of boundary-line analysis is reported in figure taken from Gerosa et al. (2009):

The main criticism formulated against this kind of approach is that the interactive effects between environmental factors are not properly taken into account, since interactions are only partially explained by the multiplicative nature of the model which simply multiplies concomitant effects, avoiding any synergistic interaction.

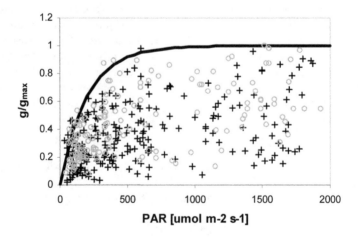

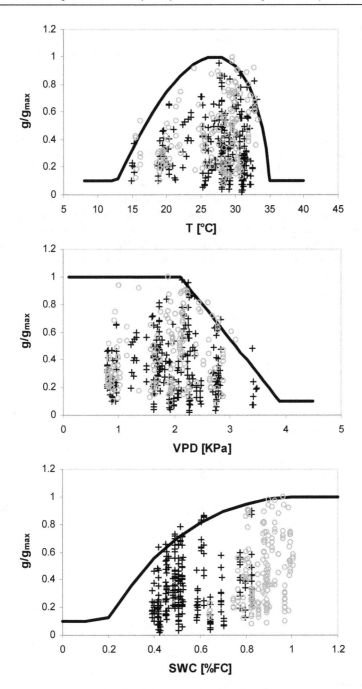

Fig. 1. Boundary-line analysis for the definition of g_s limiting function parameters (modified from Gerosa et al. 2008)

3.2 The Ball-Berry model

The Ball-Berry empirical model describes the behaviour of g_s as a function of environmental conditions and net photosynthetic rate. In its simplest form (Ball et al., 1987) the model states:

$$g_s = g_0 + a_1 An \frac{RH}{C_s} \tag{7}$$

Where g_s is the stomatal conductance to water vapour, g_0 is the stomatal conductance at the light compensation point, a_1 is a fitting parameter representing the slope of the equation, An is photosynthesis, RH is relative humidity and C_s is the molar fraction of CO_2 at the leaf surface. The model takes advantage of the feedback loop that exists between A and g_s (Farquhar at al., 1978) implying that they are interdependent. Additionally, An and g_s can respond independently to environmental variables and so they cannot be considered driving variables but rather state variables. The empirical relationship emerges from optimized vegetation behaviour that maximizes productivity (Patwardhan et al., 2006): the relationship, actually, corresponds roughly to the value of maximum surface conductance that maximizes productivity.

In order to derive g_s, the model needs to be coupled with a photosynthesis model (most often the Farquhar biochemical model) from which An is calculated. In order to derive g_s two equations must be solved simultaneously:

$$g_s = g_0 + a_1 An \frac{RH}{C_s} \tag{8}$$

$$A = (C_s - C_i)g_s \tag{9}$$

The problem is often solved by reiteration of the two equations where g_s at time t_{n+1} is computed with An at time t_n and An at time t_{n+2} is computed using g_s at time $t+1$. The reiteration approach however can give birth to oscillations in time of g_s and An due to chaotic solution in particular conditions. However, Baldocchi et al. (1994) found an analytical solution for the set of equations that bypasses this problem.

The original Ball-Berry model (Ball et al., 1987) was further implemented by Leuning (1995), considering that stomata respond to vapour pressure deficit (VPD) rather than humidity. In its modified version the equation takes the form:

$$g_s = g_0 + \frac{a_1 An}{(C_s - \Gamma)\left(1 + \dfrac{VPD_0}{VPD_s}\right)} \tag{10}$$

Where, Γ is the CO_2 compensation point, Cs and $VPDs$ are the CO_2 concentration and vapour pressure deficit at the leaf surface, and VPD_0 is an empirical coefficient.

This model encapsulates two empirical trends reported in the literature. First, through the correlation between g_s and An the equation predicts that the ratio $(C_i-\Gamma)/(C_s-\Gamma)$ is largely independent of leaf irradiance and C_s, except near the light and CO_2 compensation points. It also predicts that g_s declines linearly as VPD_s increases, in fact through the relation

$$E = 1.6 \cdot g_s \cdot VPD_s \tag{11}$$

for the transpiration rate (E), the hyperbolic function of VPD_s is equivalent to a linear decline of g_s with increasing E.

The main limitation of the Ball-Berry-Leuning (BBL) model is its failure in describing stomatal closure in drought conditions. The model has been further implemented by Dewar (2002) to take SWC in consideration by coupling the BBL model with Tardieu model for stomatal response to drought. The coupled model takes the form:

$$g_s = \frac{a_1(An + Rd)}{Ci\left(1 + \dfrac{VPD_0}{VPDs}\right)} exp\left\{-[ABA]\beta exp(\delta\Psi)\right\} \tag{12}$$

Where Rd is dark respiration, $[ABA]$ is the concentration of abscisic acid in the leaf xylem, Ψ is the leaf water potential, β is the basal sensitivity of ion diffusion to $[ABA]$ at zero leaf water potential, and δ describes the increase in the sensitivity of ion diffusion to $[ABA]$ as Ψ declines.

The model has the advantage of describing stomatal responses to both atmospheric and soil variables and has proven to reproduce a number of common water use trends reported in the literature as, for example, isohydric and anisohydric behaviour.

4. Modelling water vapour exchange between leaves and atmosphere and scaling it up to plant and ecosystem level: The *big-leaf* approach and the resistive analogy

The exchange of water vapour through stomata is a molecular diffusion process since air in the sub-stomatal cavities is motionless as well as the air in the first layer outside the stomata directly in contact with the outer leaf surface, i.e. the leaf boundary-layer,. Outside the leaf boundary-layer, it is the turbulent movement of air that removes water vapour, and this process is two orders of magnitude more efficient than the molecular diffusion. The exchange of water between the plant and the atmosphere is further complicated by the physiological control that stomatal resistance exerts on the diffusion of water vapour to the atmosphere.

Transpiration is modelled through an electric analogy (Ohm's law) introduced by Chamberlain and Chadwick (1953). Transpiration behaves analogously to an electric current, which originates from an electric potential difference and flows through a conductor of a given resistance from the high to the low potential end (Figure 2).

The driving potential of the water flux E is assumed to be the difference between the water vapour pressure in ambient air $e(T_a)$ and the water vapour pressure inside the sub-stomatal cavity $e_s(T_l)$, the latter being considered at saturation. The resistances that water vapour encounters from within the leaf to the atmosphere is given by the resistance of the stomatal openings (r_s) and the resistance of the leaf boundary laminar sub-layer (r_b). This process can be represented by the following equation:

$$E = \frac{[e_s(T_l) - e(T_a)]}{r_b + r_s} \cdot \frac{\rho c_p}{\lambda\gamma} \tag{13}$$

where T_a is air temperature (°K), T_l is leaf temperature (°K), e is water vapour pressure in the ambient air (Pa), e_s is water vapour pressure of saturated air (Pa) and the term $\rho c_p/\lambda\gamma$ is a factor to express E in mass density units (kg m^{-2} s^{-1}), equivalent to mm of water per second, being c_p the heat capacity of air at constant pressure (1005 J K^{-1} kg^{-1}), ρ the air density (kg m^{-3}), λ the vaporisation heat of water (2.5x10^6 J kg^{-1}), and $\gamma = c_p/\lambda$ the psychrometric constant (67 Pa K^{-1}). Despite the apparent difference with the well-known Penman-Monteith equation (Monteith, 1981), Eq. 13 is an equivalent formulation of this latter, as demonstrated by Gerosa et al. (2007).

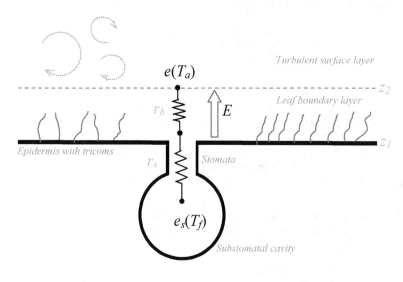

Fig. 2. Schematic picture of the transpirative process form a leaf. The symbols are explained in the text.

While the water vapour pressure deficit $[e_s(T_l)-e(T_a)]$ driving the water exchange is determined by temperature difference, the amount of water flux is regulated by the resistances along the path of the flux.

The stomatal resistance r_s, reciprocal of the stomatal conductance g_s, is obtained applying one of the stomatal prediction models presented in the previous paragraph, which are fed by meteorological and agrometeorological data.

The quasi-laminar sub-layer resistance r_b depends on the molecular properties of the diffusive substance and on the thickness of the layer. The resistance against the diffusion of a gas through air is defined as:

$$r = \int_{z_1}^{z_2} \frac{1}{D_{H_2O}} dz \qquad (14)$$

for the leaf boundary-layer the equation gives:

$$r_b = (z_2 - z_1)/D_{H2O} \qquad (15)$$

where D_{H2O} is the diffusion coefficient of water vapour in the air, z_1 and z_2 representing the lower and upper height of the leaf boundary-layer.

However, the thickness of the leaf boundary-layer depends on leaf geometry, wind intensity and atmospheric turbulence. In order to take these factors in consideration, a more practical formulation, proposed by Unsworth et al. (1984), can be used:

$$r_b = k(d / u)^{1/2} \tag{16}$$

where k is an empirical coefficient set to a value of 132 (Thom 1975), d is the downwind leaf dimension, and u is the horizontal wind speed near the leaves.

The transpiration of a whole plant, or of a vegetated surface with closed canopy, may be modelled using a similar approach referred to as the *big-leaf*. The *big-leaf* assumes the canopy vegetation as an ideal big-leaf lying at a virtual height $z=d+z_0$ above ground (Figure 3). The d parameter is the displacement height, i.e. the height of the zero-plane of the canopy, equal to $2/3$ of the canopy height, z_0 is the roughness length, i.e. the additional height above d where the wind extinguishes inside the canopy (sink for momentum), around $1/10$ of the canopy height, and $d+z_0'$ is the apparent height of water vapour source.

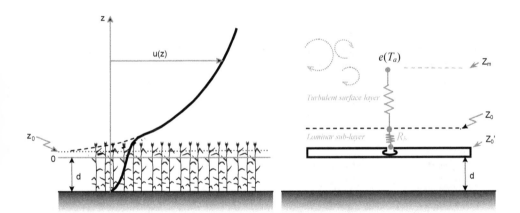

Fig. 3. The *big-leaf* approach to model water vapour exchange of a vegetated surface. Left side a real canopy; right side its big-leaf representation. The laminar sub-layer has been enlarged and the stomatal resistance is not shown. Please note the upper case notation of the resistances.

This transpiring big-leaf has a bulk stomatal resistance R_s equal to the sum of the stomatal resistances r_s of all the n leaves of the canopy. Recalling the rules of composition for parallel resistances:

$$1 / R_s = \sum_{1}^{n} 1 / r_s = n / r_s \tag{17}$$

Since the number of leaf is rarely known, a practical way of upscaling r_s is to consider the thickness of the "big-leaf" equal to the leaf area index of the canopy ($LAI = m^{-2}_{leaf} / m^{-2}_{ground}$)

i.e. the square meters of leaf area projected on each square meter of ground surface. This assumption is equivalent to stating that the light extinction coefficient of the big-leaf is equal to the light extinction of the canopy.
The transpiration rate of the "big-leaf", the whole canopy, is then obtained in a way very similar to those above developed for the leaves:

$$E = \frac{\left[e_s(T_l) - e(T_a)\right]}{R_a + R_b + R_s} \cdot \frac{\rho c_p}{\lambda \gamma}$$
(18)

It is worth noticing the upper case notation for the "bulk" resistances and the introduction of the aerodynamic resistance R_a.
The aerodynamic resistance depends on the turbulent features of the atmospheric surface layer, and it is introduced to account for the distance z_m at which the atmospheric water potential is measured above the canopy. It is formally the vertical integration of the reciprocals of the turbulent diffusion coefficients for all scalars, which in turn depends on the friction velocity u^* and the atmospheric stability. The integrated version of R_a is given by

$$R_a = \frac{1}{k \cdot u^*}\left[\ln(\frac{z_m - d}{z_0}) - \Psi_M\right]$$
(19)

where k is the von Kármán dimensionless constant (0.41), u^* is the friction velocity (m s^{-1}), a quantity indicating the turbulent characteristic of the atmosphere, and Ψ_M is the integrated form of the atmospheric stability function for momentum (non-dimensional).
The friction velocity, if not available, can be derived with the following equation:

$$u^* = \frac{k u_{zm}}{\ln(\frac{z_m - d}{z_0}) - \Psi_M}$$
(20)

where u_{zm} is the wind velocity measured at z_m, and Ψ_M is a function defined as:

$$\Psi_M = \begin{cases} 2\ln(\frac{1+y^2}{2}) \quad with \quad y = (1 - 16(z-d)/L)^{1/4} & if \ (z-d)/L < 0 \quad unstable \ condition \\ 0 & if \ (z-d)/L = 0 \quad neutral \ condition \\ -5(z-d)/L & if \ (z-d)/L > 0 \quad stable \ condition \end{cases}$$
(21)

L is the length (m) of Monin-Obukhov (1954) indicating the atmospheric stability:

$$L = \frac{-\rho c_p T_0 u_*^3}{k g H}$$
(22)

with T_0 the reference temperature (273.16 K), g the gravity acceleration (9.81 m s^{-2}) and H the sensible heat flux (W m^{-2}).
Since L is a function of u^* and H, and vice versa, concurrent determination of u^* and Ψ_M from routine weather data would normally require an iterative procedure (Holtslag and van Ulden, 1983).

If the atmospheric stability is not known as well as the sensible heat flux, and the water potential in the atmosphere is measured near the canopy, a neutral stability can be assumed by setting $\Psi_M=0$ in the u^* equation with fairly good approximation.

The laminar sub-layer resistance R_b can by computed with a general purpose formulation proposed by Hicks et al. (1987) which involves the Schmidt and Prandtl numbers, being $Sc=0.62$ for water vapour and $Pr=0.72$ respectively:

$$R_b = \frac{2}{ku^*}(Sc / Pr)^{2/3} \tag{23}$$

where k is here the von Kármán constant.

Modelling canopy transpiration using only three resistances in series might seem an oversemplification; however the approach has proven valid in different cases in predicting fast variations of water exchange over a vegetated surface following the stomatal behaviour, as well as to predict the total amount of transpired water (Grunhage et al., 2000).

To obtain a higher modelling performance, the resistive network of the "big-leaf" model can be implemented for specific needs. For example, multiple vegetation layers can be included in order to account for the transpiration of the understory vegetation below a forest, or the canopy can be decomposed in several layers, each with its own properties (De Pury and Farquhar, 1997) In such cases the models take the name of multi-layer models. Other improvements are required when multiple sources of water vapour have to be considered, for example when the evaporation from a water catchment, or evaporation from bare soil in ecosystems with sparse vegetation. .

All these models are collectively known as 1-D SVAT models (one-dimensional Soil Vegetation Atmosphere Transfer models).

In the following paragraph a multi-layer dual-source model to predict the evapotranspiration from a poplar plantation ecosystem with understory vegetation is presented.

5. Example and applications - a multi-layer model for the transpiration of a mature poplar plantation ecosystem - comparison with eddy covariance measurements

The poplar plantation used for this modelling exercise was located in the Po valley near the city of Pavia. The ecosystem was made by mature poplar trees of about 27 m height with the soil below the plant mainly covered by poplar saplings and perennial grasses. Since the canopy was completely closed, most of the evapotranspiration was due to plants transpiration i.e. evaporation from other surfaces can be considered negligible. According to Choudhury and Monteith (1988), less than 5% of the water vapour flux is due to evaporation from soil for a closed canopy. In this case study evaporation from soil was strongly limited by the absence of tillage and by the coverage of understory vegetation. Moreover the upper soil layer resulted very dry and acted as a screen against water vapour transport from wetter underlying soil layers.

The water exchange was modelled using only two water sources, both of them transpirative: the poplar crown and the understory vegetation. Thus this example model includes only two layers (Figure 4).

The model is composed of three different sub-models: one stomatal sub-model for the stomatal conductance of the transpiring plants, one soil sub-model for the soil water content, and one atmospheric sub-model to describe the water vapour exchange dynamic at canopy level following the adopted resistive network.

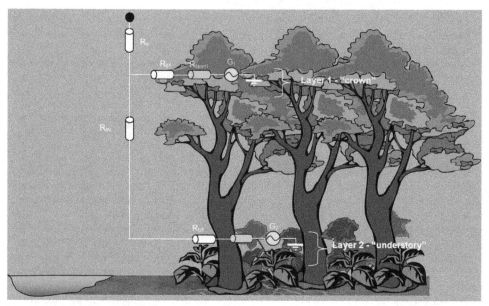

Fig. 4. A multi-layer multiple source model to estimate the water exchange between a poplar plantation ecosystem and the atmosphere.

5.1 The stomatal conductance sub-model

To describe the physiological behaviour of the bulk stomatal conductance (G_s) a Jarvis-Stewart multiplicative model was used, according to the following formulation:

$$G_s = g_{smax} \cdot [f(PHEN) \cdot f(T) \cdot f(PAR) \cdot f(VPD) \cdot f(SWC)] \tag{24}$$

where g_{smax} is the maximum stomatal conductance expressed by the poplar trees in non-limiting conditions. A maximum value of 1.87 cm s^{-1} (referred to the Projected Leaf Area) has been found in the literature for g_{smax} of poplar leaves located at 2 meter of height in Italian climatic condition (Marzuoli et al., 2009). This value has been reduced to 57% to account for the decreasing of g_{smax} with the canopy height, as proposed by Schafer et al. (2000). Thus a g_{smax} value of 0.8 cm s^{-1} was assumed for the canopy.

The phenology function $f(PHEN)$ has been assumed equal to zero when the vegetation was without leaves and equal to one after the leaf burst when the leaves were fully expanded. This was fixed to the 110th day of the year (DOY).

Compared to Eq. 2, Eq. 24 includes a limiting function based on phenology $f(PHEN)$ which grows linearly from 0 to 1 during the first 10 days after leaves emergence, and decreases linearly in the last 10 days, starting from DOY 285th, simulating leaf's senescence:

$$f(PHEN) = \begin{cases} 0 & \forall DOY \leq SGS \quad or \quad \forall DOY \geq EGS \\ 1 & \forall (SGS + DayUp) < DOY < (EGS - DayDown) \\ ((DOY - SGS) / DayUp) & \forall\ SGS < DOY < (SGS + DayUp) \\ ((EGS - DOY) / DayDown) & \forall\ (EGS - DayDown) < DOY < EGS \end{cases} \tag{25}$$

SGS and EGS are the days for the start and the end of the growing season respectively. . DayUp and DayDown are the number of days necessary to complete the new leaves expansion and to complete the leaves senescence, respectively.

The G_s dependence on light was modelled according to Eq. 3 form:

$$f(PAR) = 1 - exp^{-aPAR} \qquad (26)$$

where a represents a specie-specific coefficient (0.006 in this study) and PAR is the Photosynthetically Active Radiation expressed as μmol photons m^{-2} s^{-1}.

Eq. 4 and Eq. 5 were used for Gs dependence on temperature and VPD, respectively.

For soil water content SWC a different limiting function, from that reported by Sterwart (1988), was used. The boundary-line analysis revealed that SWC exerted its influence on g_{smax} according to the following equation:

$$f(SWC) = max\left\{0.1; min\left[1; g \cdot SWC^{(h/SWC)}\right]\right\} \qquad (27)$$

where SWC is expressed as fraction of soil field capacity while g and h are two coefficients whose values are respectively 1.0654 and 0.2951.

The bulk stomatal conductance of the understory vegetation was modelled using the same parameterization but assuming a g_{smax} value equal to 1.87 cm s^{-1}. The inherent approximation is that the understory vegetation was entirely composed of young poplar plantlets.

	Parameter	Value	Unit
	g_{max} (H$_2$O)	0.8	cm/s
f_{PHEN}	SGS	110	DOY
	EGS	285	DOY
	DayUp	10	Days
	DayDown	10	Days
f_{PAR}	a	0.006	adim.
f_T	T_{opt}	27	°C
	T_{max}	36	°C
	T_{min}	12	°C
	b	0.5625	adim.
f_{VPD}	c	3.7	KPa
	d	2.1	KPa
f_{SWC}	g	1.0654	adim.
	h	0.2951	adim.

Table 1. Values of the f limiting functions coefficients and g_{smax} for the stomatal conductance model of Populus nigra.

5.2 The soil sub-model

The water availability in the soil was modelled using a simple "bucket" model. In this paradigm the soil is considered as a bucket and the water content is assessed dynamically, step by step, via the hydrological balance between the water inputs (rains) and outputs (plant consumption) occurred in the previous time step. The model was initialised assuming the soil water saturated at the beginning of the season and assuming a root depth for soil exploitation of 3 m:

$$AWHC = (\theta_{FC} - \theta_{WP}) \cdot 1000 \cdot RootDepth = 243 \text{ mm } H_2O \text{ / m}^3 \text{ soil} \qquad (28)$$

$$AW_t = 0 = AWHC \text{ (mm)} \qquad (29)$$

where AWHC is the available water holding capability of the sandy soil between the wilting point ($\theta_{WP} = 0.114 \text{ m}^3 \text{ m}^{-3}$ for our sandy loam soil) and the field capacity ($\theta_{FC} = 0.195 \text{ m}^3 \text{ m}^{-3}$). The running equations were:

$$ET_{t-1} = F_{H20, t-1} \cdot 3600 / \lambda \text{ (mm)} \qquad (30)$$

$$AW_t = AW_{t-1} + Rain_{t-1} - ET_{t-1} \text{ (mm)} \qquad (31)$$

$$SWC_t = AW_t / AWHC \text{ (\% of FC)} \qquad (32)$$

Eq. 32 represents the water loss of plant ecosystem through the transpiration of the two layers ($F_{H20, t-1}$) in the previous time step. Since water fluxes are expressed as rates (mm s^{-1}), for an hourly time step, as in our cases, their values must be multiplied by 3600 in order to get the water consumed in one hour.

AW_t is the available water in the soil after water inputs and consumptions. The effects of runoff and groundwater level rising have been neglected due to the flatness of the ecosystem and the groundwater level which were deeper than the root exploration depth.

SWC represents the soil water content expressed as percentage of field capacity, as requested by the $f(SWC)$ function of the stomatal sub-models.

5.3 The atmospheric sub-model and the resistive network

The resistance R_a was calculated by using Eq. 19 and Eq. 21, with $z_m = 33$ m the measurement height, $h = 26.3$ m the canopy height, u^* the friction velocity, u the horizontal wind speed, L the Monin-Obhukhov length, $d = 2/3 \cdot h$ the zero-plane displacement height and $z_0 = 1/10 \cdot h$ the roughness length.

The laminar sub-layer resistances of the layers 1 and 2 (R_{b1} and R_{b2}) were both calculated using the Eq. 23 given u^*.

The stomatal resistances of the layers 1 and 2 (R_{stom1} and R_{stom2}) were calculated using the stomatal sub-model after having estimated the leaf temperatures from the air temperature T and the heat fluxes H:

$$T_l = T + H \cdot (R_a + R_{b, heat}) / (\rho \cdot c_p) \qquad (33)$$

where $R_{b,heat}$ was calculated using the Eq. 23 with $Sc = 0.67$ and $Pr = 0.71$.

Then the vapour pressure deficit $VPD = e_s(T_l) - e(T)$ was derived from the T_l for the calculation of $e_s(T_l)$ and from the air temperature T and the relative humidity RH for the actual e: $e(T) = UR \cdot e_s(T)$.

The vapour pressure of the saturated air can be calculated from the well-known Teten-Murray empirical equation:

$$e_s(T) = 0.611 \cdot exp(17.269 \cdot (T - 273) / (T - 36)) \qquad (34)$$

which gives e_s in kPa when T is expressed as °K.

The stomatal resistance of the crown R_{stom1} was obtained as the reciprocal of the stomatal conductance obtained by the Jarvis–Stewart sub-model fed with PAR, T_{leaf}, VPD and SWC_t, the latter being the soil water content calculated with the Eq. 32 .

The understory R_{stom2} was obtained in a similar way but considering a understory g_{max} (=1.87 cm s^{-1}) and the *PAR fraction* reaching the below canopy vegetation instead of the original *PAR*:

$$PAR_{fraction} = \exp(-k \cdot LAI_1) \tag{35}$$

where k is the light extinction factor within the canopy, set to 0.54, and LAI_1 is the leaf area index of the crown, assumed to be equal to 2 at maximum leaf expansion.

The in-canopy resistance R_{inc} was calculated following Erisman et al. (1994):

$$R_{inc} = (14 \cdot LAI_1 \cdot h) / u^* \tag{36}$$

where h is the canopy height and LAI_1 the leaf area index of the crown.

The stomata of the big leaves of the two layers of Figure 4 (G_1 and G_2) were assumed as water generators driven by the difference of water concentration between the leaves (χ_{sat}), assumed water saturated al leaf temperature T_l, and the air (χ_{air}):

$$G_1 = G_2 = \chi_{sat} - \chi_{air} \quad (g\ m^{-3}) \tag{37}$$

where

$$\chi_{sat} = 2.165 \cdot e_s(T_l) / T_l \quad (g\ m^{-3})$$

$$\chi_{air} = 2.165 \cdot e(UR, T) / T \quad (g\ m^{-3})$$

being 2.165 the ratio between the molar weight of water molecules M_w (18 g mol^{-1}) and the gas constant R (8.314 J mol^{-1} K^{-1}) if e and e_s are expressed in Pa (multiplied by 1000 if expressed in kPa).

Then the total water flux of the ecosystem F_{H2O} could be calculated by composing all the resistances and the generators within the modelled resistive network, following the electrical composition rules for resistances and generators in series and in parallel, and applying the scaling strategy according to the *LAI*:

$$R_1 = (R_{b1} + R_{stom1} / LAI_1) \quad (s/m) \tag{38}$$

$$R_2 = (R_{b2} + R_{stom2} / LAI_2) \quad (s/m) \tag{39}$$

$$R_3 = R_{inc} + R_2 \quad (s/m) \tag{40}$$

$$G_{eq} = G_2 - (G_2 - G_1) \cdot R_3 / (R_1 + R_3) \quad (g\ m^{-3}) \tag{41}$$

$$R_{eq} = R_1 \cdot R_3 / (R_1 + R_3) \quad (s/m) \tag{42}$$

$$F_{H2O} = G_{eq} / (R_{eq} + R_a) / 1000 \quad (kg\ m^{-2}\ s^{-1} = mm\ s^{-1}) \tag{43}$$

where LAI_2 is the leaf area index of the understory vegetation (=0.5)

5.4 Comparison with EC measurements

Concurrent measurements of λE were performed over the same ecosystem by means of eddy covariance technique with instrumentation set-up according to Gerosa et al. (2005).

The comparison between the direct λE measurements and the modelled ones allowed the evaluation of model performance.

The model performance was very good in predicting the hourly variation of λE both during the summer season (*Modeled* = 0.885 · *Measured* + 8.4389; R^2=0.85, p<0.001, n=1872) with a slight tendency to underestimate the peaks.

An example of the comparison exercise for a summer week is shown in Figure 5

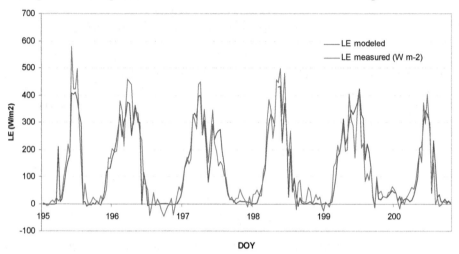

Fig. 5. Comparison between modelled and measured λE, expressed as W m^{-2} unit

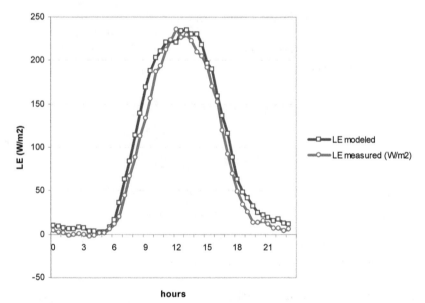

Fig. 6. Mean daily course of the modeled λE compared to the measured one. All the available hourly measurements were considered (n=3914)

For the whole year the performance was less good (*Modeled* = 0.876 · *Measured* + 21.443; R^2=0.68, p<0.001, n=3914) but still acceptable, especially in reproducing the average daily course of λE (Figure 6).

6. Conclusions

Scientific literature provides many ways (e.g. FAO) to estimate the evapotranspiration of a vegetated surface. Sometimes there is the need to predict this process at a very high-time resolution (e.g. hourly means). Hourly estimations of evapotranspiration, for example, are important in all the applications and the methodologies which couple transpiration process with carbon assimilation or air pollutants uptake by plants.

In these cases, the *big-leaf* approach, together with the resistive analogy which simulates the gas-exchange between vegetation and atmosphere, is a simple but valid example of a process-based model which includes the stomatal conductance behaviour, as well as a basic representation of the canopy features.

7. Acknowledgements

This publication was partially funded by the Catholic University's program for promotion and divulgation of scientific research.

8. References

Ball J.T., Woodrow I.E., Berry J.A., 1987. A model predicting stomatal conductance and its contribution to the control of photosynthesis under different environmental conditions. In: Biggins J, ed. *Progress in photosynthesis research*. Dordrecht: Martinus Nijhoff Publishers, 221-224

Chamberlain A.C., Chadwick R.C., 1953. Deposition of airborne radioiodine vapour. *Nucleonics* 11, 22–25

Choudhury B.J., Monteith J.L., 1988. A four-layer model for heat budget of homogeneous land surfaces. *Quarterly Journal of the Royal Meteorological Society* 114, 373–398

Damour G., Simonneau T., Cochard H., Urban L., 2010. An overview of models of stomatal conductance at leaf level. *Plant, Cell & Environment* 33, 1419-1438.

De Pury D.D.G., Farquhar G.D., 1997. Simple scaling of photosynthesis from leaves to canopies without the errors of big-leaf models. *Plant, Cell & Environment* 20, 537-557.

Dewar R.C., 2002. The Ball-Berry-Leuning and Tardieu-Davies stomatal models: synthesis and extension within a spatially aggregated picture of guard cell function. *Plant, Cell & Environment* 25: 1383-1398

Erisman J.W., Van Pul A., Wyers P., 1994. Parameterization of surface-resistance for the quantification of atmospheric deposition of acidifying pollutants and ozone. *Atmospheric Environment* 28, 2595–2607

Farquhar G.D., Dubbe D.R., Raschke K., 1978. Gain of the feedback loop involving carbon dioxide and stomata: theory and measurement. *Plant physiology* 62: 406-412

Gerosa G., Derghi F., Cieslik S., 2007. Comparison of Different Algorithms for Stomatal Ozone Flux Determination from Micrometeorological Measurements. *Water Air & Soil Pollution* 179, 309-321.

Gerosa G., Marzuoli R., Desotgiu R., Bussotti F., Ballarin-Denti A., 2008. Visible leaf injury in young trees of *Fagus sylvatica* L. and *Quercus robur* L. in relation to ozone uptake and ozone exposure. An Open-Top Chambers experiment in South Alpine environmental conditions. *Environmental Pollution* 152, 274–284.

Gerosa G., Vitale M., Finco A., Manes F., Ballarin Denti A. and Cieslik S., 2005. Ozone uptake by an evergreen Mediterranean forest (*Quercus ilex*) in Italy. Part I: Micrometeorological flux measurements and flux partitioning. *Atmospheric Environment* 39, 3255-3266.

Grünhage L., Haenel H.D., Jager H.J., 2000. The exchange of ozone between vegetation and atmosphere: micrometeorological measurement techniques and models. *Environmental Pollution* 109, 373–392.

Hicks B.B., Baldocchi, D.D., Meyers T.P., Hosker R.P., Matt D.R., 1987. A Preliminary multiple resistance routine for deriving dry deposition velocities from measured quantities. *Water, Air and Soil Pollution* 36, 311-330.

Holtslag A.A.M., van Ulden A.P., 1983. A simple scheme for daytime estimates of the surface fluxes from routine weather data. *Journal of Climate and Applied Meteorology* 22, 517–529.

Jarvis P.G., 1976. The interpretation of the variations in leaf water potential and stomatal conductance found in canopies in the field. *Philosophical Transactions of the Royal Society of London*, Series B 273, 593–610.

Körner C., Scheel J., Bauer H., 1979. Maximum leaf diffusive conductance in vascular plants. Photosynthetica 13, 45-82.

Leuning R., 1995. A critical appraisal of a combined stomatal photosynthesis model for C3 plants. *Plant, Cell & Environment.* 18, 339–355.

Marzuoli R., Gerosa G., Desotgiu R., Bussotti F., Ballarin-Denti A., 2008. Ozone fluxes and foliar injury development in the ozone-sensitive poplar clone Oxford *(Populus maximowiczii x Populus berolinensis)*: a dose–response analysis. Tree Physiology 29, 67-76.

Monin A.S., Obukhov A.M., 1954. Basic laws of turbulent mixing in the atmosphere near the ground. *Translation in Aerophysics of Air Pollution* (In: Fay, J.A., Hoult D.P. (Eds.), AIAA, New York, 1969, pp. 90–119). Akademija Nauk CCCP, Leningrad, TrudyGeofizich eskowo Instituta 151(24), 163–187.

Monteith J.L., 1981. Evaporation and surface temperature. *Quarterly Journal of the Royal Meteorological Society* 107, 1–27.

Patwardhan S., Pavlick, R. Kleidon A., 2006. Does the empirical Ball-Berry law of stomatal conductance emerge from maximization of productivity? *American Geophysical Union*, Fall Meeting 2006, abstract #H51C-0498.

Schafer K.V.R., Oren R., Tenhunen J.D., 2000. The effect of tree height on crown level stomatal conductance. *Plant, Cell & Environment* 23, 4 365-375.

Stewart J.B. (1988) Modelling surface conductance of pine forest. *Agricultural and Forest Meteorology* 43, 19–35.

Thom A.S., 1975. Momentum, mass and heat exchange of plant communities. In Vegetation and Atmosphere. Ed. J.L. Monteith. Academic Press, London.

Unsworth M.H., Heagle A.S., Heck W.W., 1984. Gas Exchange in open field chambers – I. Measurement and analysis of atmospheric resistance to gas exchange. Atmospheric Environment 18, 373–380.

Webb R.A., 1972. Use of the boundary line in the analysis of biological data. *Journal of Horticultural Science* 47, 309-319.

9

A Distributed Benchmarking Framework for Actual ET Models

Yann Chemin
International Water Management Institute
Sri Lanka

1. Introduction

With the various types of actual ET models being developed in the last 20 years, it becomes necessary to inter-compare methods. Most of already published ETa models comparisons address few number of models, and small to medium areas (Chemin et al., 2010; Gao & Long, 2008; García et al., 2007; Suleiman et al., 2008; Timmermans et al., 2007). With the large amount of remote sensing data covering the Earth, and the daily information available for the past ten years (i.e. Aqua/Terra-MODIS) for each pixel location, it becomes paramount to have a more complete comparison, in space and time.

To address this new experimental requirement, a distributed computing framework was designed, and created. The design architecture was built from original satellite datasets to various levels of processing until reaching the requirement of various ETa models input dataset. Each input product is computed once and reused in all ETa models requiring such input. This permits standardization of inputs as much as possible to zero-in variations of models to the models internals/specificities.

2. Theoretical points of observation

2.1 Net radiation and soil heat flux

In the two-source energy balance approach, like TSEB and SEBS differ from the single-source concept of SEBAL and METRIC in the sense that the radiation and energy balances have separate formulations for either bare soil or canopy. The energy balance at any instantaneous moment is expressed by equation Eq. 1:

$$Rn = G + H + LE \qquad (1)$$

Where Rn is Net Radiation, G is soil heat flux, H is sensible heat flux and LE is latent heat of vaporization. This is what is appearing in single-source models like SEBAL and METRIC. Single source models concentrate on identifying Rn and G from astronomical and semi-empirical equations respectively, while H is being iteratively solved based on thermodynamically exceptional geographical locations, often referred in literature (Bastiaanssen, 1995) as *wet* and *dry* pixels, also the technique to identify them is referred in more recent literature as *end-members selection/identification* (Timmermans et al., 2007).

In two-source models, it is separated into bare soil and canopy energy balances as in Eq. 2 and 3, respectively:

$$Rns = G + Hs + LEs \tag{2}$$

Where Rns is the net radiation of bare soil surface, Hs is the sensible heat flux from bare soil, LEs is the latent heat of vaporization from soil surface.

$$Rnc = Hc + LEc \tag{3}$$

Where Rnc is the net radiation from canopy of crop, Hc is the sensible heat flux from canopy, LEc is the latent heat of vaporization of crop. Once the elements of those two equations are found, the fraction of vegetation cover (fc) is used to combine them into the area of a satellite remote sensing pixel, which is inherently a mixel of bare soil and canopy.

The Net Radiation is partitioned according to the formulation commonly used in two-sources model (Eq. 4 and 5), where the soil partition of Rn is an LAI-based extinction coefficient (Choudhury, 1989) with a coefficient C ranging from 0.3 to 0.7 (Friedl, 2002), depending on the arrangement of the canopy elements. Friedl (2002) mentions that a canopy with spherical (random) leaf angle distribution would lead to a C value of 0.5.

$$Rns = Rn\, e^{\frac{-C\,LAI}{Cos(sunza)}} \tag{4}$$

$$Rnc = Rn - Rns \tag{5}$$

Where LAI is the leaf area index, sunza is the sun zenith angle. Friedl (2002) mentions that he derived his soil heat flux formulation from his previous work (Friedl, 1996). It takes the already available soil fraction of net radiation and the cosine of the sun zenith angle (Eq. 6). A coefficient is then multiplied to those whereby soil type and moisture conditions are taken into consideration after (Choudhury et al., 1987).

$$G = Kg\, Rns\, Cos(sunza) \tag{6}$$

Where Kg is the soil type and moisture condition coefficient in the soil heat flux. The Fraction of Vegetation cover is necessary to split the two-sources of heat transfer studied in such models. They are the soil surface (bare soil) and the vegetation canopy surface. The fraction of vegetation cover from Jia et al. (2003) quoting Baret et al. (1995) is developed as in Eq. 7:

$$fc = 1 - [\frac{(NDVI - NDVI_{min})}{(NDVI_{min} - NDVI_{max})}]^K \tag{7}$$

with K being taken as 0.4631 in Jia et al. (2003) and NDVImin at LAI=0 and NDVImax at LAI = +INF. As can be seen, a very large weight of potential deviation from the expected result is resting in the proper assessment of fc (Eq. 7). There are also uncertainties in the LAI raster input (Yang, Huang, Tan, Stroeve, Shabanov, Knyazikhin, Nemani & Myneni, 2006; Yang, Tan, Huang, Rautiainen, Shabanov, Wang, Privette, Huemmrich, Fensholt, Sandholt, Weiss, Ahl, Gower, Nemani, Knyazikhin & Myneni, 2006).

The soil heat flux computed for Bastiaanssen (1995), is what could be called a *partial contribution* of soil heat flux to the energy balance of the pixel, as the semi-empirical relationship is proportional to various elements of thermodynamic forcing within each pixel (Eq. 8).

$$G = \frac{Rn}{Albedo}\, T_c\, (0.0032(\frac{Albedo}{r_0}) + 0.0062(\frac{Albedo}{r_0})^2)\, (1 - 0.978NDVI^4) \tag{8}$$

with T_c the temperature in Celsius and r_0 the Albedo to apparent Albedo correction ranging 0.9 to 1.1 depending on the time of the day.

SEBS uses a two-source Albedo anchors stretching equation multiplied by the soil fraction of the pixel to extract a percentage of the net radiation as soil heat flux (Eq. 9).

$$G = Rn \left(Albedo_{dark} + (1 - f_c) \left(Albedo_{bright} - Albedo_{dark} \right) \right) \tag{9}$$

Generic values are $Albedo_{dark} = 0.05$ and $Albedo_{bright} = 0.35$, while adjustements are made when concentrating on a specific land use, eventually.

2.2 Monin-Obukhov Similarity Theory

The Monin-Obukhov Similarity Theory (Monin & Obukhov, 1954) is being used in single source and two-source energy balance models. It is interesting to note that Monin & Obukhov (1954), in the development of their Monin-Obukhov Similarity Theory (MOST) considered the friction velocity to be about 5% of the geostrophic wind velocity having an average speed of 10m/s results in the friction velocity being around 0.5 m/s, and with the Coriolis parameter $l = 10^{-4}s^{-1}$ and a tolerance of 20%, an estimate of the height of the surface layer is found at h=50m, that is also the DisALEXI blending height for air temperature (Norman et al., 2003).

The dynamic velocity within this layer can be considered near to constant and the effect of Coriolis Force neglected (Monin & Obukhov, 1954). Under those conditions of neutral stratification the processes of turbulent mixing in the surface layer can be described by the logarithmic model of the boundary layer (Eq. 10).

$$
\begin{aligned}
L &= \frac{-1004\,\rho u^3 T}{kgH} \\
most_x &= (1 - 16\frac{h}{L})^{\frac{1}{4}} \\
\psi_h &= 2\,log(\frac{1 + most_x^2}{2}) \\
\psi_m &= 2\,log(\frac{1 + most_x}{2}) + log(\frac{1 + most_x^2}{2}) - 2\,atan(most_x) + 0.5\pi
\end{aligned}
\tag{10}
$$

with ψ_m, ψ_h the diabatic correction of momentum and heat through their changes of states, $most_x$ a MOST internal parameter, L the Monin-Obukhov Length (MOL), k is the von Karman constant, g the gravity acceleration, u is the wind speed, ρ is the air density, T is the temperature and h is the height of interest (measurement height of the wind speed, roughness length, etc.).

Constraints to MOST as found in Bastiaanssen (1995) are of two types, first avoiding the latent heat flux input to be nil as its input location is in the denominator of the MOL equation (Equation Eq. 11), the second constraint is when the MOL is becoming positive, to force ψ_m and ψ_h to a ranged negative value (Bastiaanssen, 1995).

$$
\begin{aligned}
if\,(H &= 0.0) : L = -1000.0 \\
if\,(L &> 0.0) : \psi_h = \psi_m = -5\frac{2}{L}
\end{aligned}
\tag{11}
$$

It turns out that Su (2002), extending the reach of his SEBS model to the GCM community has included a dual model for the convective processes within the Atmospheric Boundary Layer (ABL). Su (2002) followed the observations of Brutsaert (1999) that the ABL lower layer is either stable, either unstable and that the thickness of this lower layer is α = 10-15% of the ABL height, which is about β = 100-150 times the surface roughness. SEBS takes the highest from both as its estimation of h_{st}, the height of ABL sublayer separation. If the reference height is lower than that, then the lower sublayer model is run, otherwise the upper sublayer models is used.

The cutline between the two sublayers of the ABL permits SEBS to process the lower layer (Atmospheric Surface Layer, ASL) under the MOST paradigm (Eq. 12), whether it proves unstable (generally in the day) or stable (generally in the night). The momentum and heat functions for the upper sublayer of the ABL where flow is laminar (free convection) can then be merged with the ASL by what Su (2002) calls a Bulk ABL Similarity (BAS) stability correction set of functions called here ζ_m and ζ_h for momentum and heat respectively (Eq. 13).

$$if(\frac{-z_0}{L} < 0.0) : \zeta_m = -2.2\,alog(1 + \frac{z_0}{L})$$

$$if(z_0 < \frac{\alpha}{\beta}h_i) : \zeta_m = -alog(\alpha) + \psi_m(\frac{-\alpha h_i}{L}) - \psi_m(-\frac{z_0}{L}) \tag{12}$$

$$if(z_0 \geq \frac{\alpha}{\beta}h_i) : \zeta_m = alog(\frac{h_i}{\beta z_0}) + \psi_m(\frac{-\beta z_0}{L}) - \psi_m(-\frac{z_0}{L})$$

$$if(\frac{-z_0}{L} < 0.0) : \zeta_h = -7.6\,alog(1 + \frac{z_0}{L})$$

$$if(z_0 < \frac{\alpha}{\beta}h_i) : \zeta_h = -alog(\alpha) + \psi_h(\frac{-\alpha h_i}{L}) - \psi_h(-\frac{z_0}{L}) \tag{13}$$

$$if(z_0 \geq \frac{\alpha}{\beta}h_i) : \zeta_h = alog(\frac{h_i}{\beta z_0}) + \psi_h(\frac{-\beta z_0}{L}) - \psi_h(-\frac{z_0}{L})$$

with z_0 the surface roughness for momentum, h_i the height of ABL or Planetary Boundary Layer (PBL). The formulation of ψ_m and ψ_h functions are inherited from Beljaars & Holtslag (1991) and include either correction weights inside the standard equations in some cases (unstable conditions of ψ_m and ψ_h), either a polynomial with exponential in other cases (stable conditions of ψ_m and ψ_h). However, Beljaars & Holtslag (1991) stated categorically that the data described are characteristic for grassland and agricultural land *with sufficient water supply*.

2.3 Roughness height

Allen et al. (2005) mentions that METRIC and SEBAL do not require knowledge of crop type (no satellite based crop classification is needed). SEBAL relies on a type of semi-empirical equation relating NDVI to the roughness length (also called roughness height) for momentum and heat (Eq. 14).

$$z_{0m} = e^{a+b\,NDVI} \tag{14}$$

Among many others, Chandrapala & Wimalasuriya (2003) proposed $a = -5.5$ and $b = 5.8$ for Sri Lanka using AVHRR NDVI images (sensor response curves, atmospheric correction and pixel size are all influencing NDVI response). Bastiaanssen (1995) preferred using extrema conditions to define a and b.

SEBS takes a ground truth added to mapping point of view and uses a look up table to translate land use raster maps into roughness length.

Land Cover	NDVI	z_{0m}
Vegetation	$NDVI_{max}$	$\frac{h_v}{7}$
Desert	0.02	0.002

Table 1. Boundary conditions on z_{0m} from NDVI satellite data after Bastiaanssen (1995)

While SEBAL and METRIC use a fixed conversion rate between roughness length for momentum and roughness length for heat, Su (2002) is introducing in SEBS the use of the exponential of kB^{-1} (Eq. 15).

$$z_{0m} = 0.136\, h_{vegetation}$$
$$z_{0h} = \frac{z_{0m}}{e^{kB^{-1}}} \tag{15}$$

with $h_{vegetation}$ the height of the vegetation relating to the roughness length for momentum z_{0m} from Brutsaert (1982), z_{0h} the roughness length for heat, Su (2002) refers to the work of Massman (1999) on the combined *von Karman constant - sublayer Stanton number* (kB^{-1}), where B^{-1} is defined by Gieske (2007) as in Eq. 16:

$$B^{-1} = St_k^{-1} - C_d^{-\frac{1}{2}}$$
$$\frac{St_k^{-1}}{u_*} = \frac{\rho C_p \Delta T}{H} \tag{16}$$

C_d is the drag coefficient, St_k^{-1} is the roughness Stanton number, u_* is the friction velocity, ρ the air density, C_p the specific heat and ΔT the temperature difference.

2.4 Aerodynamic roughness for heat

The aerodynamic resistance (roughness) for heat r_{ah} is an input to the sensible heat flux and is often a source of concentration in ET models based on energy balance. For logical reasons, as the parameterization of r_{ah} needs prior knowledge of the state of the sensible heat flux to enable knowledge of the MOL to parameterize ψ_m and ψ_h the diabatic correction of momentum and heat through their changes of states. In turn, ψ_m and ψ_h, offset the logarithmic relation of the observation height to the respective roughness lengths (z_{0m} and z_{0h}) being the driving force to curve the relation to the wind shear profile.

SEBS resistance at the wet limiting case re_{wet} (Eq. 17) is using the MOL as L_{wet} configured to use all the energy available for evaporation, which in turn is used in conjunction with the reference height (z_{ref}) either in the computation of the ψ_h or χ_h whether ASL or BAS models are at work.

$$re_{wet} = \frac{alog(\frac{z_{ref}}{L_{wet}} - [\psi|\chi]_h)}{ku^*} \tag{17}$$

2.5 Ground to air temperature difference

Allen et al. (2005) mentions that METRIC is a variant of the important model SEBAL and that it has been extended to provide tighter integration with ground-based reference ET. SEBAL formulation for the ground to air temperature difference (dT) is estimated (Eq. 18) as an affine function with two extreme conditions found in the satellite image processed (Bastiaanssen,

1995).

$$pixel_{cold} \mapsto dT = 0$$

$$pixel_{hot} \mapsto dT = \frac{(Rn - G) \times r_{ah}}{\rho C_p} \tag{18}$$

$$dT = a + b \times T$$

with r_{ah} the aerodynamic resistance for heat, $pixel_{cold}$ and $pixel_{hot}$ the end-members defined in Bastiaanssen (1995); Bastiaanssen et al. (1998); Timmermans et al. (2007) and that are the *signet ring* of the model.

METRIC formulation (Eq. 19) includes the reference ET paradigm found in Allen et al. (1998), the $Kc \times ETo$ crop ET, also called ETc, being translated into METRIC as $k \times ET_r$ (Allen et al., 2005), practically using Alfalfa at full growth as anchor point in their Idaho study. Some extra metorological data being available when using METRIC, permits a daily surface soil water balance to be run to enforce conditions on the *dry/hot* pixel energy balance and effective dT. Selection of the extreme pixels is focused on cropped area as much as possible.

$$pixel_{cold} \mapsto dT = \frac{(Rn - G - k \times ET_r) \times r_{ah}}{\rho C_p}$$

$$pixel_{hot} \mapsto dT = \frac{(Rn - G) \times r_{ah}}{\rho C_p} \tag{19}$$

$$dT = a + b \times T$$

SEBS (Su, 2002) is computing dT (Eq. 20) from a surface skin virtual temperature (T_0) and PBL virtual temperature (T_{pbl}).

$$T_v = \frac{log(h_{pbl} - h_{disp})}{log(h_u - h_{disp})}$$

$$T_{pbl} = \frac{T_s \times (1 - f_c) + T_v \times f_c}{\frac{1 - DEM}{44331.0}^{1.5029}}$$

$$T_0 = \frac{T_c}{\frac{1 - DEM}{44331.0}^{1.5029}} \tag{20}$$

$$dT = T_0 - T_{pbl}$$

Where T_s is the soil temperature, T_v is the (virtual) vegetation canopy temperature, f_c is the fraction of vegetation cover, DEM is the elevation and T_c is the satellite sensed temperature in Celsius, h_u is the wind speed measurement height. h_{disp} is the displacement height being 0.65 of the canopy height in SEBS, Monin & Obukhov (1954) mention that for observations made at height superior to 1 meter, the displacement height can be nullified. The blending height (h_{pbl}) is given by an external mean or if no data is available, default value of 200m is used (same as in SEBAL) or 1000m.

2.6 Actual ET
Single sources models (METRIC and SEBAL) have promoted a particular way of closing the energy-balance (Eq. 21), using a ratio of fluxes called the evaporative fraction (Λ; Eq. 21).

$$\Lambda = \frac{Rn - G - H}{Rn - G} \tag{21}$$

This instantaneous ratio (could be called an efficiency, since it is unitless) is multiplied with a potential value of ET in order to provide with an Actual ET (equation Eq. 22).

$$ET_a = \Lambda \, ET_{potential} \tag{22}$$

METRIC (Allen et al., 2007) evaporative fraction is used with the potential ET based on Penman-Monteith method as published in Allen et al. (1998) in order to produce an actual ET estimation. In contrast SEBAL (Bastiaanssen, 1995; Bastiaanssen et al., 1998) evaporative fraction is used with the potential ET computed from exo-atmospheric solar radiation, a single-way atmospheric transmissivity, and the reflected proportion from Albedo.

SEBS (Su, 2002) is computing an instantaneous latent heat flux (LE) from the evaporative fraction and the subtraction of net radiation and soil heat flux (Eq. 23). However, using what is called in SEBS the *relative evaporation*, the pixel value of H the sensible heat flux is allowed to vary only between H_{wet} and H_{dry}.

$$H_{dry} = Rn - G$$

$$H_{wet} = \frac{(Rn - G) - \frac{(\frac{\rho C_p}{r_{e\_wet}} e_{sat})}{\gamma}}{\frac{1 + slope}{\gamma}}$$

$$Enforce : H_{wet} \leq H \leq H_{dry} \tag{23}$$

$$E_{relative} = 1 - \frac{H - H_{wet}}{H_{dry} - H_{wet}}$$

$$\Lambda = E_{relative} \frac{Rn - G - H_{wet}}{Rn - G}$$

H_{wet} is derived from the Priestley-Taylor equation (Priestley & Taylor, 1972), from which γ and *slope* come from, r_{e_wet} is the resistance at the *wet* limiting case. $E_{relative}$ is the relative evaporation.

SSEB (Senay et al., 2007) is evaluating a regional approximation of the evaporative fraction Λ as a ranging of satellite-based temperature products (Eq. 24) and uses the reference ET found in Allen et al. (1998) as a mean to compute the actual ET.

$$\Lambda = \frac{T_{hot} - T}{T_{hot} - T_{cold}} \tag{24}$$

3. Methodology

3.1 Conceptual design and processing flow

The distributed framework is a Linux system based on GDAL library (GDAL, 2011) and C programming, enhanced with a distributed language called OpenMP (OpenMP, 2011), used essentially for data distribution as seen in (Chemin, 2010).

As the conceptual architecture of the framework (Fig. 1) signifies, there are several layers of processing involved. Initially, downloaded satellite imagery is located in a single directory (referred as *RS data* in Fig. 1).

During the pre-processing phase, a parsing agent will select images of the same kind and same date and stitch them together, then reproject them to the projection system selected. Upon completion, a second agent will perform a relatively conservative quality check and assign null value to failed pixels according to the satellite imagery information available.

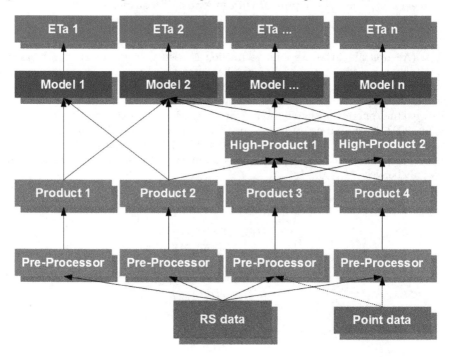

Fig. 1. Architecture Concept of the Framework

Once these two steps are performed, the raster maps are tagged as *products*. Those products will be shared in between any of the ET models that require such type of input.
Some ET models require some higher-level input raster maps, by this, we define *higher-level product* as a raster that requires at least one *product* as defined above as precursor to its creation.

3.2 Meteorological data

Meteorological data (referred as *Point data* in Fig. 1) is encoded with Fourier Transforms (FT) in function of cumulative day of year from the beginning of the satellite imagery data set. This insures both faithfulness of the data and high-portability as well as an elegant way to summarize a complex and variable non-spatial dataset.
Actual state of the research in meteorological data time-series encoding for this framework is to develop an array of geo-tagged FT. Also under consideration are Wavelet Transforms (WT), for reasons of time tagging. This array will be used to interpolate on-the-fly from an appropriate number of neighbours and with an interpolation algorithm suiting best the operator requirement. The reason for this, is that it transfers the load from storage requirements (which can be heavy for daily meteorological raster datasets) to thread computing that is benefiting from distributed speed-up as only a marginal number of equations will be added by such process compared to the ET models processing load.

4. Implementation

Practically (Fig. 2), MODIS datasets are grouped by products and by day and batch processed each in one core of the computer in parallel. This involves format changing, merging tiles, reprojecting, renaming outputs according to the nomenclature of the processing system. The tools involved in that step are either standard Linux Shell tools, either part of GDAL (2011) standard tools (i.e. *gdalwarp* and *gdal_translate*). Both of these tools are still essentially sequential programs at this time, thus, they are being sent to each core in a distributed manner through the Shell with a check loop to ensure that there is at all time the same number of programs running as there are cores/threads available in the CPU architecture. It becomes clear that for each new leap in number of cores in future commercial offerings, the framework will automatically increase its processing capacity to the new enlarged number of cores/threads available, thus also reducing by the same factor the time needed to process a given number of satellite images.

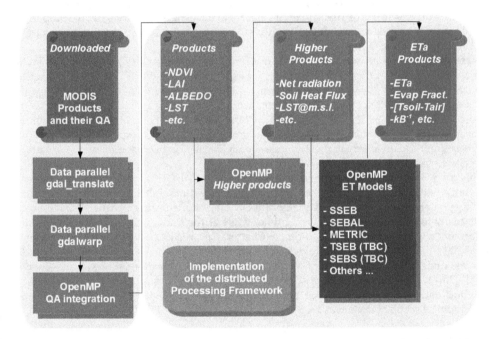

Fig. 2. Architecture Implementation of the Framework

Models that are already inside the framework account to SSEB from Senay et al. (2007), METRIC from Allen et al. (2007), SEBAL from Bastiaanssen et al. (1998) using the work from Alexandridis et al. (2009), in progress are SEBS from Su (2002) and TSEB from both Kustas & Norman (1999) and Norman et al. (1995).

One of the many candidate for inclusion was the Two-Source Algorithm (TSA) from Yunhao et al. (2005). After extensive calculus and referring to the mathematic academia, it became clear that both temperature equations below when combined to extract $T_{vegetation}$ and T_{soil} (Eq. 25) have a large amount of solutions even within the constraining dimension of surface skin temperature range. The available information to possibly close the equations unknown parameters are relating to satellite T after Price (1984) and the satellite-based emissivity (ϵ), partitioned into $\epsilon_{vegetation} = 0.93$ & $\epsilon_{soil} = 0.97$ quoting Sui et al. (1997). Unless missing information or other sources of solution dimension constraints are published, the proposition of the model is so far considered impossible and not included in the framework until such condition is being met through publication.

$$T = f_c\, T_{vegetation} + (1 - f_c)\, T_{soil}$$
$$\epsilon\sigma T^4 = f_c\, \epsilon_{vegetation}\sigma T_{vegetation}^4 + (1 - f_c)\, \epsilon_{soil}\sigma T_{soil}^4 \qquad (25)$$
$$\epsilon = f_c\, \epsilon_{vegetation} + (1 - f_c)\, \epsilon_{soil}$$

With ϵ the satellite-based emissivity, $\epsilon_{vegetation}$ & ϵ_{soil} assumed fixed emissivity values for vegetation and bare soil (satellite response dependent), σ the Stephan-Boltzmann constant, T the satellite-based land surface temperature, and $T_{vegetation}$ & T_{soil} the pixel vegetation and soil fractions temperatures. The first equation answers to the geographical two-source proportion within the pixel, while the second equation answers to the two-source flux merging according to Yunhao et al. (2005).

Reference ET models included are Allen et al. (1998) from Cannata (2006), Priestley and Taylor (Priestley & Taylor, 1972) and Hargreaves (Hargreaves et al., 1985), Modified Hargreaves (Droogers & Allen, 2002), Hargreaves-Samani (Hargreaves & Samani, 1985). Only the reference ET from Allen et al. (1998) is being used as a precursor of SSEB (Senay et al., 2007) and METRIC (Allen et al., 2007) actual ET. It was found preponderant to have a minimum group of reference ET models available as baseline for all the work, especially when looking into geographical areas where meteorological data has always been dominant in agricultural literature.

Some models requiring operator intervention (SEBAL, METRIC) have add there internals modified with specially designed heuristics acting as operators. Initial developments were not looking into heuristics but stochastic algorithms. Some efforts using a genetic algorithm were eventually too expensive in processing time, while at the same time end-member selection information were becoming more common (Chandrapala & Wimalasuriya, 2003; Timmermans et al., 2007). Thus heuristics were designed and implemented on a regional basis, initially studied under the Greek conditions for the purpose of Alexandridis et al. (2009) and Chemin et al. (2010). Eventually, the heuristics are extended to fit data sources, continent/climate combinations and model types on an adhoc basis as new regions are included into the geographical scope of research.

5. Initial results

In the case of SEBAL heuristic, the convergence reached 82% of the images processed for the Australian Murray-Darling Basin (1 Million Km^2), enabling the automatic processing of 3635 MODIS multi-tiles images within a single day of computing. Fig. 3 is the output from SEBAL with such heuristic for some irrigated areas in Australia, the total area being processed amounts to more than 5 Billions pixels of ETa values, being multiplied by as many temporary rasters and original data as required for each of the ET models. The Australian irrigation system (less than 100,000 ha) has a sharp, contrasted and well-defined pattern of water depletion, characteristic of continental dry climate with high water supply control for defined periods of the year where crops are in the field.

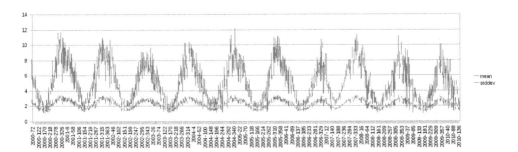

Fig. 3. Daily RS-based ETa (mm/day) in an Australian irrigation system (2000-2010)

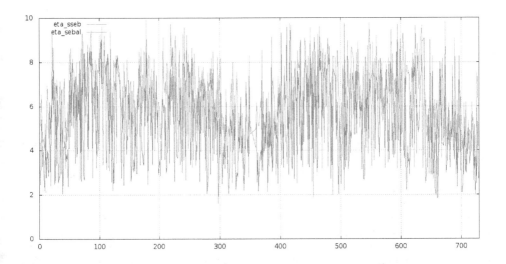

Fig. 4. Daily ETa (mm/day) averages for Sri Lanka (2003-2004)

Looking into the matter of comparing ETa results from different ETa models, Fig. 4 is the averaged ETa output from two models (SEBAL and SSEB) over the tropical island of Sri Lanka in 2003 and 2004. It turns out that the relatively small island of Sri Lanka has an average ETa that is changing much more on a day to day basis than our previous example in Australia. Scale, climate, topography yield exposure to ocean events frequently, having drastic impact on thermodynamics of the island surface as the Fig. 5 also confirms. Changes between models of actual ET from SEBAL and SSEB are relatively constant throughout the RS modeling period. Actual ET from SSEB is in the upper range of SEBAL's one. The work of de Silva (1999) in the dry zone of Sri Lanka and the work of Hemakumara et al. (2003) in the wet zone of Sri Lanka are falling within the expected results found here. Likewise the average evaporative fractions found for Sri Lanka in Fig. 5 are especially leveraging the larger dry zone area of the island with value in the range of 0.3 to 0.5.

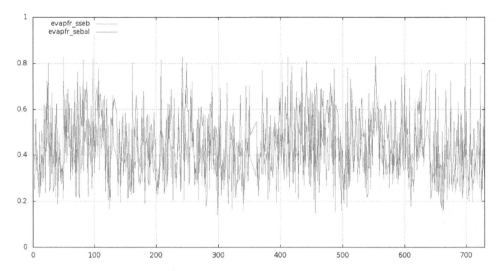

Fig. 5. Instantaneous Evaporative Fractions for Sri Lanka (2003-2004)

6. Conclusion

Challenges to experimentally compare ET models are immense, the theoretical points of comparison are sometimes clear, sometimes rather difficult to pinpoint. To try and address this situation, a framework for benchmarking ET actual models has been designed. Its implementation has embedded parallel data distribution at the base of each parts of the framework to remove the resistance of the data size to process large areas, high frequency and large time period with commonly available computers.

Future work includes the finalization of SEBS (Su, 2002) and TSEB (Kustas & Norman, 1999) integration in the framework, looking for other ETa model candidates to add to existing ones. Also there is a need for designing and creating statistical tools to cross-compare several depths and layers of ETa models processing datasets. Finally, the use of OpenMPI (OpenMPI, 2011) is envisaged for concurrently running several ET models diagnostics in different multi-core machines or OpenCL (Khronos.org, 2011) kernel-based data distributed language to process all analysis as one large computation on a Graphical Processing Unit (GPU).

7. References

Alexandridis, T. K., Cherif, I., Chemin, Y., Silleos, G. N., Stavrinos, E. & Zalidis, G. C. (2009). Integrated methodology for estimating water use in mediterranean agricultural areas, *Remote Sensing* 1(3): 445–465.
URL: *http://www.mdpi.com/2072-4292/1/3/445/*

Allen, R. G., Peirera, L. S., Raes, D. & Smith, M. (1998). *Crop evapotranspiration - Guidelines for computing crop water requirements - FAO Irrigation and Drainage Paper 56*, FAO - Food and Agriculture Organization of the United Nations.
URL: *http://www.fao.org/docrep/x0490e/x0490e00.htm*

Allen, R. G., Tasumi, M., Morse, A., Kramber, W. J. & Bastiaanssen, W. G. M. (2005). Computing and mapping evapotranspiration, *in* U. Aswathanarayana (ed.), *Advances in Water Science Methodologies*, Taylor & Francis, pp. 73–85.

Allen, R. G., Tasumi, M. & Trezza, R. (2007). Satellite-based energy balance for mapping evapotranspiration with internalized calibration (metric) model, *Journal of Irrigation and Drainage Engineering* 133(6): 380–394.
URL: *http://ascelibrary.org/iro/resource/1/jidedh/v133/i4/p380*

Baret, F., Clevers, J. G. P. W. & Steven, M. D. (1995). The robustness of canopy gap fraction estimates from red and near-infrared reflectances: A comparison of approaches, *Remote Sensing of Environment* 54(2): 141–151.
URL: *http://www.sciencedirect.com/science/article/B6V6V-3YYMS35-10/2/dcb889861d7a7 68d454934f074fe487a*

Bastiaanssen, W. G. M. (1995). *Regionalization of surface flux densities and moisture indicators in composite terrain. A remote sensing approach under clear skies in mediterranean climates*, PhD thesis, Agricultural University of Wageningen.
URL: *http://library.wur.nl/WebQuery/clc/918192*

Bastiaanssen, W. G. M., Pelgrum, H., Wang, J., Ma, Y., Moreno, J. F., Roerink, G. J. & van der Wal, T. (1998). A remote sensing surface energy balance algorithm for land (sebal).: Part 2: Validation, *Journal of Hydrology* 212-213: 213–229.
URL: *http://www.sciencedirect.com/science/article/B6V6C-4CYFRND-H/2/6bc33f78398c9 c74ffabe611b3ed1b6b*

Beljaars, A. C. M. & Holtslag, A. A. M. (1991). Flux parameterization over land surfaces for atmospheric models, *Journal of Applied Meteorology* 30(3): 327–341.
URL: *http://journals.ametsoc.org/doi/abs/10.1175/1520-0450%281991%29030%3C0327% 3AFPOLSF%3E2.0.CO%3B2*

Brutsaert, W. (1982). *Evaporation in the Atmosphere - Theory, History and Applications*, D. Reidel Publishing Company.
URL: *http://www.springer.com/978-90-277-1247-9*

Brutsaert, W. (1999). Aspects of atmospheric boundary layer similarity under free-convective conditions, *Reviews of geophysics* 37(4): 439–451.
URL: *http://www.agu.org/journals/ABS/1999/1999RG900013.shtml*

Cannata, M. (2006). *GIS embedded approach for Free & Open Source Hydrological Modelling*, PhD thesis, Department of Geodesy and Geomatics, Polytechnic of Milan, Italy.
URL: *http://istgis.ist.supsi.ch:8001/geomatica/index.php?id=1*

Chandrapala, L. & Wimalasuriya, M. (2003). Satellite measurements supplemented with meteorological data to operationally estimate evaporation in sri lanka, *Agricultural Water Management* 58(2): 89–107.
URL: *http://www.sciencedirect.com/science/article/B6T3X-47PPC5B-1/2/d0844a2f9391840 f54c00ff0da926a75*

Chemin, Y. H. (2010). *Remote Sensing Raster Programming*, Lulu Eds.
URL: *http://www.lulu.com/product/paperback/remote-sensing-raster-programming/16217224*

Chemin, Y. H., Alexandridis, T. K. & Cherif, I. (2010). Grass image processing environment - application to evapotranspiration direct readout, *OSGEO Journal* 6: 27–31.
URL: *http://www.osgeo.org/ojs/index.php/journal/article/viewFile/131/132*

Choudhury, B. (1989). Estimating evaporation and carbon assimilation using infrared temperature data: vistas in modeling, *in* G. Asrar (ed.), *Theory and applications of remote sensing*, New York Wiley, pp. 628–690.

Choudhury, B., Idso, S. & Reginato, R. (1987). Analysis of an empirical model for soil heat flux under a growing wheat crop for estimating evaporation by an infrared-temperature based energy balance equation, *Agricultural and Forest Meteorology* 39(4): 283–297.
URL: *http://www.sciencedirect.com/science/article/B6V8W-4894NGY-9B/2/5668c2204c443 52f9fca7d9aef38216c*

de Silva, R. (1999). A comparison of different models of estimating actual evapotranspiration from potential evapotranspiration in the dry zone of sri lanka, *Sabaragamuwa University Journal* 2: 87–100.
URL: *http://www.sab.ac.lk/journal/1999/1999A10.pdf*

Droogers, P. & Allen, R. G. (2002). Estimating reference evapotranspiration under inaccurate data conditions, *Irrigation and Drainage Systems* 16: 33–45.
URL: *http://dx.doi.org/10.1023/A:1015508322413*

Friedl, M. A. (1996). Relationships among remotely sensed data, surface energy balance, and area-averaged fluxes over partially vegetated land surfaces, *Journal of Applied Meteorology* 35(11): 2091–2103.
URL: *http://journals.ametsoc.org/doi/abs/10.1175/1520-0450%281996%29035%3C2091% 3ARARSDS%3E2.0.CO%3B2*

Friedl, M. A. (2002). Forward and inverse modeling of land surface energy balance using surface temperature measurements, *Remote Sensing of Environment* 79(2-3): 344–354.
URL: *http://www.sciencedirect.com/science/article/B6V6V-44R1BH4-K/2/57215e156bb3b7 681a6460684503a761*

Gao, Y. & Long, D. (2008). Intercomparison of remote sensing-based models for estimation of evapotranspiration and accuracy assessment based on swat, *Hydrological Processes* 22: 4850–4869.
URL: *http://onlinelibrary.wiley.com/doi/10.1002/hyp.7104/abstract*

García, M., Villagarcía, L., Contreras, S., Domingo, F. & Puigdefábregas, J. (2007). Comparison of three operative models for estimating the surface water deficit using aster reflective and thermal data, *Sensors* 7(6): 860–883.
URL: *http://www.mdpi.com/1424-8220/7/6/860/*

GDAL (2011). Gdal - geospatial data abstraction library.
URL: *http://www.gdal.org*

Gieske, A. (2007). Numerical modeling of heat and water vapor transport through the interfacial boundary layer into a turbulent atmosphere, *in* B. J. Geurts, H. Clercx & W. Uijttewaal (eds), *Particle-Laden Flow*, Vol. 11 of *ERCOFTAC Series*, Springer Netherlands, pp. 71–83. 10.1007/978-1-4020-6218-6_6.
URL: *http://dx.doi.org/10.1007/978-1-4020-6218-6_6*

Hargreaves, G. H. & Samani, Z. A. (1985). Reference crop evapotranspiration from temperature, *Applied Engineering in Agriculture* 1: 96–99.

Hargreaves, G. L., Hargreaves, G. H. & Riley, J. P. (1985). Agricultural benefits for senegal river basin, *Journal of Irrigation and Drainage Engineering* 111: 113–124.

Hemakumara, H., Chandrapala, L. & Moene, A. (2003). Evapotranspiration fluxes over mixed vegetation areas measured from large aperture scintillometer, *Agricultural Water Management* 58: 109–122.

Jia, L., Su, Z., van den Hurk, B., Menenti, M., Moene, A., De Bruin, H. A. R., Yrisarry, J. J. B., Ibanez, M. & Cuesta, A. (2003). Estimation of sensible heat flux using the surface energy balance system (sebs) and atsr measurements, *Physics and Chemistry of the Earth* 28(1-3): 75–88. Applications of Quantitative Remote Sensing to Hydrology.

URL: *http://www.sciencedirect.com/science/article/B6X1W-483SMNY-5/2/735667f1cf3fef ea27e001736c7d728a*

Khronos.org (2011). Opencl: Open source computing language.
URL: *http://www.khronos.org/opencl/*

Kustas, W. P. & Norman, J. M. (1999). Evaluation of soil and vegetation heat flux predictions using a simple two-source model with radiometric temperatures for partial canopy cover, *Agricultural and Forest Meteorology* 94(1): 13–29.
URL: *http://www.sciencedirect.com/science/article/B6V8W-3W7XBGD-2/2/8c44f505c27c3 9f01073e096dcef8b8d*

Massman, W. J. (1999). A model study of kbh-1 for vegetated surfaces using 'localized near-field' lagrangian theory, *Journal of Hydrology* 223(1-2): 27–43.
URL: *http://www.sciencedirect.com/science/article/B6V6C-3XG1T11-3/2/b77eb802830b880 f3946171e0adfb12a*

Monin, A. & Obukhov, A. (1954). Basic laws of turbulent mixing in the surface layer of the atmosphere, *Tr. Akad. Nauk. SSSR Geophiz. Inst.* 24: 163–187.

Norman, J. M., Anderson, M., Kustas, W. P., French, A. N., Mecikalski, J., Torn, R., Diak, G. R., Schmugge, T. J. & Tanner, B. C. W. (2003). Remote sensing of energy fluxes at 101m pixel resolution, *Water Resources Research* 39(8): 1221–1229.

Norman, J. M., Kustas, W. P. & Humes, K. S. (1995). Source approach for estimating soil and vegetation energy fluxes in observations of directional radiometric surface temperature, *Agricultural and Forest Meteorology* 77(3-4): 263–293.
URL: *http://www.sciencedirect.com/science/article/B6V8W-4031C00-9/2/a0a99cd55406a97 5e6f7b6ab446006f0*

OpenMP (2011). The openmp api specification for parallel programming.
URL: *http://www.openmp.org*

OpenMPI (2011). Open mpi: Open source high performance computing.
URL: *http://www.open-mpi.org*

Price, J. C. (1984). Land surface temperature measurements from the split window channels of the noaa 7 advanced very high resolution radiometer, *Journal of Geophysical Research* 89(D5): 7231–7237.
URL: *www.agu.org/journals/ABS/JD089iD05p07231.shtml*

Priestley, C. H. B. & Taylor, R. J. (1972). On the assessment of surface heat flux and evaporation using large-scale parameters, *Monthly Weather Review* 100(2): 81–92.
URL: *http://journals.ametsoc.org/doi/abs/10.1175/1520-0493%281972%29100%3C0081% 3AOTAOSH%3E2.3.CO%3B2*

Senay, G. B., Budde, M., Verdin, J. P. & Melesse, A. M. (2007). A coupled remote sensing and simplified surface energy balance approach to estimate actual evapotranspiration from irrigated fields, *Sensors* 7(6): 979–1000.
URL: *http://www.mdpi.com/1424-8220/7/6/979/*

Su, Z. (2002). The surface energy balance system (sebs) for estimation of turbulent heat fluxes, *Hydrology and Earth System Sciences* 6(1): 85–100.
URL: *http://www.hydrol-earth-syst-sci.net/6/85/2002/*

Sui, H. Z., Tian, G. L. & Li, F. Q. (1997). Two-layer model for monitoring drought using remote sensing, *Journal of Remote Sensing* 1: 220–224.

Suleiman, A., Al-Bakri, J., Duqqah, M. & Crago, R. (2008). Intercomparison of evapotranspiration estimates at the different ecological zones in jordan, *Journal of*

Hydrometeorology 9(5): 903–919.
    URL: *http://journals.ametsoc.org/doi/abs/10.1175/2008JHM920.1*

Timmermans, W. J., Kustas, W. P., Anderson, M. C. & French, A. N. (2007). An intercomparison of the surface energy balance algorithm for land (sebal) and the two-source energy balance (tseb) modeling schemes, *Remote Sensing of Environment* 108(4): 369 – 384.
    URL: *http://www.sciencedirect.com/science/article/B6V6V-4MV19YK-3/2/02bab62a1b4 dc9f9298946f723c196b3*

Yang, W., Huang, D., Tan, B., Stroeve, J. C., Shabanov, N. V., Knyazikhin, Y., Nemani, R. R. & Myneni, R. B. (2006). *IEEE Transactions on Geoscience and Remote Sensing* 44: 1829–1842.

Yang, W., Tan, B., Huang, D., Rautiainen, M., Shabanov, N. V., Wang, Y., Privette, J. L., Huemmrich, K. F., Fensholt, R., Sandholt, I., Weiss, M., Ahl, D. E., Gower, S. T., Nemani, R. R., Knyazikhin, Y. & Myneni, R. B. (2006). *IEEE Transactions on Geoscience and Remote Sensing* 44: 1885–1898.

Yunhao, C., Xiaobing, L., Jing, L., Peijun, S. & Wen, D. (2005). Estimation of daily evapotranspiration using a two-layer remote sensing model, *International Journal of Remote Sensing* 26(8): 1755–1762.
    URL: *http://www.informaworld.com/10.1080/01431160512331314074*

10

Operational Remote Sensing of ET and Challenges

Ayse Irmak[1], Richard G. Allen[2], Jeppe Kjaersgaard[2], Justin Huntington[3],
Baburao Kamble[4], Ricardo Trezza[2] and Ian Ratcliffe[1]
[1]*School of Natural Resources University of Nebraska–Lincoln, HARH, Lincoln NE*
[2]*University of Idaho, Kimberly, ID*
[3]*Desert Research Institute, Raggio Parkway, Reno, NV*
[4]*University of Nebraska-Lincoln, Lincoln, NE*
USA

1. Introduction

Satellite imagery now provides a dependable basis for computational models that determine evapotranspiration (ET) by surface energy balance (EB). These models are now routinely applied as part of water and water resources management operations of state and federal agencies. They are also an integral component of research programs in land and climate processes. The very strong benefit of satellite-based models is the quantification of ET over large areas. This has enabled the estimation of ET from individual fields among populations of fields (Tasumi et al. 2005) and has greatly propelled field specific management of water systems and water rights as well as mitigation efforts under water scarcity. The more dependable and universal satellite-based models employ a surface energy balance (EB) where ET is computed as a residual of surface energy. This determination requires a thermal imager onboard the satellite. Thermal imagers are expensive to construct and more a required for future water resources work. Future moderate resolution satellites similar to Landsat need to be equipped with moderately high resolution thermal imagers to provide greater opportunity to estimate spatial distribution of actual ET in time. Integrated ET is enormously valuable for monitoring effects of water shortage, water transfer, irrigation performance, and even impacts of crop type and variety and irrigation type on ET. Allen (2010b) showed that the current 16-day overpass return time of a single Landsat satellite is often insufficient to produce annual ET products due to impacts of clouds. An analysis of a 25 year record of Landsat imagery in southern Idaho showed the likelihood of producing annual ET products for any given year to increase by a factor of NINE times (from 5% probability to 45% probability) when two Landsat systems were in operation rather than one (Allen 2010b).

Satellite-based ET products are now being used in water transfers, to enforce water regulations, to improve development and calibration of ground-water models, where ET is a needed input for estimating recharge, to manage streamflow for endangered species management, to estimate water consumption by invasive riparian and desert species, to estimate ground-water consumption from at-risk aquifers, for quantification of native

American water rights, to assess impacts of land-use change on wetland health, and to monitor changes in water consumption as agricultural land is transformed into residential uses (Bastiaanssen et al., 2005, Allen et al., 2005, Allen et al. 2007b).

The more widely used and operational remote sensing models tend to use a 'CIMEC' approach ("calibration using inverse modeling of extreme conditions") to calibrate around uncertainties and biases in satellite based energy balance components. Biases in EB components can be substantial, and include bias in atmospheric transmissivity, absolute surface temperature, estimated aerodynamic temperature, surface albedo, aerodynamic roughness, and air temperature fields. Current CIMEC models include SEBAL (Bastiaanssen et al. 1998a, 2005), METRIC (Allen et al., 2007a) and SEBI-SEBS (Su 2002) and the process frees these models from systematic bias in the surface temperature and surface reflectance retrievals. Other models, such as the TSEB model (Kustas and Norman 1996), use absolute temperature and assumed air temperature fields, and so can be more susceptible to biases in these fields, and often require multiple times per day imagery. Consequently, coarser resolution satellites must be used where downscaling using finer resolution reflectance information is required.

Creating 'maps' of ET that are useful in management and in quantifying and managing water resources requires the computation of ET over monthly and longer periods such as growing seasons or annual periods. Successful creation of an ET 'snapshot' on a satellite overpass day is only part of the required process. At least half the total effort in producing a quantitative ET product involves the interpolation (or extrapolation) of ET information between image dates. This interpolation involves treatment of clouded areas of images, accounting for evaporation from wetting events occurring prior to or following overpass dates, and applying a grid of daily reference ET with the relative ET computed for an image, or a direct Penman-Monteith type of calculation, over the image domain for periods between images to account for day to day variation in weather. The particular methodology for estimating these spatial variables substantially impacts the quality and accuracy of the final ET product.

2. Model overview

Satellite based models can be separated into the following classes, building on Kalma et al. (2008):

- Surface Energy Balance
  - Full energy balance for the satellite image: $\lambda E = R_n - G - H$
  - Water stress index based on surface temperature and vegetation amounts
  - Application of a continuous Land Surface Model (LSM) that is partly initialized and advanced, in time, using satellite imagery
- Statistical methods using differences between surface and air temperature
- Simplified correlations or relationships between surface temperature extremes in an image and endpoints of anticipated ET
- Vegetation-based relative ET that is multiplied by a weather-based reference ET

where λE is latent heat flux density, representing the energy 'consumed' by the evaporation of water, R_n is net radiation flux density, G is ground heat flux density and H is sensible heat flux density to the air.

Except for the LSM applications, none of the listed energy balance methods, in and of themselves, go beyond the creation of a 'snapshot' of ET for the specific satellite image date. Large periods of time exist between snapshots when evaporative demands and water availability (from wetting events) cause ET to vary widely, necessitating the coupling of hydrologically based surface process models to fill in the gaps. The surface process models employed in between satellite image dates can be as simple as a daily soil-surface evaporation model based on a crop coefficient approach (for example, the FAO-56 model of Allen et al. 1998) or can involve more complex plant-air-water models such as SWAT (Arnold et al. 1994), SWAP (van Dam 2000), HYDRUS (Šimůnek et al. 2008), Daisy (Abrahamsen and Hansen 2000) etc. that are run on hourly to daily timesteps.

2.1 Problems with use of absolute surface temperature

Error in surface temperature (T_s) retrievals from many satellite systems can range from 3 – 5 K (Kalma et al. 2008) due to uncertainty in atmospheric attenuation and sourcing, surface emissivity, view angle, and shadowing. Hook and Prata (2001) suggested that finely tuned T_s retrievals from modern satellites could be as accurate as 0.5 K. Because near surface temperature gradients used in energy balance models are often on the order of only 1 to 5 K, even this amount of error, coupled with large uncertainties in the air temperature fields, makes the use of models based on differences in absolute estimates of surface and air temperature unwieldy.

Cleugh et al. (2007) summarized challenges in using near surface temperature gradients (dT) based on absolute estimates of T_s and air temperature, T_{air}, attributing uncertainties and biases to error in T_s and T_{air}, uncertainties in surface emissivity, differences between radiometrically derived T_s and the aerodynamically equivalent T_s required as a sourcing endpoint to dT.

The most critical factor in the physically based remote sensing algorithms is the solution of the equation for sensible heat flux density:

$$H = \rho_a c_p \frac{T_{aero} - T_a}{r_{ah}} \tag{1}$$

where ρ_a is the density of air (kg m^{-3}), c_p is the specific heat of air (J kg^{-1} K^{-1}), r_{ah} is the aerodynamic resistance to heat transfer (s m^{-1}), T_{aero} is the surface aerodynamic temperature, and T_a is the air temperature either measured at standard screen height or the potential temperature in the mixed layer (K) (Brutsaert et al., 1993). The aerodynamic resistance to heat transfer is affected by wind speed, atmospheric stability, and surface roughness (Brutsaert, 1982). The simplicity of Eq. (1) is deceptive in that T_{aero} cannot be measured by remote sensing. Remote sensing techniques measure the radiometric surface temperature T_s which is not the same as the aerodynamic temperature. The two temperatures commonly differ by 1 to 5 °C, depending on canopy density and height, canopy dryness, wind speed, and sun angle (Kustas et al., 1994, Qualls and Brutsaert, 1996, Qualls and Hopson, 1998). Unfortunately, an uncertainty of 1 °C in $T_{aero} - T_a$ can result in a 50 W m^{-2} uncertainty in H (Campbell and Norman, 1998) which is approximately equivalent to an evaporation rate of 1 mm day^{-1}. Although many investigators have attempted to solve this problem by adjusting r_{ah} or by using an additional resistance term, no generally applicable method has been developed.

Campbell and Norman (1998) concluded that a practical method for using satellite surface temperature measurements should have at least three qualities: (i) accommodate the difference between aerodynamic temperature and radiometric surface temperature, (ii) not require measurement of near-surface air temperature, and (iii) rely more on differences in surface temperature over time or space rather than absolute surface temperatures to minimize the influence of atmospheric corrections and uncertainties in surface emissivity.

2.2 CIMEC Models (SEBAL and METRIC)

The SEBAL and METRIC models employ a similar inverse calibration process that meets these three requirements with limited use of ground-based data (Bastiaanssen et al., 1998a,b, Allen et al., 2007a). These models overcome the problem of inferring T_{aero} from T_s and the need for near-surface air temperature measurements by directly estimating the temperature difference between two near surface air temperatures, T_1 and T_2, assigned to two arbitrary levels z_1 and z_2 without having to explicitly solve for absolute aerodynamic or air temperature at any given height. The establishment of the temperature difference is done via inversion of the function for H at two known evaporative conditions in the model using the CIMIC technique. The temperature difference for a dry or nearly dry condition, represented by a bare, dry soil surface is obtained via $H=R_n - G- \lambda E$ (Bastiaanssen et al., 1998a):

$$T_1 - T_2 = \Delta T_a = \frac{H\, r_{ah_{1-2}}}{\rho_a c_p} \tag{2}$$

where $r_{ah,1-2}$ is the aerodynamic resistance to heat transfer between two heights above the surface, z_1 and z_2. At the other extreme, for a wet surface, essentially all available energy $R_n - G$ is used for evaporation λE. At that extreme, the classical SEBAL approach assumes that H ≈ 0, in order to keep requirements for high quality ground data to a minimum, so that $\Delta T_a \approx 0$. Allen et al. (2001, 2007a) have used reference crop evapotranspiration, representing well-watered alfalfa, to represent λE for the cooler population of pixels in satellite images of irrigated fields in the METRIC approach, so as to better capture effects of regional advection of H and dry air, which can be substantial in irrigated desert. METRIC calculates $H = R_n - G - k_1 \lambda ET_r$ at these pixels, where ET_r is alfalfa reference ET computed at the image time using weather data from a local automated weather station, and ΔT_a from Eq. (2), where $k_1 \sim 1.05$. In typical SEBAL and METRIC applications, z_1 and z_2 are taken as 0.1 and 2 m above the zero plane displacement height (d). z_1 is taken as 0.1 m above the zero plane to insure that T_1 is established at a height that is generally greater than $d + z_{oh}$ (z_{oh} is roughness length for heat transfer). Aerodynamic resistance, r_{ah}, is computed for between z_1 and z_2 and does not require the inclusion and thus estimation of z_{oh}, but only z_{om}, the roughness length for momentum transfer that is normally estimated from vegetation indices and land cover type. H is then calculated in the SEBAL and METRIC CIMEC-based models as:

$$H = \rho_a c_p \frac{\Delta T_a}{r_{ah_{1-2}}} \tag{3}$$

One can argue that the establishment of ΔT_a over a vertical distance that is elevated above $d + z_{oh}$ places the r_{ah} and established ΔT_a in a blended boundary layer that combines influences of sparse vegetation and exposed soil, thereby reducing the need for two source modeling and problems associated with differences between radiative temperature and aerodynamic temperature and problems associated with estimating z_{oh} and specific air temperature associated with the specific surface.

Evaporative cooling creates a landscape having high ΔT_a associated with high H and high radiometric temperature and low ΔT_a with low H and low radiometric temperature. For example, moist irrigated fields and riparian systems have much lower ΔT_a and much lower T_s than dry rangelands. Allen et al. (2007a) argued, and field measurements in Egypt and Niger (Bastiaanssen et al., 1998b), China (Wang et al., 1998), USA (Franks and Beven, 1997), and Kenya (Farah, 2001) have shown the relationship between T_s and ΔT_a to be highly linear between the two calibration points

$$\Delta T_a = c_1 T_s - c_2 \qquad (4)$$

where c_1 and c_2 are empirical coefficients valid for one particular moment (the time and date of an image) and landscape. By using the minimum and maximum values for ΔT_a as calculated for the nearly wettest and driest (i.e., coldest and warmest) pixel(s), the extremes of H are used, in the CIMEC process to find coefficients c_1 and c_2. The empirical Eq. (4) meets the third quality stated by Campbell and Norman (1998) that one should rely on differences in radiometric surface temperature over space rather than absolute surface temperatures to minimize the influence of atmospheric corrections and uncertainties in surface emissivity.

Equation (3) has two unknowns: ΔT_a and the aerodynamic resistance to heat transfer $r_{ah,1-2}$ between the z_1 and z_2 heights, which is affected by wind speed, atmospheric stability, and surface roughness (Brutsaert, 1982). Several algorithms take one or more field measurements of wind speed and consider these as spatially constant over representative parts of the landscape (e.g. Hall et al., 1992; Kalma and Jupp, 1990; Rosema, 1990). This assumption is only valid for uniform homogeneous surfaces. For heterogeneous landscapes a unique wind speed near the ground surface is required for each pixel. One way to meet this requirement is to consider the wind speed spatially constant at a blending height about 200 m above ground level, where wind speed is presumed to not be substantially affected by local surface heterogeneities. The wind speed at blending height is predicted by upward extrapolation of near-surface wind speed measured at an automated weather station using a logarithmic wind profile. The wind speed at each pixel is obtained by a similar downward extrapolation using estimated surface momentum roughness z_{0m} determined for each pixel.

Allen et al. (2007a) have noted that the inverted value for ΔT_a is highly tied to the value used for wind speed in its CIMEC determination. Therefore, they cautioned against the use of a spatial wind speed field at some blending height across an image with a single ΔT_a function. The application of the image-specific ΔT_a function with a blending height wind speed in a distant part of the image that is, for example, double that of the wind used to determine coefficients c_1 and c_2 can estimate higher H than is possible based on energy availability. In those situations, the 'calibrated' ΔT_a would be about half as much to compensate for the larger wind speed. Therefore, if wind fields at the blending height (200 m) are used, then fields of ΔT_a calibrations are also needed, which is prohibitive. The single ΔT_a function of SEBAL and METRIC, coupled with a single wind speed at blending height, transcends these problems. Gowda et al., (2008) presented a summary of remote sensing based energy balance algorithms for mapping ET that complements that by Kalma et al. (2008).

Aerodynamic Transport. The value for $r_{ah,1,2}$ is calculated between the two heights z1 and z2 in SEBAL and METRIC. The value for $r_{ah,1,2}$ is strongly influenced by buoyancy within the boundary layer driven by the rate of sensible heat flux. Because both $r_{ah,1,2}$ and H are unknown at each pixel, an iterative solution is required. During the first iteration, $r_{ah,1,2}$ is computed assuming neutral stability:

$$r_{ah_{1-2}} = \frac{\ln\left(\dfrac{z_2}{z_1}\right)}{u_* \, k} \tag{5}$$

where z_1 and z_2 are heights above the zero plane displacement of the vegetation where the endpoints of dT are defined, u_* is friction velocity (m s^{-1}), and k is von Karman's constant (0.41). Friction velocity u_* is computed during the first iteration using the logarithmic wind law for neutral atmospheric conditions:

$$u^* = \frac{k \, u_{200}}{\ln\left(\dfrac{200}{z_{om}}\right)} \tag{6}$$

where u_{200} is the wind speed (m s^{-1}) at a blending height assumed to be 200 m, and z_{om} is the momentum roughness length (m). z_{om} is a measure of the form drag and skin friction for the layer of air that interacts with the surface. u_* is computed for each pixel inside the process model using a specific roughness length for each pixel, but with u_{200} assumed to be constant over all pixels of the image since it is defined as occurring at a "blending height" unaffected by surface features. Eq. (5) and (6) support the use of a temperature gradient defined between two heights that are both above the surface. This allows one to estimate $r_{ah,1-2}$ without having to estimate a second aerodynamic roughness for sensible heat transfer (z_{oh}), since height z_1 is defined to be at an elevation above z_{oh}. This is an advantage, because z_{oh} can be difficult to estimate for sparse vegetation.

The wind speed at an assumed blending height (200 m) above the weather station, u_{200}, is calculated as:

$$u_{200} = \frac{u_w \ln\left(\dfrac{200}{z_{omw}}\right)}{\ln\left(\dfrac{z_x}{z_{omw}}\right)} \tag{7}$$

where u_w is wind speed measured at a weather station at z_x height above the surface and z_{omw} is the roughness length for the weather station surface, similar to Allen and Wright (1997). All units for z are the same. The value for u_{200} is assumed constant for the satellite image. This assumption *is required* for the use of a constant relation between dT and T_s to be extended across the image (Allen 2007a).

The effects of mountainous terrain and elevation on wind speed are complicated and difficult to quantify (Oke, 1987). In METRIC, z_{om} or wind speed for image pixels in mountains are adjusted using a suite of algorithms to account for the following impacts (Allen and Trezza, 2011):

- Terrain roughness – the standard deviation of elevation within a 1.5 km radius is used to estimate an additive to zom to account for vortex and channeling impacts of turbulence
- Elevation effect on velocity – the relative elevation within a 1.5 km radius is used to estimate a relative increase in wind speed, based on slope.
- Reduction of wind speed on leeward slopes – when the general wind direction aloft can be estimated in mountainous terrain, then a reduction factor is made to wind speed on leeward slopes, using relative elevation and amount of slope as factors.

These algorithms have been developed for western Oregon and are being tested in Idaho, Nevada and Montana and are described in an article in preparation (Allen and Trezza, 2011). Allen and Trezza (2011) also refined the estimation of diffuse radiation on steep mountainous slopes.

Iterative solution for $r_{ah,1-2}$. During subsequent iterations for the solution for H, a corrected value for u_* is computed as:

$$u_* = \frac{u_{200}k}{ln\left(\dfrac{200}{z_{0m}}\right) - \Psi_{m(200m)}} \tag{8}$$

where $\psi_{m(200m)}$ is the stability correction for momentum transport at 200 meters. A corrected value for $r_{ah,1-2}$ is computed each iteration as:

$$r_{ah,1,2} = \frac{ln\left(\dfrac{z_2}{z_1}\right) - \Psi_{h(z_2)} + \Psi_{h(z_1)}}{u_* \times k} \tag{9}$$

where $\psi_{h(z2)}$ and $\psi_{h(z1)}$ are the stability corrections for heat transport at z_2 and z_1 heights (Paulson 1970 and Webb 1970) that are updated each iteration.

Stability Correction functions. The Monin-Obukhov length (L) defines the stability conditions of the atmosphere in the iterative process. L is the height at which forces of buoyancy (or stability) and mechanical mixing are equal, and is calculated as a function of heat and momentum fluxes:

$$L = -\frac{\rho_{air} c_p u_*^3 T_s}{kgH} \tag{10}$$

where g is gravitational acceleration (= 9.807 m s⁻²) and units for terms cancel to m for L. Values of the integrated stability corrections for momentum and heat transport (ψ_m and ψ_h) are computed using formulations by Paulson (1970) and Webb (1970), depending on the sign of L. When L < 0, the lower atmospheric boundary layer is unstable and when L > 0, the boundary layer is stable. For L<0:

$$\Psi_{m(200m)} = 2ln\left(\frac{1 + x_{(200m)}}{2}\right) + ln\left(\frac{1 + x_{(200m)}^2}{2}\right) - 2ARCTAN\left(x_{(200m)}\right) + 0.5\pi \tag{11}$$

$$\Psi_{h(2m)} = 2ln\left(\frac{1 + x_{(2m)}^2}{2}\right) \tag{12a}$$

$$\Psi_{h(0.1m)} = 2ln\left(\frac{1 + x_{(0.1m)}^2}{2}\right) \tag{12b}$$

where

$$x_{(200m)} = \left(1 - 16\frac{200}{L}\right)^{0.25} \tag{13a}$$

$$x_{(2m)} = \left(1 - 16\frac{2}{L}\right)^{0.25} \tag{13b}$$

$$x_{(0.1m)} = \left(1 - 16\frac{0.1}{L}\right)^{0.25} \tag{14}$$

Values for $x_{(200m)}$, $x_{(2m)}$, and $x_{(0.1m)}$ have no meaning when $L \geq 0$ and their values are set to 1.0. For $L > 0$ (stable conditions):

$$\Psi_{m(200m)} = -5\left(\frac{2}{L}\right) \tag{15}$$

$$\Psi_{h(2m)} = -5\left(\frac{2}{L}\right) \tag{16a}$$

$$\Psi_{h(0.1m)} = -5\left(\frac{0.1}{L}\right) \tag{16b}$$

When $L = 0$, the stability values are set to 0. Equation (15) uses a value of 2 m rather than 200 m for z because it is assumed that under stable conditions, the height of the stable, inertial boundary layer is on the order of only a few meters. Using a larger value than 2 m for z can cause numerical instability in the model. For neutral conditions, $L = 0$, $H = 0$ and ψ_m and $\psi_h = 0$.

2.2.1 The use of inverse modeling at extreme conditions during calibration (CIMEC)
In METRIC, the satellite-based energy balance is internally calibrated at two extreme conditions (dry and wet) using locally available weather data. The auto-calibration is done for each image using alfalfa-based reference ET (ET_r) computed from hourly weather data. Accuracy and dependability of the ET_r estimate has been established by lysimetric and other studies in which we have high confidence (ASCE-EWRI, 2005). The internal calibration of the sensible heat computation within SEBAL and METRIC and the use of the indexed temperature gradient eliminate the need for atmospheric correction of surface temperature (T_s) and reflectance (albedo) measurements using radiative transfer models (Tasumi et al., 2005b). The internal calibration also reduces impacts of biases in estimation of aerodynamic stability correction and surface roughness.

The calibration of the sensible heat process equations, and in essence the entire energy balance, to ET_r corrects the surface energy balance for lingering systematic computational biases associated with empirical functions used to estimate some components and uncertainties in other estimates as summarized by Allen et al. (2005), including:

- atmospheric correction
- albedo calculation
- net radiation calculation
- surface temperature from the satellite thermal band
- air temperature gradient function used in sensible heat flux calculation
- aerodynamic resistance including stability functions
- soil heat flux function
- wind speed field

This list of biases plagues essentially all surface energy balance computations that utilize satellite imagery as the primary spatial information resource. Most polar orbiting satellites orbit about 700 km above the earth's surface, yet the transport of vapor and sensible heat from land surfaces is strongly impacted by aerodynamic processes including wind speed, turbulence and buoyancy, all of which are essentially invisible to satellites. In addition, precise quantification of albedo, net radiation and soil heat flux is uncertain and potentially biased. Therefore, even though best efforts are made to estimate each of these parameters as accurately and as unbiased as possible, some biases do occur and calibration to ET_r helps to compensate for this by introducing a bias correction into the calculation of H. The end result is that biases inherent to R_n, G, and subcomponents of H are essentially cancelled by the subtraction of a bias-canceling estimate for H. The result is an ET map having values ranging between near zero and near ET_r, for images having a range of bare or nearly bare soil and full vegetation cover.

2.3 Calculation of evapotranspiration

ET at the instant of the satellite image is calculated for each pixel by dividing LE from $LE = R_n - G - H$ by latent heat of vaporization:

$$ET_{inst} = 3600\frac{LE}{\lambda \rho_w} \tag{17}$$

where ET_{inst} is instantaneous ET (mm hr^{-1}), 3600 converts from seconds to hours, ρ_w is the density of water [~1000 kg m^{-3}], and λ is the latent heat of vaporization (J kg^{-1}) representing the heat absorbed when a kilogram of water evaporates and is computed as:

$$\lambda = [2.501 - 0.00236(T_s - 273.15)] \times 10^6 \tag{18}$$

The reference ET fraction (ET_rF) is calculated as the ratio of the computed instantaneous ET (ET_{inst}) from each pixel to the reference ET (ET_r) computed from weather data:

$$ET_rF = \frac{ET_{inst}}{ET_r} \tag{19}$$

where ET_r is the estimated instantaneous rate (interpolated from hourly data) (mm hr^{-1}) for the standardized 0.5 m tall alfalfa reference at the time of the image. Generally only one or two weather stations are required to estimate ET_r for a Landsat image that measures 180 km x 180 km, as discussed later. ET_rF is the same as the well-known crop coefficient, K_c, when used with an alfalfa reference basis, and is used to extrapolate ET from the image time to 24-hour or longer periods.

One should generally expect ET_rF values to range from 0 to about 1.0 (Wright, 1982; Jensen et al., 1990). At a completely dry pixel, $ET = 0$ and therefore $ET_rF = 0$. A pixel in a well established field of alfalfa or corn can occasionally have an ET slightly greater than ET_r and therefore $ET_rF > 1$, perhaps up to 1.1 if it has been recently wetted by irrigation or precipitation. However, ET_r generally represents an upper bound on ET for large expanses of well-watered vegetation. Negative values for ET_rF can occur in METRIC due to systematic errors caused by various assumptions made earlier in the energy balance process and due to random error components so that error should oscillate about $ET_rF = 0$ for completely dry pixels. In calculation of ET_rF in Equation (19), each pixel retains a unique value for ET_{inst} that is derived from a common value for ET_r derived from the representative weather station data.

24-Hour Evapotranspiration (ET_{24}). Daily values of ET (ET_{24}) are generally more useful than the instantaneous ET that is derived from the satellite image. In the METRIC process, ET_{24} is estimated by assuming that the instantaneous ET_rF computed at image time is the same as the average ET_rF over the 24-hour average. The consistency of ET_rF over a day has been demonstrated by various studies, including Romero (2004), Allen et al., (2007a) and Collazzi et al., (2006).

The assumption of constant ET_rF during a day appears to be generally valid for agricultural crops that have been developed to maximize photosynthesis and thus stomatal conductance. In addition, the advantage of the use of ET_rF is to account for the increase in 24-hour ET that can occur under advective conditions. The impacts of advection are represented well by the Penman-Monteith equation. However, the ET_rF may decrease during afternoon for some native vegetation under water short conditions where plants endeavor to conserve soil water through stomatal control. In addition, by definition, when the vegetation under study is the same as or similar to the vegetation for the surrounding region and experiences similar water inputs (natural rainfall, only), then (by definition) no advection can occur. This is because as much sensible heat energy is generated by the surface under study as is generated by the region. Therefore, the net advection of energy is nearly zero. Therefore, under these conditions, the estimation by ET_r that accounts for impacts of advection to a *wet* surface do not occur, and the use of ET_rF to estimate 24-hour ET may not be valid. Instead, the use of evaporative fraction, EF, that is used with SEBAL applications may be a better time-transfer approach for rainfed systems. Various schemes of using EF for rainfed portions of Landsat images and ET_rF for irrigated, riparian or wetland portions were explored by Kjaersgaard and Allen (2010). When used, the EF is calculated as:

$$EF = \frac{ET_{inst}}{R_n - G} \tag{20}$$

where ET_{inst} and R_n and G have the same units and represent the same period of time. Finally, the ET_{24} (mm/day) is computed for each image pixel in SEBAL as:

$$ET_{24} = (EF)\left(R_{n\_24}\right) \tag{21}$$

and in METRIC as:

$$ET_{24} = C_{rad} \left(ET_r F \right) \left(ET_{r\_24} \right) \tag{22}$$

where $ET_r F$ (or EF) is assumed equal to the $ET_r F$ (or EF) determined at the satellite overpass time, ET_{r-24} is the cumulative 24-hour ET_r for the day of the image and C_{rad} is a correction term used in sloping terrain to correct for variation in 24-hr vs. instantaneous energy availability. C_{rad} is calculated for each image and pixel as:

$$C_{rad} = \frac{R_{so(inst)Horizontal}}{R_{so(inst)Pixel}} \cdot \frac{R_{so(24)Pixel}}{R_{so(24)Horizontal}} \tag{23}$$

where R_{so} is clear-sky solar radiation (W m^{-1}), the "$(inst)$" subscript denotes conditions at the satellite image time, "(24)" represents the 24-hour total, the "$Pixel$" subscript denotes slope and aspect conditions at a specific pixel, and the "$Horizontal$" subscript denotes values calculated for a horizontal surface representing the conditions impacting ET_r at the weather station. For applications to horizontal areas, $C_{rad} = 1.0$.

The 24 hour R_{so} for horizontal surfaces and for sloping pixels is calculated as:

$$R_{so(24)} = \int_0^{24} R_{so\_i} \tag{24}$$

where R_{so_i} is instantaneous clear sky solar radiation at time i of the day, calculated by an equation that accounts for effects of slope and aspect. In METRIC, $ET_{r\ 24}$ is calculated by summing hourly ET_r values over the day of the image.

After ET and $ET_r F$ have been determined using the energy balance, and the application of the single dT function, then, when interpolating between satellite images, a full grid for ET_r is used for the extrapolation over time, to account for both spatial and temporal variation in ET_r. The ET_r grid is generally made on a 3 or 5 km base using as many quality-controlled weather stations located within and in the vicinity of the study area as available. Depending on data availability and the density of the weather stations various gridding methods including krieging, inverse-distance, and splining can be used.

Seasonal Evapotranspiration (ET_{seasonal}). Monthly and seasonal evapotranspiration "maps" are often desired for quantifying total water consumption from agriculture. These maps can be derived from a series of $ET_r F$ images by interpolating $ET_r F$ on a pixel by pixel basis between processed images and multiplying, on a daily basis, by the ET_r for each day. The interpolation of $ET_r F$ between image dates is not unlike the construction of a seasonal K_c curve (Allen et al., 1998), where interpolation is done between discrete values for K_c.

The METRIC approach assumes that the ET for the entire area of interest changes in proportion to change in ET_r at the weather station. This is a generally valid assumption and is similar to the assumptions used in the conventional application of $K_c \times ET_r$. This approach is effective in estimating ET for both clear and cloudy days in between the clear-sky satellite image dates. Tasumi et al., (2005a) showed that the $ET_r F$ was consistent between clear and cloudy days using lysimeter measurements at Kimberly, Idaho. ET_r is computed at a specific weather station location and therefore may not represent the actual condition at each pixel. However, because ET_r is used only as an index of the relative change in weather, specific information at each pixel is retained through the $ET_r F$.

Cumulative ET for any period, for example, month, season or year is calculated as:

$$ET_{period} = \sum_{i=m}^{n} \left[\left(ET_r F_i \right) \left(ET_{r24i} \right) \right] \tag{25}$$

where ET_{period} is the cumulative ET for a period beginning on day m and ending on day n, $ET_r F_i$ is the interpolated $ET_r F$ for day i, and ET_{r24i} is the 24-hour ET_r for day i. Units for ET_{period} will be in mm when ET_{r24} is in mm d^{-1}. The interpolation between values for $ET_r F$ is best made using a curvilinear interpolation function, for example a spline function, to better fit the typical curvilinearity of crop coefficients during a growing season (Wright, 1982). Generally one satellite image per month is sufficient to construct an accurate $ET_r F$ curve for purposes of estimating seasonal ET (Allen et al., 2007a). During periods of rapid vegetation change, a more frequent image interval may be desirable. Examples of splining $ET_r F$ to estimate daily and monthly ET are given in Allen et al. (2007a) and Singh et al. (2008).

If a specific pixel must be masked out of an image because of cloud cover, then a subsequent image date must be used during the interpolation and the estimated $ET_r F$ or K_c curve will have reduced accuracy.

Average $ET_r F$ over a period. An average $ET_r F$ for the period can be calculated as:

$$ET_r F_{period} = \frac{\sum_{i=m}^{n} \left[\left(ET_r F_i \right) \left(ET_{r24i} \right) \right]}{\sum_{i=m}^{n} ET_{r24i}} \tag{26}$$

Moderately high resolution satellites such as Landsat provide the opportunity to view evapotranspiration on a field by field basis, which can be valuable for water rights management, irrigation scheduling, and discrimination of ET among crop types (Allen et al., 2007b). The downside of using high resolution imagery is less frequent image acquisition. In the case of Landsat, the return interval is 16 days. As a result, monthly ET estimates are based on only one or two satellite image snapshots per month. In the case of clouds, intervals of 48 days between images can occur. This can be rectified by combining multiple Landsats (5 with 7) or by using data fusion techniques, where a more frequent, but more coarse system like MODIS is used as a carrier of information during periods without quality Landsat images (Gao et al., 2006, Anderson et al., 2010).

2.4 Reflectance based ET methods

Reflectance based ET methods typically estimate relative fractions of reference ET ($ET_r F$, synonymous with the crop coefficient) based on some sort of vegetation index, for example, the normalized difference vegetation index, NDVI, and multiply the $ET_r F$ by daily computed reference ET_r (Groeneveld et al., 2007). NDVI approaches don't directly or indirectly account for evaporation from soil, so they have difficulty in estimating evaporation associated with both irrigation and precipitation wetting events, unless operated with a daily evaporation process model. The VI-based methods are therefore largely blind to the treatment of both irrigation and precipitation events, except on an average basis. In contrast, thermally based models detect the presence of evaporation from soil, during the snapshot, at least, via evaporative cooling. VI-based methods also do not pick up on acute water stress caused by drought or lack of irrigation, which is often a primary reason for quantifying ET. These models can be run with a background daily evaporation process model, similar to the EB-based models, to estimate evaporation from precipitation between satellite overpass dates.

2.5 Challenges with snapshot models

The SEBAL, METRIC, and other EB models, that can be applied at the relatively high spatial resolution of Landsat and similar satellites, despite their different relative strengths and weaknesses, all suffer from the inability to capture evaporation signals from episodic precipitation and irrigation events occurring between overpass dates. In the case of irrigation events, which are typically unknown to the processer in terms of timing and location, the random nature of these events in time can be somewhat accommodated via the use of multiple overpass dates during the irrigation season (Allen et al. 2007a). In this manner, the ET retrieval for a specific field may be biased high when the overpass follows an irrigation event, but may be biased low when the overpass just precedes an irrigation event. Allen et al. (2007a) suggested that monthly overpass dates over a seven month growing season, for example, can largely compensate for the impact of irrigation wetting on individual fields, especially when it is total growing season ET that is of most interest. The variance of the error in ET estimate caused by unknown irrigation events should tend to decrease with the square root of the number of images processed during the irrigation season.

The impact by precipitation events is a larger problem in converting the 'snapshot' ET images from energy balance models or other methods into monthly and longer period ET. Precipitation timing and magnitudes tend to be less random in time and have much larger variance in depth per wetting event than with irrigation. Because of this, the use of snapshot ET models to construct monthly and seasonal ET maps is more likely to be biased high (if a number of images happen to be 'wet' following a recent precipitation event) or low (if images happen to be 'dry', with precipitation occurring between images). The latter may often be the case since the most desired images for processing are cloud free.

One important use of ET maps is in the estimation of ground water recharge (Allen et al., 2007b). Ground water recharge is often uncertain due to uncertainty in both precipitation and ET, and is usually computed using the difference between P and ET, with adjustment for runoff. It is therefore important to maintain congruency between ET and P data sets or 'maps'. Lack of congruency can cause very large error in estimated recharge, especially in the more arid regions.

3. Adjusting for background evaporation

Often a Landsat or other image is processed on a date where antecedent rainfall has caused the evaporation from bare soil to exceed that for the surrounding monthly period. Often, for input to water balance applications, it is desirable that the final ET image represent the average evaporation conditions for the month. In that case, one approach is to adjust the 'background' evaporation of the processed image to better reflect that for the month or other period that it is to ultimately represent. This period may be a time period that is half way between other adjacent images.

An example of a sequence of Landsat images processed using the METRIC surface energy balance model for the south-western portion of the Nebraska Panhandle (Kjaersgaard and Allen, 2010a) is shown in Figure 1 along with daily precipitation from the Scottsbluff High Plains Regional Climate Center (HPRCC) weather station. The August 13 image date was preceded by a wet period and followed by a very dry period, thus the evaporation from non-irrigated areas at the satellite image date is not representative for the month.

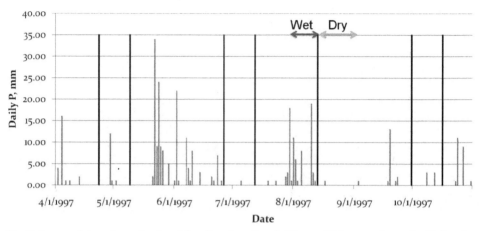

Fig. 1. Image dates of nearly cloud free Landsat 5 path 33 row 31 images from the Nebraskan Panhandle in 1997 (black vertical bars) and precipitation recorded at the Scottsbluff HPRCC weather station (red bars). After Kjaersgaard and Allen (2010).

In making the adjustment for background evaporation, the background evaporation on the overpass date is subtracted out of the image and the average background evaporation is substituted in. Full adjustment is made for areas of completely bare soil, represented by NDVI = NDVI$_{bare soil}$, with no adjustment to areas having full ground covered by vegetation, represented by NDVI = NDVI$_{full cover}$, and with linear adjustment in between.

The following methodology is taken from a white paper developed by the University of Idaho during 2008 and 2009 (Allen 2008, rev. 2010). The ET$_r$F of the Landsat image is first adjusted to a 'basal' condition, where the evaporation estimate is free of rainfall induced evaporation, but still may contain any irrigation induced evaporation:

$$\left(ET_rF_i\right)_b = ET_rF_i - \left(ET_rF_{background}\right)_i \left(\frac{NDVI_{full cover} - NDVI_i}{NDVI_{full cover} - NDVI_{bare soil}}\right) \tag{27}$$

where (ET$_r$F$_{background}$)$_i$ is the background evaporation on the image date (i) for bare soil, computed using a gridded FAO-56 two-stage evaporation model of Allen et al. (1998) with modification to account for 'flash' evaporation from the soil skin (Allen 2010a) or some other soil evaporation model such as Hydrus or DAISY. The soil evaporation model is on a daily timestep using spatially distributed precipitation, reference ET, and soil properties. (ET$_r$F$_i$)$_b$ is the resulting 'basal' ET image for a particular image date, representing a condition having NDVI amount of vegetation and a relatively dry soil surface. This parameter represents the foundation for later adjustment to represent the longer period.

3.1 Adjustment for cases of riparian vegetation

For riparian vegetation and similar systems, where soil water stress is not likely to occur due to the frequent presence of shallow ground water, an adjusted ET_rF is computed for the image date that reflects background evaporation averaged over the surrounding period in proportion to the amount of ground cover represented by NDVI:

$$\left(ET_rF_i\right)_{adjusted} = \left(ET_rF_i\right)_b + \overline{\left(ET_rF_{background}\right)} \left(\frac{NDVI_{full\,cover} - NDVI_i}{NDVI_{full\,cover} - NDVI_{bare\,soil}} \right) \tag{28}$$

where $\overline{\left(ET_rF_{background}\right)}$ is the average evaporation from bare soil due to precipitation over the averaging period (e.g., one month), calculated as:

$$\overline{\left(ET_rF_{background}\right)} = \frac{\sum_{1}^{n}\left(ET_rF_{background}\right)_i}{n} \tag{29}$$

Equations 5 and 6 can be combined as:

$$\left(ET_rF_i\right)_{adjusted} = ET_rF_i + \left(\overline{\left(ET_rF_{background}\right)} - \left(ET_rF_{background}\right)_i\right)\left(\frac{NDVI_{full\,cover} - NDVI_i}{NDVI_{full\,cover} - NDVI_{bare\,soil}} \right) \tag{30}$$

with limits $NDVI_{bare\,soil} \leq NDVI_i \leq NDVI_{full\,cover}$.

The outcome of this adjustment is to preserve any significant evaporation stemming from irrigation or ground-water and any transpiration stemming from vegetation, with adjustment only for evaporation stemming from precipitation to account for differences between the image date and that of the surrounding time period. In other words, if the initial ET_rF_i, prior to adjustment is high due to evaporation from irrigation or from high ground-water condition, much of that evaporation would remain in the adjusted ET_rF_i estimate.

3.2 Adjustments for non-riparian vegetation

The following refinement to Eq. 30 is made for application to non-riparian vegetation, to account for those situations where, during long periods (i.e., months), soil moisture may have become limited enough that even transpiration of vegetation has been reduced due to moisture stress. If the Landsat image is processed during that period of moisture stress, then the ET_rF value for vegetated or partially vegetated areas will be lower than the potential (nonstressed) value. This can happen, for example, during early spring when winter wheat may go through stress prior to irrigation or a rainy period, or in desert and other dry systems.

This causes a problem in that the method of Eq. 8 attempts to 'preserve' the ET_rF of the vegetated portion of a pixel that was computed by METRIC on the image date. However, when a rain event occurs following the image date, not only will the ET_rF of exposed soil increase, but any stressed vegetation will equally 'recover' from moisture stress and the ET_rF of the vegetation fraction of the surface will increase. This situation may occur for rangeland and dryland agricultural systems. It is therefore assumed that the ET_rF of nonstressed vegetation will be at least as high as the ET_rF of bare soil over the same time period, since it should have equal access to shallow water. An exception would be if the vegetation were sufficiently stressed to not recover transpiration potential. However, this amount of stress should be evidenced by a reduced NDVI. A minimum limit is therefore placed, using the background ET_rF $\left(\overline{ET_rF_{background}} \right)$ for the period.

To derive a modified Eq. 8, it is useful to first isolate the 'transpiration' portion of the ET_rF. On the satellite image date, the bulk ET_rF computed by METRIC for a pixel, is decomposed to:

$$ET_rF_i = (1 - f_c)\left(ET_rF_{background} \right)_i + f_c \left(ET_rF_{transpiration} \right)_i \tag{31}$$

where $ET_rF_{transpiration}$ is the apparent transpiration from the fraction of ground covered by vegetation, f_c. The f_c is estimated as $1 - f_s$, where f_s is the fraction of bare soil, and for consistency with equations 30, f_s is estimated as:

$$f_s = \left(\frac{NDVI_{fullcover} - NDVI_i}{NDVI_{fullcover} - NDVI_{baresoil}} \right) \tag{32a}$$

so that:

$$f_c = 1 - \left(\frac{NDVI_{fullcover} - NDVI_i}{NDVI_{fullcover} - NDVI_{baresoil}} \right) \tag{32b}$$

Eq. 31 is not used as is, since ET_rF_i comes from the energy balance-based ET image (i.e., from METRIC, etc.). However, one can rearrange Eq. 31 to solve for $ET_rF_{transpiration}$:

$$f_c \left(ET_rF_{transpiration} \right)_i = ET_rF_i - (1 - f_c)\left(ET_rF_{background} \right)_i \tag{33}$$

Now, if $ET_rF_{transpiration}$ is limited to the maximum of the $ET_rF_{transpiration}$ on the day of the image, or the $\left(\overline{ET_rF_{background}} \right)$ for the period, then:

$$\left(ET_rF_{transpiration} \right)_{adjusted} = \max\left[\left(ET_rF_{transpiration} \right)_i , \left(\overline{ET_rF_{background}} \right) \right] \tag{34}$$

Then the new ET_rF adjusted value becomes:

$$\left(ET_rF_i \right)_{adjusted} = (1 - f_c)\left(\overline{ET_rF_{background}} \right) + f_c \left(ET_rF_{transpiration} \right)_{adjusted}$$

or

$$\left(ET_rF_i \right)_{adjusted} = (1 - f_c)\left(\overline{ET_rF_{background}} \right) + f_c \max\left[\left(ET_rF_{transpiration} \right)_i , \left(\overline{ET_rF_{background}} \right) \right] \tag{35}$$

where $\left(\overline{ET_rF_{background}} \right)$ is the average evaporation from bare soil due to precipitation over the averaging period (e.g., one month) and $ET_rF_{transpiration}$ is the original transpiration computed from Eq. 33. Eq. 33 and 35 can be combined so that:

$$\left(ET_rF_i\right)_{adjusted} = \left(1 - f_c\right)\overline{\left(ET_rF_{background}\right)} +$$
$$+\max\left[\left(ET_rF_i - \left(1 - f_c\right)\left(ET_rF_{background}\right)_i\right)_i , f_c\overline{\left(ET_rF_{background}\right)}\right]$$

(36)

Only areas with bare soil or partial vegetation cover are adjusted. Pixels having full vegetation cover, defined as when NDVI > 0.75, are not adjusted. An example of an image date where the adjustment increased the ET_rF for bare soil and partially vegetated areas is shown in Figure 2. Figure 3 shows an example of an image date where the ET_rF from bare soil and partial vegetation cover was decreased by the adjustment.

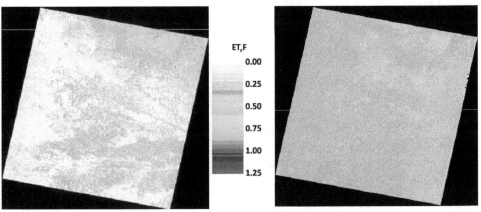

Fig. 2. ET_rF in western Nebraska from May 9 1997 before (left) and after (right) adjustment for background evaporation representing the time period (~month) represented by that image. After Kjaersgaard and Allen (2010).

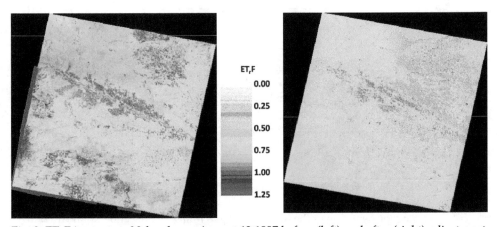

Fig. 3. ET_rF in western Nebraska on August 13 1997 before (left) and after (right) adjustment to reflect soil evaporation occurring over the time period (~ 1 month) represented by that image. Note that irrigated fields with full vegetation cover having a substantial transpiration component were not affected by the adjustment. After Kjaersgaard and Allen (2010).

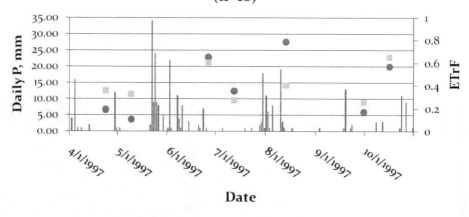

Fig. 4. Average ET$_r$F from ten rangeland locations in western Nebraska before and after adjustment. Also shown is the precipitation from the Scottsbluff HPRCC weather station (after Kjaersgaard and Allen 2010).

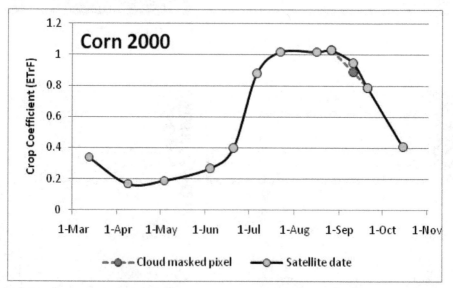

Fig. 5. Schematic representation of the linear cloud gap filling and the cubic spline used to interpolate between image dates for a corn crop. The green points represent image dates and the black line is the splined interpolation between points; the red point represents the value of ET$_r$F that is interpolated linearly from the two adjacent image dates had the field had cloud cover on September 10.

Average ET$_r$F on image dates before and after adjustment for background evaporation is shown in Figure 4 from ten rangeland locations in western Nebraska. For some image dates, such as early and late in the season, the adjusted ET$_r$F values are "wetter" than that represented by the original image. Similarly, for other images dates, such as in the middle of the growing season, the images were "drier". The adjustment for one image in August reduced the estimated ET for the month of August by nearly 50%, which is considerable.

It is noted, that the images no longer represent the ET from the satellite overpass dates after the adjustment for background evaporation. The images are merely an intermediate product that is used as the input into an interpolation procedure when producing ET estimates for monthly or longer time periods.

4. Dealing with clouded parts of images

Satellite images often have clouds in portions of the images. ET$_r$F cannot be directly estimated for these areas using surface energy balance because cloud temperature masks surface temperature and cloud albedo masks surface albedo. Generally ET$_r$F for clouded areas must be filled in before application of further integration processes so that those processes can be uniformly applied to an entire image. The alternative is to directly interpolate ET$_r$F between adjacent (in time) image dates or to run some type of daily ET process model that is based on gridded weather data.

In METRIC applications (Allen et al. 2007b), ET$_r$F for clouded areas of images is usually filled in prior to interpolating ET$_r$F for days between image dates (and multiplying by gridded ET$_r$ for each day to obtain daily ET images). A linear interpolation, as shown in Figure 5, is used to fill in ET$_r$F for clouded portions of images rather than curvilinear interpolation that is used to interpolate ET$_r$F between nonclouded image portions because some periods between cloud-free pixel locations can be as long as several months. Often, the change in crop vegetation amount and thus ET$_r$F is uncertain during that period. Thus, the use of curvilinear interpolation can become speculative.

Image processing code can be created to conduct the 'filling' of cloud masked portions of images. The code used with METRIC accommodates up to eight image dates and corresponding ET$_r$F, with conditionals used to select the appropriate set of images to interpolate between, depending on the number of consecutive images that happen to be cloud masked for any specific location. Missing (clouded) ET$_r$F for end-member images (those at the start or end of the growing season) must be estimated by extrapolation of the nearest (in time) image having valid ET$_r$F, or alternatively, for end-member images, a 'synthetic' image can be created, based on daily soil water balance or other methods, to be used to substitute for cloud-masked areas. Often, the availability of images for early spring is limited due to clouds. In these cases, the ET$_r$F values in the synthetic image are based on a soil-water balance–weather data model, such as the FAO-56 evaporation model or Hydrus or DAISY, applied over the month of April, for example, to provide an improved estimate of ET$_r$F over the early season. The synthetic image(s) are strategically placed, date-wise, so that the cloud-filling process and the subsequent cubic spline process used to interpolate final ET$_r$F has end-points early enough in the year to provide ET$_r$F for all days of interest during the growing period.

Examples of cloud masking for a METRIC application in western Nebraska are shown in Figure 6. Black portions within each image are the areas masked for clouds. ET$_r$F for cloud masked areas was filled in for individual Landsat dates prior to splining ET$_r$F between images. The cloud mask gap filling and interpolation of ET between image dates entails interpolating the ET$_r$F for the missing area from the previous and following images that have ET$_r$F for that location.

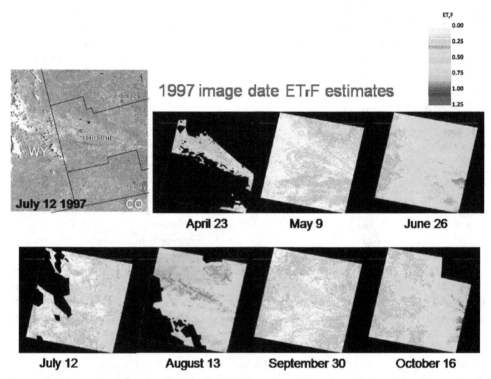

Fig. 6. Maps of cloud masked ET$_r$F from seven 1997 images dates. The geographical extent of the North Platte and South Platte Natural Resource Districts boundaries and principal cities is shown on the image in the top left corner (after Kjaersgaard and Allen 2010).

In current METRIC applications, gaps in the ET$_r$F maps occurring as a result of the cloud masking are filled in using linear time-weighted interpolation of ET$_r$F values from the previous image and the nearest following satellite image date having a valid ET$_r$F estimate, adjusted for vegetation development. The NDVI is used to indicate change in vegetation amount from one image date to the next. The principle is sketched in Figure 7, where a location in the two nearest images (i-1 and i+1) happen to be clouded. During the gap filling, the interpolated values for the clouded and cloud-shadowed areas are adjusted for differences in residual soil moisture between the image dates occurring as a result of heterogeneities in precipitation (such as by local summer showers) in inverse proportion to NDVI and by adding an interpolated 'basal' ET$_r$F from the previous and following satellite image dates. This procedure is needed to remove artifacts of this precipitation-derived evapotranspiration that are unique to specific image dates but that may not be representative of the image date that is to be represented by the ET$_r$F from the previous and the following images. A comparison between cloud gap filling without and with adjustment for background evaporation is shown in Figure 8. An additional example from Singh et al. (2008) is shown in Figure 9 for central Nebraska, where filled in areas that were clouded are difficult to detect due to the adjustment for background evaporation via a daily process model.

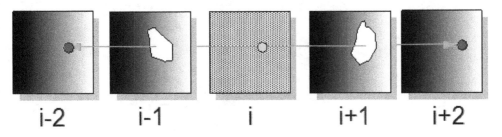

Fig. 7. Principle of cloud gap filling. "i" is the image having cloud masked areas to be filled; "i-1" and "i-2" are the two earlier images than image I; "i+1" and "i+2" are the two following images.

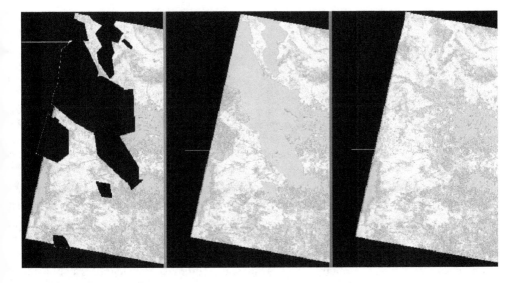

Fig. 8. Maps of ET_rF from Landsat 5, July 12 1997, in western Nebraska after cloud masking (left) (black indicate areas removed during cloud masking or background); and after cloud gap filling without (center) and with (right) adjustment for vegetation amount and background evaporation from antecedent rainfall. The August 13 image from which part of the ET_rF data was borrowed was quite wet from precipitation, and thus had high ET_rF for low-vegetated areas, and therefore created substantially overestimated ET_rF for July 12 in the filled areas (center). After Kjaersgaard and Allen (2010).

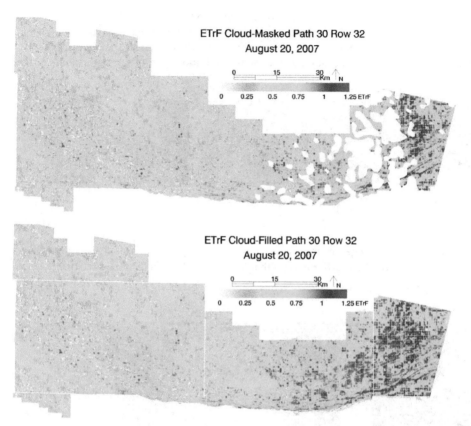

Fig. 9. ETrF product for August 20, 2007 over the Central Platte Natural Resources District, Nebraska, with clouded areas masked (top) and filled (bottom) using a procedure that adjusted for background evaporation from antecedent precipitation events (after Singh et al., 2008).

5. Other remaining challenges with operational models for spatial ET

In addition to challenges in producing daily time series of spatial ET, as described in the previous section, other challenges remaining with all models, snapshot and process models alike include the following. These were described by Allen et al., (2010) and include estimation of aerodynamic roughness at 30 m scale; aerodynamic roughness and wind speed variation in complex terrain and in tall, narrow vegetation systems such as riparian systems; and estimation of hemispherical reflectance from bi-direction reflectance in deep vegetation canopies from nadir-looking satellites such as Landsat. Other remaining challenges include estimation of soil heat and aerodynamic sensible heat fluxes in sparse desert systems and in playa and estimation of ET over 24-hour periods using one-time of day observation (for example ~1000 solar time for Landsat) based on energy balance, especially where substantial stomatal control exists (desert and forest). METRIC capitalizes on using weather-based reference ET to make this transfer over time, which has been shown to work well for irrigated crops, especially in advective environments (Allen et al. 2007a). However, the evaporative fraction, as used in early SEBAL (Bastiaanssen et al. 1998a) and other models may perform best for rainfed systems where, by definition, advection can not exist. Therefore, a mixture of ET_rF and EF may be optimal, based on land-use class.

6. Conclusions

Satellite-based models for determining evapotranspiration (ET) are now routinely applied as part of water and water resources management operations of state and federal agencies. The very strong benefit of satellite-based models is the quantification of ET over large areas. Strengths and weaknesses of common EB models often dictate their use. The more widely used and operational remote sensing models tend to use a 'CIMEC' approach ("calibration using inverse modeling of extreme conditions") to calibrate around uncertainties and biases in satellite based energy balance components. Creating 'maps' of ET that are useful in management and in quantifying and managing water resources requires the computation of ET over monthly and longer periods such as growing seasons or annual periods. This requires accounting for increases in ET from precipitation events in between images. An approach for estimating the impacts on ET from wetting events in between images has been described. This method is empirical and can be improved in the future with more complex, surface conductance types of process models, such as used in Land surface models (LSM's). Interpolation processes involve treatment of clouded areas of images, accounting for evaporation from wetting events occurring prior to or following overpass dates, and applying a grid of daily reference ET with the relative ET computed for an image, or a direct Penman-Monteith type of calculation. These approaches constitute a big step forward in computing seasonal ET over large areas with relatively high spatial (field-scale) definition, where impacts of intervening wetting events and cloud occurrence are addressed.

7. References

Abrahamsen, P. and S. Hansen . 2000. Daisy: an open soil-crop-atmosphere system model. *Environmental Modelling and Software*, 15(3):313-330.

Allen, R.G. 2008, rev. 2010. Procedures for adjusting METRIC-derived ETrF Images for Background Evaporation from Precipitation Events prior to Cloudfilling and Interpretation of ET between Image Dates. Internal memo., University of Idaho. 11 pages. Version 7, last revised April 2010.

Allen, R.G., 2010a. Modification to the FAO-56 Soil Surface Evaporation Algorithm to Account for Skin Evaporation during Small Precipitation Events. Memorandum prepared for the UI Remote Sensing Group, revised May 16, 2010, August 2010, 9 pages.

Allen, R.G. 2010b. Assessment of the probability of being able to produce Landsat resolution images of annual (or growing season) evapotranspiration in southern Idaho – and effect of the number of satellites. Memorandum prepared for the Landsat Science Team. 5 p.

Allen, R.G. and Wright, J.L. (1997). Translating Wind Measurements from Weather Stations to Agricultural Crops. *J. Hydrologic Engineering*, ASCE 2(1): 26-35.

Allen, R.G. and Trezza, R. (2011). New aerodynamic functions for application of METRIC in mountainous areas. Internal memo. University of Idaho, Kimberly, ID. 12 p.

Allen, R. G., Pereira, L., Raes, D., and Smith, M. 1998. *Crop Evapotranspiration*, Food and Agriculture Organization of the United Nations, Rome, It. ISBN 92-5-104219-5. 300 p.

Allen, R.G., W.B.M. Bastiaanssen, M. Tasumi, and A. Morse. 2001. Evapotranspiration on the watershed scale using the SEBAL model and LandSat Images. Paper Number 01-2224, ASAE, Annual International Meeting, Sacramento, California, July 30-August 1, 2001.

Allen, R.G., M.Tasumi, A.T. Morse, and R. Trezza. 2005. A Landsat-based Energy Balance and Evapotranspiration Model in Western US Water Rights Regulation and Planning. *J. Irrigation and Drainage Systems.* 19: 251-268.

Allen, R.G., M. Tasumi and R. Trezza. 2007a. Satellite-based energy balance for mapping evapotranspiration with internalized calibration (METRIC) – Model. ASCE *J. Irrigation and Drainage Engineering* 133(4):380-394.

Allen, R.G., M. Tasumi, A.T. Morse, R. Trezza, W. Kramber, I. Lorite and C.W. Robison. 2007b. Satellite-based energy balance for mapping evapotranspiration with internalized calibration (METRIC) – Applications. ASCE *J. Irrigation and Drainage Engineering* 133(4):395-406.

Allen,R.G., J. Kjaersgaard, R. Trezza, A. Oliveira, C. Robison, and I. Lorite-Torres. 2010. Refining components of a satellite based surface energy balance model to complex land use systems. Proceedings of the Remote Sensing and Hydrology Symposium, Jackson Hole, Wyo, IAHS. Oct. 2010. 3 p.

Anderson, M.C., Kustas, W.P., Dulaney, W., Feng, G., Summer, D., 2010. Integration of Multi-Scale Thermal Satellite Imagery for Evaluation of Daily Evapotranspiration at the Sub-Field Scale. Remote Sensing and Hydrology Symposium 2010. Jackson Hole, Wyoming, USA. p. 62.

Arnold, J.G., J.R. Williams, R. Srinivasan, K.W. King and R.H. Griggs. 1994. SWAT--Soil and Water Assessment Tool--User Manual. Agricultural Research Service, Grassland, Soil and Water Research Lab, US Department of Agriculture.

ASCE – EWRI. (2005). *The ASCE Standardized reference evapotranspiration equation.* ASCE-EWRI Standardization of Reference Evapotranspiration Task Comm. Report, ASCE Bookstore, ISBN 078440805, Stock Number 40805, 216 pages.

Bastiaanssen, W.G.M., M. Menenti, R.A. Feddes, and A.A. M. Holtslag. 1998a. A remote sensing surface energy balance algortithm for land (SEBAL). Part 1: Formulation. *J. of Hydrology* 198-212.

Bastiaanssen, W.G.M., H. Pelgrum, J. Wang, Y. Ma, J.F. Moreno, G.J. Roerink, R.A. Roebeling, and T. van der Wal. 1998b. A remote sensing surface energy balance algortithm for land (SEBAL). Part 2: Validation. *J. of Hydrology* 212-213: 213-229.

Bastiaanssen , W.G.M., E.J.M. Noordman , H. Pelgrum, G. Davids, B.P. Thoreson and R.G. Allen. 2005. SEBAL model with remotely sensed data to improve water resources management under actual field conditions. *J. Irrig. Drain. Engrg,* ASCE 131(1): 85-93.

Brutsaert, W. 1982. *Evaporation into the atmosphere.* Reidel, Dordrecht, The Netherlands.

Brutsaert, W., A.Y. Hsu, and T.J. Schmugge. 1993. Parameterization of surface heat fluxes above a forest with satellite thermal sensing and boundary layer soundings. *J. Appl. Met.* 32: 909-917.

Campbell, G.S. and J.M. Norman. 1998. *An introduction to environmental biophysics.* Sec. Edition. Springer, New York.

Cleugh, H.A., R. Leuning, Q. Mu and S.W. Running. 2007. Regional evaporation estimates from flux tower and MODIS satellite data. *Remote Sens. Environ.* 106:285-304.

Colaizzi, P.D., S. R. Evett, T. A. Howell, J. A. Tolk. 2006. Comparison of Five Models to Scale Daily Evapotranspiration from One-Time-of-Day Measurements. Trans. ASABE. 49(5):1409-1417.

Farah, H.O. 2001. Estimation of regional evaporation under different weather conditions from satellite and meteorological data. A case study in the Naivasha Basin, Kenya. Doctoral Thesis Wageningen University and ITC.

Franks, S.W. and K.J. Beven. 1997. Estimation of evapotranspiration at the landscape scale: a fuzzy disaggregation approach. *Water Resour. Res.* 33:2929-2938.

Gao, F., Masek, J., Schwaller, M., Hall, F., 2006. On the Blending of the Landsat and Modis Surface Reflectance: Predicting Daily Landsat Surface Reflectance. IEEE Trans on Geosci. and Remote Sens. 44, 2207-2218.

Gowda, P.H., J.L. Chávez, P.D. Colaizzi, S.R. Evett, T.A. Howell, and J.A. Tolk. 2008. ET mapping for agricultural water management: present status and challenges. *Irrig. Sci.* 26:223-237

Groeneveld, D.P., W.J. baugh, J.S. Sanderson and D.J. Cooper. 2007. Annual groundwater evapotranspiration mapped from single satellite scenes. *J. Hydrol.* 344:146-156.

Hall, F.G., K.F. Huemmrich, S.J. Goetz, P.J. Sellers, and J.E. Nickeson. 1992. Satellite remote sensing of the surface energy balance: success, failures and unresolved issues in FIFE. *J. Geophys. Res.* 97(D17):19061-19090.

Hook, S. and A.J. Prata. 2001. Land surface termpature measured by ASTER-First results. *Geophys. Res. Abs.,* 26th Gen. Assemb. 3:71.

Jensen, M.E., Burman, R.D. and Allen, R.G. (eds.) (1990). *Evapotranspiration and Irrigation Water Requirements,* ASCE Manuals and Reports on Engineering Practice No. 70. ISBN 0-87262-763-2, 332 p. ASCE, Reston, VA.

Kalma, J.D. and D.L.B. Jupp. 1990. Estimating evaporation from pasture using infrared thermography: evaluation of a one-layer resistance model. *Agr. and Forest Met.* 51:223-246.

Kalma, J.D., T.R. McVicar, and M.F. McCabe. 2008. Estimating land surface evaporation: a review of methods using remotely sensed surface temperature data. *Surv. Geophys* 29:421-469.

Kjaersagaard, J. and R.G. Allen. 2010. Remote Sensing Technology to Produce Consumptive Water Use Maps for the Nebraska Panhandle. Final completion report submitted to the University of Nebraska. 60 pages.

Kustas, W. P., and J. M. Norman. 1996. Use of remote sensing for evapotranspiration monitoring over land surfaces. *Hydrol. Sci. J.* 41(4): 495-516.

Kustas, W. P., Moran, M. S., Humes, K. S., Stannard, D. I., Pinter, J., Hipps, L., and Goodrich, D. C. 1994. Surface energy balance estimates at local and regional scales using optical remote sensing from an aircraft platform and atmospheric data collected over semiarid rangelands. *Water Resources Research,* 30(5): 1241-1259.

Oke, T.R, (1987). *Boundary Layer Climates.* 2nd Ed., Methuen, London, 435 pp, ISBN 0-415-04319-0.

Paulson, C.A. (1970). The mathematical representation of wind speed and temperature profiles in the unstable atmospheric surface layer. *Appl. Meteorol.* 9:857-861.

Qualls, R., and Brutsaert, W. (1996). "Effect of vegetation density on the parameterization of scalar roughness to estimate spatially distributed sensible heat fluxes." *Water Resources Research*, 32(3): 645-652.

Qualls, R., and Hopson, T. (1998). "Combined use of vegetation density, friction velocity, and solar elevation to parameterize the scalar roughness for sensible heat." *J. Atmospheric Sciences*, 55: 1198-1208.

Romero, M.G. (2004) *Daily evapotranspiration estimation by means of evaporative fraction and reference ET fraction.* Ph.D. Diss., Utah State Univ., Logan, Utah.

Rosema, A. 1990. Comparison of meteosat-based rainfall and evapotranspiration mapping of Sahel region. *Rem. Sens. Env.* 46: 27-44.

Šimůnek, J., M.T. van Genuchten, and M. Šejna. 2008. Development and applications of the HYDRUS and STANMOD software packages and related codes. *Vadose Zone J.* 7:587-600.

Singh, R.K., A. Irmak, S. Irmak and D.L. Martin. 2008. Application of SEBAL Model for Mapping Evapotranspiration and Estimating Surface Energy Fluxes in South-Central Nebraska. J. Irrigation and Drainage Engineering 134(3):273-285.

Su, Z. 2002. The surface energy balance system (SEBS) for estimation of turbulent fluxes. *Hydrol. Earth Systems Sci.* 6(1): 85-99.

Tasumi, M., R. G. Allen, R. Trezza, J. L. Wright. 2005. Satellite-based energy balance to assess within-population variance of crop coefficient curves, *J. Irrig. and Drain. Engrg,* ASCE 131(1): 94-109.

Van Dam, J.C., 2000. Field-scale water flow and solute transport: SWAP model concepts, parameter estimation and case studies. Proefschrift Wageningen Universiteit.

Wang, J., W.G.M Bastiaanssen, Y. Ma, and H. Pelgrum. 1998. Aggregation of land surface parameters in the oasis-desert systems of Northwest China. *Hydr. Processes* 12:2133-2147.

Webb, E.K. (1970). Profile relationships: the log-linear range, and extension to strong stability. *Quart. J. Roy. Meteorol. Soc.* 96:67-90.

Wright, J.L. (1982) New Evapotranspiration Crop Coefficients. *J. of Irrig. and Drain. Div.* (ASCE), 108:57-74.

11

Adaptability of Woody Plants in Aridic Conditions

Viera Paganová and Zuzana Jureková
Slovak University of Agriculture in Nitra
Slovak Republic

1. Introduction

Ecological conditions and sources such as water, temperature, solar radiation, and carbon dioxide concentration are factors that limit plant growth, development, and reproduction. Deviations from the optimal values of these factors can cause stress. Plants are subjected to multiple abiotic and biotic stresses that adversely influence plant survival and growth by inducing physiological dysfunctions (Kozlowski & Pallardy, 2002). On the other hand, plants use different strategies for survival that are important for their distribution throughout various regions. Plants differ widely in their ability to adjust to a changing environment and the associated stress (Itail et al., 2002), including the ability to cope with drought (Kozlowski & Pallardy 1997).

Water deficiency is the most significant stress factor for plant growth and reproduction. Drought is mostly associated with the dieback of trees within various regions and throughout the world (Mc Dowel et al., 2008). However, physiological mechanisms of woody plant survival have not yet been described. According to Passioura (2002a), all mechanisms that support physiological functions of plants under conditions of limited water availability are mechanisms of stress resistance. These mechanisms have developed over a long period of time as part of plant adaptability. According to Jones (1993), there are three mechanisms for plant drought resistance. The first mechanism consists of avoiding water deficit and involves the limitation of transpiration and maximisation of root uptake. The second mechanism involves the tolerance to water deficit (Passioura, 2002b; Gielen et al., 2008), and the third mechanism optimises the utilisation of water (Jones 2004).

Plant water stress is the result of a disproportionate balance between the amount of received and released water through various interactions with plant growth, development, and biomass production. The interactions are modified by genetic properties of the specimen and by the character and degree of plant adaptation. The amount of water that a plant can receive depends on the water supply in the soil and on eco-physiological characteristics of plant roots. The transport of water enables for a potential water gradient between the atmosphere and soil, and depends on the hydraulic resistance of the root and stem vascular system. Another component of the water regime of plants – release through transpiration – is a function of the physiological availability and mobility of the water. Plant regulation of the stomata opening and transpiration depend on the pressure potential and other influencing factors. Maintenance of a positive pressure potential is therefore conditional for the survival of plants under drought. The water regime of plants is therefore an ensemble of the physical and physiological rules of the water transport within the soil-plant-atmosphere continuum.

2. Distribution of the wild pear *Pyrus pyraster* (L.) Burgsd. and service tree *Sorbus domestica* L. in Slovakia

The wild pear and service tree are members of the rare woody plants in Slovakia. The wild pear often grows in the scattered vegetation of the landscape, but also on the forest margins mainly in communities of oak stands. The service tree appears mostly in the rural landscape, and mainly in vineyards and fruit orchards. In many European countries, wild pear and service tree are often sought after by landscape designers and foresters, because both species have aesthetic influence in the landscape, a good growth rate, and provide valuable timber.

The vertical distribution of wild pear has been documented mainly at lower altitudes up to 400 m (Hofmann, 1993; Schmitt, 1998; Wilhelm, 1998). The highest location found was at an altitude of 754 m in bundesland Süd-Niedersachsen und Nordhessen (Schmitt, 1998).

In Slovakia, wild pear grows in the lowlands to sub-mountain areas, up to an approximate altitude of 950 m (Peniašteková, 1992), and in some cases up to an altitude of 1163 m (Blattný & Šťastný, 1959). A detailed study on the environmental conditions of stands where wild pear naturally occurs was conducted in 1994-1999 (Paganová, 2003). The basic data were obtained from 64 locations (Fig. 1).

Stands with wild pear were located mostly on grazing lands, meadows, and in the scattered woodlands. Wild pear populations were also found along a dry stream channel (locality 8) and in a thin forest (location 21). Wild pear often grows on the forest edge (locations 30, 34, 41, and 48), or on former grazing land that gradually changed to woodlands (location 42, 55, and 56).

The majority of stands with wild pear (80%) were found at altitudes up to 500 m. The lowest location in Slovakia was at an altitude of 100 m (12 Solnička), and the highest analysed stand was at an altitude of 800 m (19 Jezersko) (Paganová, 2003).

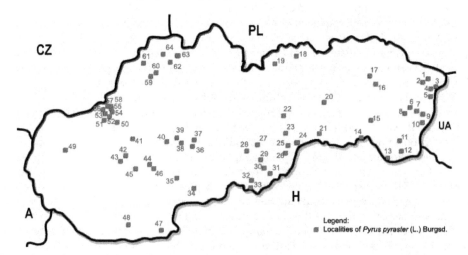

Fig. 1. Distribution of the wild pear *(Pyrus pyraster)* populations in the territory of Slovakia (Paganová, 2003).

The service tree is one of the rare autochthonous woody plants in the entire area of natural distribution. The area of natural distribution of the service tree reaches the northern part of Asia Minor and Africa as well as the northern border crosses of North Rhine-Westphalia, Lower Saxony, Saxony-Anhlat, and Thüringen, Bavaria. The northernmost occurrence is located in the Federal Republic of Germany at approximate latitude 51° of north width (Haeupeler & Schönfelder, 1988), and then continues to South Moravia and Slovakia, Hungary, Romania, and Crimea Mt.

According to Májovský (1992), the service tree has higher demands for light and high temperatures. In Slovakia, it is cultivated in the uplands on sunny south and southwest exposed stands. The vertical distribution of this woody plant occurs at an altitude of 109 m (Benčať, 1995) or 175 m (Michalko, 1961) up to an altitude of 610 m (Michalko, 1961; Benčať, 1995).

In 1996-2000, the environmental conditions of 24 locations of the service tree were analysed (Paganová, 2008). In Slovakia, the service tree appears in the southern regions in warmer stands at lower altitudes (Fig. 2). The distribution of the analysed stands containing the service tree confirmed its occurrence mainly at lower altitudes. The lowest stand with the service tree was found at an altitude of 200 m (location 23 - Vinné), and the highest stand was found at an altitude of 490 m (location 1 - Predpoloma) (Fig. 2).

The majority of the analysed stands containing the service tree (50%) were located in an open landscape near vineyards and fruit orchards (location 10, 11, 12, 13, 14, 15, 16, 17, 18, 19, 21, and 22). The service tree was frequently (46% of analysed stands) found in abandoned fruit orchards or on grazing lands as well (location 1, 2, 3, 4, 5, 6, 7, 19, and 20). Only a few plants were found in woodlands (location 23 and 24) and one stand of service tree (location 8) was located in an oak forest.

According to the analysis of the vertical distribution (Fig. 3), the service tree grows mainly on uplands in Slovakia. Approximately 66% of the analysed stands were found at an

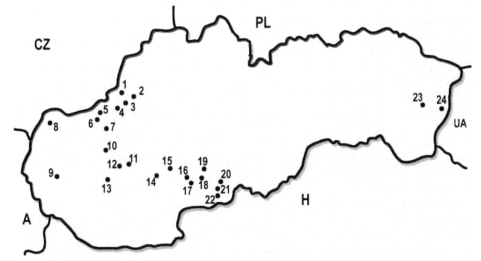

Fig. 2. Location of the stands with a higher number of service trees (*Sorbus domestica*) in Slovakia (Paganová, 2008).

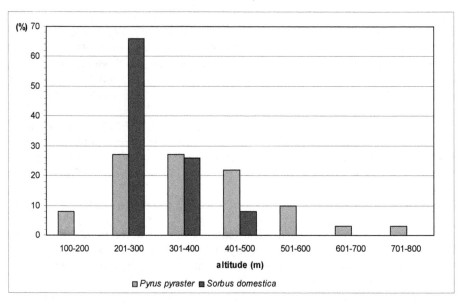

Fig. 3. Distribution of wild pear (*Pyrus pyraster*) and service tree (*Sorbus domestica*) in Slovakia according to stand altitude.

altitude of 201–300 m, and 26% of the stands were at an altitude of 301–400 m. One location was at altitudes of 450 m and 490 m. Compared to the wild pear (Fig. 3), the service tree is absent from the lowlands, and the occurrence of this woody plant at altitudes above 400 m is rare. According to Kárpáti (1960), the service tree is frequently found within communities of oak forests at lower altitudes and on fertile soils in communities of *Lithospermo-Quercetum, Melico (uniflorae)-Quercetum petraeae,* and others. Michalko (1961) confirmed the findings by Kárpáti, but according to his opinion, the service tree grows at higher altitudes only in extreme communities of *Corneto-Quercetum (pubescentis* and *petraeae)*, and can even be found in the beech woodland *Corneto-Fagetum* and in relict pinewoods growing on limestone and dolomite parent rock.

Similar to the data mentioned above, very similar findings regarding the range of altitudinal distribution were found in Switzerland. In this country, the service tree was found within an altitude of 384 m in the Basel region and 675 m in the Schaffhausen region (Brütsch & Rotach, 1993). In the southeast section of the Wiener Wald in the area of Merkenstein, the service tree has been found up to an altitude of 550 m (Steiner, 1995). At the northern border of its natural distribution in Germany in the region of Sachsen-Anhalt, the service tree is distributed from 140 m to 310 m, and predominantly within an altitude of 161-240 m (Steffens, 2000). On the Plateau of Lorraine, the service tree appears in forest crops at an altitude of 200-400 m (Wilhelm, 1998).

In southern regions of the natural distribution, the service tree grows at higher altitudes than in Slovakia. For example, in Spain, it grows at altitudes up to 1400 m, in Greece up to 1350 m, in Turkey up to 1300 m, and in southern Bulgaria from 300 to 800 m (Kausch, 2000). In southern Italy (Mt. Vesuvius), the service tree grows from the banks of the sea up to an altitude of 800 m (Bignami, 2000).

3. Ecological characteristics of the stands with *Pyrus pyraster* (L.) Burgsd. and *Sorbus domestica* L.

The wild pear is considered to be a light-demanding woody plant, which prefers warm stands with a sufficient amount of sunlight (Namvar & Spethman 1986; Hofmann, 1993; Wagner, 1995; Kleinschmit & Svolba 1998; Schmitt, 1998; Rittershoffer, 1998; Roloff, 1998; Wilhelm, 1998). Hofmann (1993) previously created a diagram for the occurrence of 300 wild pear plants according to the stand exposure. The plants were predominantly in locations with south and southwest exposures. In support of these findings, Roloff (1998) also found that the most frequent occurrence of wild pear plants was on slopes with a south or west exposure.

The service tree is explicitly regarded as a light-demanding woody plant (Michalko, 1961; Májovský, 1992; Brütsch & Rotach, 1993; Pagan, 1996; Wilhelm, 1998). In Slovakia, 96% of stands with the service tree were found in the open landscape with solitary trees. Two stands were on the margin of woodlands with a few service trees in the crop, and only one location was an oak forest, with service trees found in the upper tree canopy or slightly above it. In all of these stands, the individual service trees grew under nearly full light without competition from other woody plants (Paganová, 2008). The service tree is intolerant to shading at an early age, and similar to the wild pear, will die quickly without a minimum light supply (Wilhelm, 1998).

Based on data obtained on the distribution of 507 wild pear plants in Slovakia, the majority of stands (80%) had a south, southeast, or southwest exposure. However, a limited number of stands (14%) containing wild pears were also found to have west, east, or northwest exposures, and four locations (6%) were on a plain stand (Fig. 4).

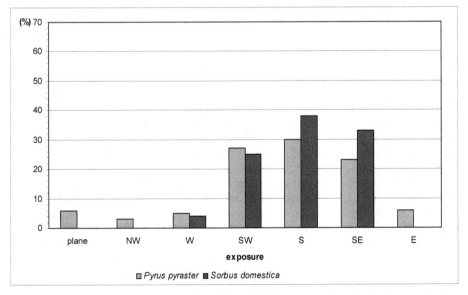

Fig. 4. The distribution of locations with wild pear (*Pyrus pyraster*) and service tree (*Sorbus domestica*) in Slovakia according to stand exposure.

In comparison to the wild pear, a majority of stands with service trees (38%) had a southern exposure, and many locations also had southeast (33%) and southwest (25%) exposures. One location (4%) had a western exposure. According to measurements by Geiger (1961), slopes with northern exposure obtain just half of the absolute total light emission as slopes with southern exposure. The prevalent distribution of service trees in southern-exposed stands supports the hypothesis regarding their high demand of light and warm climate. In Slovakia, none of the analysed locations had northern exposure. In Switzerland, 74% of the locations with service trees had southern exposure (Brütsch & Rotach, 1993).

The ecological-climatic amplitude of the wild pear locations in Slovakia is relatively wide. In stands with wild pear plants, the conditions range from plain and fold climates, to a mountain climate (Paganová, 2003). Stands were classified to climate-geographic types and subtypes according to Tarábek (1980) and Špánik et al. (1999).

Within the analysed scale of the wild pear stands in Slovakia, the average January temperatures range from -1.4°C to -5.8°C and the average July temperatures range from 13.5°C to 20.4°C. The annual sum of precipitation reaches values ranging from 570 mm to 900 mm. The majority of the pear locations (53%) fall within the climate-geographic type of mountain climate (Fig. 5), which is humid or very humid with rare temperature inversion. These stands were most frequently found in the warm and moderately warm subtypes of the mountain climate at an altitude of 250-550 m. The average January temperatures range from -1.4°C to -5.0°C, the average July temperatures range from 17.0°C to 20.4°C, and the annual sum of precipitation for these stands ranges from 580 mm to 790 mm.

Stands within the warm subtype of the fold climate are observed quite frequently. The fold climate is semi-humid to semi-arid with a remarkable inversion of temperatures. These locations are at altitudes of 210-450 m. The average January temperatures range from -2.0°C to -4.0°C, the average July temperatures range from 18.0°C to 19.2°C, and the annual sum of precipitation ranges from 628 mm to 765 mm in the respective locations.

The lowest number of wild pear locations was documented for stands in the warm or mostly warm subtypes of the plane climate, which is arid and semi-arid. Locations were registered at altitudes of 120-400 m. The average January temperatures in these stands range from -1.5°C to -3.3°C, and the average July temperatures range from 17.2°C to 20.1°C. The annual sum of precipitation is 570-700 mm.

The climate–geographic characteristics of stands with service trees are slightly different than stands containing wild pears (Fig. 5). The climate with the highest number of locations (42%) belongs to the mostly warm subtype of the plane climate, which is arid or semi-humid with a mild inversion of air temperatures. These stands are at an altitude of 200-300 m. The average annual temperature ranges from 8.3°C to 9.0°C, and the annual sum of precipitation is 610–650 mm.

A high number of stands (33%) were classified in the mountain climate with a mild inversion of air temperatures (the climate is rather humid). These stands are at an altitude of 220-490 m. The average annual temperature is 8.3°C and the annual sum of precipitation ranges from 650 to 620 mm.

The climate with the fewest locations of service trees was the fold climate (25%), which has a markedly high inversion of air temperatures (with an arid or even humid climate). The stand altitude ranges from 250 m to 380 m, the annual average temperature ranges between 8.1–8.5°C, and the annual sum of precipitation is 620–700 mm.

According to a recent climatic evaluation (Škvarenina et al., 2004), the annual average temperature of the majority of locations with service trees is above 8°C, with the exception

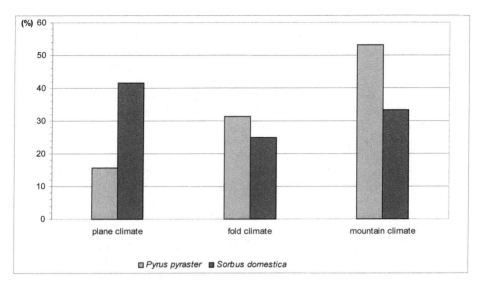

Fig. 5. Wild pear (*Pyrus pyraster*) and service tree (*Sorbus domestica*) stand classification in Slovakia according to climate-geographic types.

of two of the analysed locations (1 and 5), where the annual average temperature ranges from 7.5°C to 7.7°C. The average annual sum of precipitation for the majority of the stands is 610–700 mm, with the exception of the two mentioned locations, where this parameter reaches 790 mm and 750 mm, respectively. The potential evapotranspiration amount in the majority of the analysed stands was 600–750 mm during one year. Considering that the annual average sum of precipitation is 610–700 mm, it is possible that the service tree has to obtain enough moisture during the main growing season predominantly from water resources in soil. The deficit of rain during the summer occurs in the majority of the stands with this woody plant.

Warm and arid (southeast, south, southwest, and even west) stand exposures play an important role in the formation of the arid microclimate and mezzo-climate of the mentioned locations. According to climate classification in Slovakia (Špánik et al., 1999), the analysed locations with service trees were in warm and even moderately warm as well as semi-humid and even semi-arid climates. Compared to the wild pear, the service tree prefers stands at lower altitudes and is prevalent in warm and arid climates. The wild pear has wider ecological amplitude and grows at higher altitudes in stands with different water regimes and climate extremes.

Based on the brief pedology characteristics of our experimental plots, we can hypothesize that these soils are very well fertile (Chernozem, Fluvi-mollic soils, Cambisols, and Orthic Luvisols) or well supplied with nutrients (Luvisols, Pararendzinas, and Fluvisols). In addition, browned Rendzinas can be considered as relatively favourable soils.

According to the ecological scheme of Ellenberg (1978), the wild pear is a woody plant with broad ecological amplitude that grows in nearly all soil types, with the exception of extreme acidic soils. Rittershoffer (1998) found that mildly acidic or mildly alkaline soils were optimal for wild pear growth. According to information from Westfalen-Lippe (Germany), the wild pear prefers soils developed on limestone or on the rich nutrient parent rocks (80% of stands) (Schmitt, 1998). In the area of Süd-Niedersachsen und Nordhessen (Germany), the wild pear frequently grows in shallow rendzinas from mussel limestone or from lime sandstone, and very rarely appears on deeper brown forest soils. More than 92% of all natural stands were on rich basic rocks (Hofmann, 1993). In the forest on the Plateau Lorraine, the wild pear grows mainly on parent rock of the mussel limestone and on keuper sediments. There are deep terra fusca soils and shallow Rendzinas (Wilhelm, 1998). The colective data from different areas of the natural distribution indicate that suitable growth conditions for the wild pear are mainly basic and rich nutrient soils with occasional water deficits.

In Slovakia, the wild pear grows on fertile soils (Chernozem, Fluvi-mollic soils, Cambisols, and Orthic Luvisols), or soils well supplied with nutrients (Albic Luvisol, Pararendzina, and Fluvisol). In addition, in some stands it grows on soils that are rich in minerals but under conditions of unbalanced soil chemistry with little fertility (Paganová, 2003).

In general, Fluvi-mollic and Cambisol soils have a sufficient water supply. At lower altitudes, the water deficit appears mainly in Rendzinas. The water deficit in Luvisols is usually a result of a lower amount of precipitation and higher evaporation. Orthic Luvisols have a lower water supply, and therefore the possibility of their aridization is higher. In addition, a fluctuating water regime appears within the Planosols and Fluvisols (Šály, 1988).

According to the ecological scheme by Ellenberg (1978), the wild pear has optimal growth conditions on fresh basic soils (its potential optimum). Another more frequent existence optimum of this woody plant is near the xeric forest limit, where the wild pear grows in arid soils rich in bases as well as in moderately acidic soils. Some authors have placed the wild pear among xerophytic woody plants according to its lower demands on soil humidity (Bouček, 1954). When under competition with some woody plants, it grows on its synecological optimum in extreme arid stands - rocky hilltops, stands of the xeric forest limit close to steppe communities, and in the sparse xerophytic oak woodlands (Rittershoffer, 1998). However, another existence optimum of this woody plant occurs in the hydric forest boundary in stands of the hardwood floodplain forests, where wild pear growth is limited by inundation (Rittershoffer, 1998). Based on these findings, the wild pear is a flexible woody plant with tolerance to a large range of soil humidity.

In Slovakia, stands with the service tree have favourable physical characteristics, good saturation, and are very fertile (Orthic-Luvisols and Cambisols), or have soils that are well supplied with nutrients (Rendzinas). However, under conditions of unbalanced soil chemistry, there is little fertility, and the pH of this soil is moderately acidic, neutral, or moderately alkaline. Cambisols generally have a sufficient water supply, and Orthic-Luvisols have a lower water supply with the possibility of aridization. Water deficiency can appear in Rendzinas as a result of the water penetration, so the water supply in this soil is usually low (Šály, 1988).

According to Wilhelm (1998), the service tree grows on mussel limestone and on keuper sediments on the Plateau Lorraine. These soils are well or very well supplied with nutrients. On slopes based with mussel limestones are deep terra fusca soils, and in the upper parts of the slopes are shallow Rendzinas. On the keuper, there are abundant, deep Vertic Cambisols with water deficiency during summer.

In east Austria, the service tree grows in the oak forest communities and is considered to be a woody plant of the uplands with less demands on soil humidity, but with quite high demands on the nutrient content of the soils (Kirisits, 1992). In the Wiener Wald (Steiner, 1995), the service tree appears on limestone and dolomite parent rock with prevalence in semi-humid and arid Rendzinas. In addition, the service trees in Switzerland are found mainly in arid soil with less skeleton that is rich in bases (Landolt, 1977; Brütsch & Rotach, 1993), as determined from detailed studies of the service tree stands in Canton Genf, which refer to the medium deep and deep skeletal Cambisols and Luvisols with slower water penetration and possible water logging. In the Bassel region, the service tree also grows on Rendzinas or Lithosols, which are shallow and extreme skeletal soils that have a very low water capacity. In the Schaffhausen, approximately 92% of the service tree plants grow on limestones, and the rest of the stands grow on gravels of the high terrace that belong to Riss. In the deeper strata, there are limestones that are part of the morena and gravels. Various soils, even acidic soils, can appear randomly on small areas of the parent rocks . These data document quite a broad range of soil conditions for the stands containing service trees, and tolerance of the taxon to periodic or rare occurrences of water deficit in the soils is evident. On some stands within the area of its natural distribution, the service tree grows under conditions of a soil drought.

4. Potential adaptability of the analysed woody plants to progressive drought

Drought can be considered in meteorological, agricultural, hydrological, and socio-economic terms (Wilhite & Glantz, 1985). Meteorological drought reflects one of the primary causes of drought. It is usually defined as precipitation less than a long-term average (defined as normal) over a specific period of time. Agricultural drought is expressed in terms of the moisture availability at a particular time during the growing season for a particular crop. Hydrological drought is usually expressed as a deficiency in surface and subsurface suppliers, and refers to a period when stream flows are unable to supply the established users under a given water management system. Socio-economic definitions of drought relate to the supply and demand of specific goods. Importantly, humans can create a drought situation through land-use choices or an excess demand for water (Wilhite & Glantz, 1985).

According to Škvarenina et al. (2009a), drought is a temporary aberration that differs from aridity, which is restricted to low rainfall regions and is a permanent feature of the climate. The altitude and topography are significant climate-differentiating factors. In Slovakia, a considerably broken topography plays an important role in the variability of climate conditions. The increase in altitude causes changes in solar radiation as well as thermal and water balance of the land (Škvarenina et al., 2009a). Vertical differentiation of the climate conditions has a significant influence on species structure of the natural vegetation. The biogenocenoses can be classified into nine vegetation stages described by Zlatník (1976) based on altitude, exposure, and topography, which are named after woody plants that are dominant in the area.

Škvarenina et al. (2009a) analysed trends in the occurrence of dry and wet periods in altitudinal vegetation stages in Slovakia between 1951 and 2005. The authors considered relative evapotranspiration (E/E_0), which is defined as the rate of the actual evapotranspiration (E) to potential evapotranspiration (E_0), as an excellent measure of water sufficiency for vegetation. According to their findings, the smallest annual values of (E/E_0) were recorded in the Danube lowland (1st Oak vegetation stage) with relatively high totals of potential evapotranspiration (E_0) above 700 mm and with annual precipitation totals below (P) 550 mm. The lowest value of the relative evapotranspiration (approximately 60%) was recorded in the lowest areas of Slovakia with an altitude up to 200 m. Relative evapotranspiration reached higher values towards higher vegetation stages (above 90% in the 4th Beech vegetation stage at altitudes above 650 m).

In addition to relative evapotranspiration, the drought index (E_0/P) has also been used to describe the relationship between the energy and precipitation (P) inputs within particular vegetation stages. Warm forest-steppe stands in Slovakia with oak communities have drought index values (E_0/P) of approximately 1. The predominant areas of Slovak forests are stands with drought index values up to 0.3. Moreover, the vegetation stages with $E_0/P <$ 0.3 are within the mountain climate (Škvarenina et al., 2009b).

In Slovakia, wild pear stands are distributed from lowlands up to an altitude of 800 m. Specimens also appear in 1st (oak) and 2nd (beech-oak) vegetation stages with a water deficit during the growing season. The stands in these vegetation stages are classified as a territory with a dry (arid) climate according to the relative evapotranspiration and drought index. On the other hand, the wild pear is also distributed in stands at higher altitudes in the 4th (beech) and 5th (fir-beech) vegetation stages, which have a higher humidity (higher relative evapotranspiration). This type of distribution shows that the wild pear is tolerant to different conditions of water sufficiency.

The service tree is predominantly distributed in the 1st (oak), 2nd (beech oak), and 3rd (oak-beech) vegetation stages in Slovakia, avoids lowland stands, and appears mainly on slope terrain of the forest steppe stands. This taxon often grows in conditions of warm oak communities with an arid climate. At higher altitudes, the service tree most likely avoids the consequences of a strong beech competition. In the Slovak lowlands, the absence of the service tree is most likely due to the higher underground water level and the intensive agricultural utilization of the land.

According to Škvarenina et al. (2009a), a markedly severe drought between 1951 and 2005 was only identified in the Danube Lowland (1st Oak vegetation stage) and in the Záhorská lowland (2nd Beech-oak vegetation stage) of Slovakia. Considering the natural distribution of the wild pear and tolerance to a wider range of water supply, this woody plant has the potential to adapt to the decreasing humidity of the Danube Lowland. The service tree has similar qualities and the potential to grow in arid conditions; however, this taxon is mainly found on the slopes of forest-steppe stands.

According to a drought analysis of the Slovak territory conducted on the climatic data obtained from 1960-1990, agricultural regions become more sensitive to conditions of climate change upon drought occurrence (Šiška & Takáč, 2009). The authors used two indices for spatial evaluation of drought conditions in Slovakia: the climatic index of drought and the evapotranspiration deficit. The climatic index of drought (K) was applied for the entire growing season (GS10 period) and $K_{GS10} = \Delta E$, where E_0 is the potential evapotranspiration during GS10 and R is the rainfall during GS10. The evapotranspiration deficit ΔE during the growing season was calculated as $\Delta E_{GS10} = E_0 - E$, where E_0 is the potential evapotranspiration during the main growing season (GS10) and E is the actual evapotranspiration during the main growing season.

Two very dry and hot regions were classified in Slovakia, the Danubian and east Slovakian lowlands, which represent maize production areas with a water deficit that exceeds 250 mm during the growing season. These evapotranspiration deficit values will most likely be present in river valleys up to altitudes of 300 m as well (Šiška & Takáč, 2009).

The findings described here support the hypothesis that a higher frequency of drought occurs in agroclimatic regions of the Slovak Republic. In the future, it is important to elaborate on several concepts of the stabilization of agricultural production against water deficit and soil aridity. With the exception of breeding programs that focus on developing new crop varieties that can tolerate the changed climatic conditions and development of integrated irrigation systems, there are also possibilities for landscape stabilization using non-forest woodlands. These types of woodlands should be established with woody plants that are tolerant to water deficit and that are adaptable to dynamic changes of water regimes. The taxa analysed here, including the wild pear and service tree, belong among the prospective woody plant species that are suitable for planting in regions potentially endangered by droughts.

The described research focused on an analysis of the physiological parameters of two woody plant species (wild pear and service tree) under conditions of a regulated water regime and water stress. The aims of the study were to verify the adaptive potential of both taxa to drought, and to obtain information on the mechanisms used by these woody plants under conditions of water deficit.

5. Interspecific differences of the selected physiological parameters of woody plants

Woody plants make different ecological adjustments to water deficit, and can modify their physiological functions and anatomical structures for adaptation. Adaptability is a rather complex quality, and the explicit function of a typical plant response to water deficit is very difficult to define. Therefore, we established experiments that regulated the water regime of juvenile (two-year old) wild pear and service tree plants under semi-controlled conditions.

The plants were planted in pots (content 2 L) with mixed peat substrate enriched with clay (content of clay 20 kg.m^{-3}; pH 5.5-6.0; fertilizer 1.0 kg.m^{-3}). The potted plants were placed under a polypropylene cover with 60% shading. The plants were regularly watered and maintained on 60% of the full substrate saturation for 28 days. In the phenological stage of shoot elongation (at the beginning of June), the plants of both taxa were divided in two variants according to a differentiated water regime. Variant "stress" was supplied with water at 40% of full substrate saturation and "control" at 60% of full substrate saturation. The model of the differentiated water regime was maintained for 126 days (to the end of September). Sampling was performed at 14 day periods for both conditions.

The size of the leaf area (A) and leaf water content (LWC) were measured, and a determination of fresh weight (F_W) and dry weight (D_W) was done gravimetrically. The size of leaf area (A) was calculated from leaf scans using ImageJ software (http://rsbweb.nih.gov/ij/). The LWC and specific leaf area (SLA) were calculated according to the methods described by Larcher (2003). For metabolic characteristics, the total chlorophyll and carotenoid content were determined according to the methods described by Šesták & Čatský (1966).

Data were analysed from three growing seasons in 2008-2010 for each taxon under two variations of water regimes (40% and 60% substrate saturation). The relationship between SLA and LWC of the plants under stress and control conditions as well as changes in the assimilatory pigments during water stress were also analysed. A statistical assessment of these parameters was conducted by regression analysis using the statistical software Statgraphics Centurion XV (StatPoint Technologies, USA). A $P < 0.05$ was consisted statistically significant.

5.1 The influence of water stress on the production of leaf dry mass

The different reactions of the analysed taxa (wild pear and service tree) to water stress were confirmed by the dry mass (DM) measurements taken under controlled and stress experimental conditions (Table 1). Under control conditions, the increment of leaf dry mass of the wild pear was 14.67 mg p^{-1} d^{-1} and the increment of leaf dry mass of the service tree was 18.37 mg p^{-1} d^{-1}. Under conditions of water deficit (stress), the increment of the leaf dry mass of wild pear plants was 12.78 mg p^{-1} d^{-1} and the increment of leaf dry mass for the service tree plants decreased to 3.04 mg p^{-1} d^{-1}. The impact of water stress on the wild pear was less significant, and this plant is probably more tolerant to drought. Importantly, the relationship to photosynthesis economy depends on the leaf structure. The wild pear is a typical taxon of sunny and arid stands, and contains heterobaric leaves. Parenchyma (or sclerenchyma) cells without chloroplasts accompany the vascular system, and similar to ribs, lead to the top (adaxial) or bottom (abaxial) epidermis (Essau, 1977; Fahn, 1990; Terashima, 1992). The tips (ribs) of the vascular bundles divide leaf mesophyll hermetically into compartments that are reciprocally isolated against gas exchange. In the compartments, the intercellular space is relatively small with low chlorophyll content. The compartments are similar to "open windows", which transmit visible light into the internal layers of the mesophyll (Liakoura et al., 2009). Heterobaric leaf structures are also significant because they allow for easier transport of water to the epidermis due to increased hydraulic conductivity. One predominant factor that limits plant transpiration is leaf area. The reduction of leaf area during water deficit is typical for plants from arid stands. Several authors (Reich et al., 2003, Wright et al., 2004; Niclas & Cobb, 2008) have confirmed the narrow relationship between leaf structure and function. Our comparison of the leaf area ratio to dry weight of the leaves (SLA) of the analysed species confirmed the interspecific differences (Table 1). Wild pear leaves with higher values of SLA were thinner than leaves of the service tree under control conditions. The leaf water content per unit of dry weight in pear leaves was higher than service tree leaves. In experiments with fast growing woody plants, Dijkstra (1989) confirmed the thinner leaves of these species as well as the presence of larger vacuoles in the cells, which accumulate a larger amount of water per unit of dry mass. In our experiments with wild pear, the values of SLA decreased after 70 days under both conditions (stress and control), and the pear leaves became xeromorphous. There were no significant differences in SLA values of the pear leaves after 70 days under the differentiated water regime or due to water stress (Table 1).

The different functional qualities of the leaves can be effected by 1) changes in the leaf structure, and 2) different compositions of the leaf, including sclerenchyma elements and organic compounds (lignins and phenols), which increase leaf dry mass as described by Mooney & Gulmon (1982) and Lin & Harnly (2008).

Interspecific differences in the reaction to water deficit were not confirmed in the analysed taxa of this study. However, at the beginning of the experiments and after 70 days of cultivation, the values of LWC of the wild pear and service tree plants were different, and these values did not change under conditions of water stress (Table 1).

Based on our analysis of the relationship between SLA and LWC, both of the analysed taxa maintained higher LWC with increasing values of the specific leaf area, regardless of the level of substrate saturation (Fig. 6, 7, 8, and 9). In addition, a significant linear correlation was observed between SLA and LWC under control and stress conditions without interspecific differences.

Physiological characteristics	Taxon							
	Pyrus pyraster				Sorbus domestica			
	control		stress		Control		stress	
	0 day	70 day	0 day	70 day	0 day	70 day	0 day	70 day
Size of the leaf area A (mm²)	23 312	34 570	23 312	32 578	59 499	78 713	59 499	51 210
Specific leaf area SLA (mm².mg⁻¹)	19.13	15.14	19.13	15.80	16.57	16.71	16.57	16.06
Leaf dry weight DW_1 (mg)	1 176	2 203	1 176	2 071	3 488	4 774	3 488	3 275
Leaf water content LWC (%)	66.3	57.0	66.3	57.6	45.4	52.2	45.4	51.5
Chlorophyll content (mg.mm⁻²)	515.7 .10⁻⁶	679.2 .10⁻⁶	515.7 .10⁻⁶	779.7 .10⁻⁶	333.9 .10⁻⁶	470.5 .10⁻⁶	333.9 .10⁻⁶	452.0 .10⁻⁶
Carotenoid content (mg.mm⁻²)	110.2 .10⁻⁶	138.4 .10⁻⁶	110.2 .10⁻⁶	147.4 .10⁻⁶	76.3 .10⁻⁶	105.4 .10⁻⁶	76.3 .10⁻⁶	101.6 .10⁻⁶

Table 1. Physiological characteristics of leaves taken from 2-year old potted plants of wild pear (*Pyrus pyraster*) and service tree (*Sorbus domestica*) grown in conditions of differentiated water regime - control (60% of the full substrate saturation) and stress (40% of the full substrate saturation) conditions.

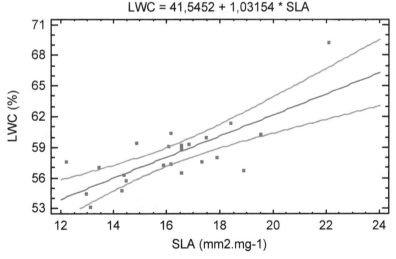

Plot of Fitted Model for Pyrus pyraster with 40% saturation of the substrate
LWC = 41,5452 + 1,03154 * SLA

Fig. 6. Positive linear correlation between SLA (mm².mg⁻¹) and LWC (%) of wild pear (*Pyrus pyraster*) leaves under conditions of water stress. Correlation coefficient (r) = 0.760432, p value = 0.0000.

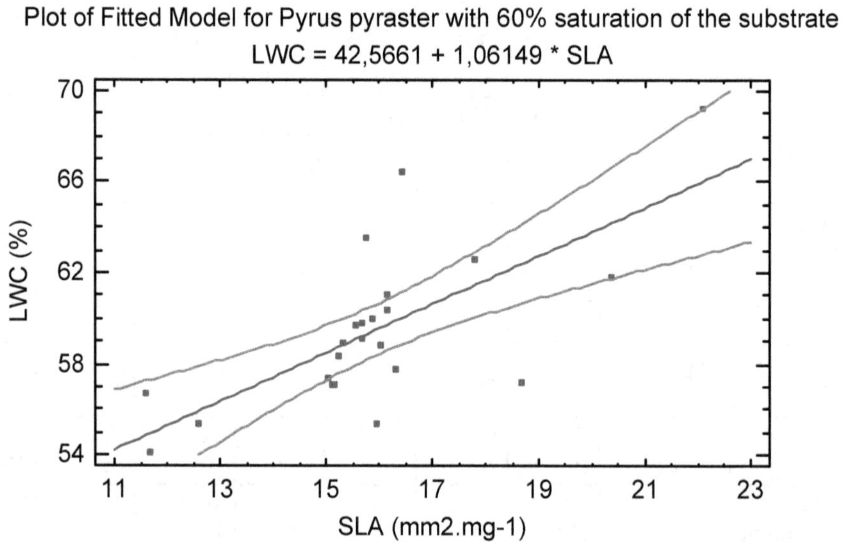

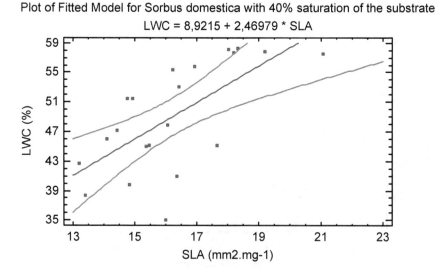

Fig. 7. Positive linear correlation between SLA (mm2.mg-1) and LWC (%) of wild pear (*Pyrus pyraster*) leaves under control conditions. Correlation coefficient (r) = 0.704177, p value = 0.0002.

Fig. 8. Positive linear correlation between SLA(mm2.mg-1) and LWC (%) parameters of service tree (*Sorbus domestica*) leaves under water stress. Correlation coefficient (r) = 0.669898, p value = 0.0009.

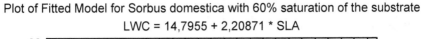

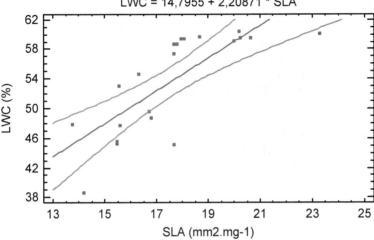

Fig. 9. Positive linear correlation between SLA (mm².mg⁻¹) and LWC (%) parameters of service tree (*Sorbus domestica*) leaves under control conditions. Correlation coefficient (r) = 0.76925, p value = 0.0000.

5.2 Changes in the assimilatory pigment content in leaves under conditions of water stress

The content of assimilatory pigments is an important factor that has a significant influence on thermal characteristics of the leaves. Leaves with lower chlorophyll content have higher reflexion, and the leaf surface temperature can have relatively lower values than the temperature of leaves with a higher content of assimilatory pigments. In addition, leaves with a higher content of carotenoids should have a relatively higher resistance against water stress. On the other hand, the ability of a plant to maintain a higher content of assimilatory pigments during stress can be very important for the functional activity of the leaves. Our analysis confirmed a different content profile of assimilatory pigments (chlorophyll a and chlorophyll b), β-carotene, and neoxantine in the leaves of the wild pear and service tree. There was a significant positive linear correlation between carotenoid and chlorophyll content in the leaves of both analysed taxa, regardless of the level of water saturation of the substrate (Table 2). This relationship is illustrated in Figure 10 for the wild pear plants at 40% substrate saturation. The results of the regression analysis for the wild pear under the control condition as well as for the service tree under both conditions are shown in Table 2. The SLA values of the service tree leaves did not change significantly under the differentiated water regime or under conditions of water stress (Table 1). The values of SLA in wild pear leaves decreased during the differentiated water regime under both conditions (control and stress). The decrease of SLA was most likely influenced by the specific quality of the taxon, which produces so called "summer leaves" during twig elongation. Two-year old plants of the service tree created leaves on terminal shoots only, and the values of SLA were not significantly changed in both variants of the water regime (control and stress) within the analysed period of time. During summer, the chlorophyll content in leaves of the wild pear increased under control and water stress conditions. The chlorophyll content in

taxon/substrate saturation	wild pear/40	wild pear/60	service tree/40	service tree/60
correlation coefficient r	0.973681	0.964724	0.982228	0.974045
p value	0.0002	0.0004	0.0005	0.0001

Table 2. Results of a simple regression between total chlorophyll content and carotenoid content in leaves of the analysed taxa wild pear (*Pyrus pyraster*) and service tree (*Sorbus domestica*) under two conditions of substrate saturation. Legend: 40 – conditions of water stress (40% substrate saturation); 60 – control conditions (60 % of substrate saturation).

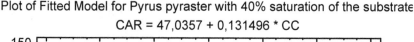

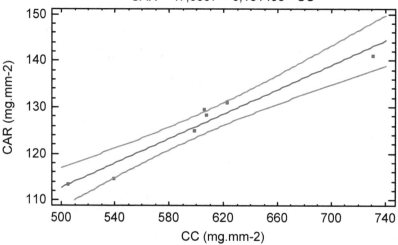

Fig. 10. Positive linear regression between total chlorophyll content (CC) and carotenoid content (CAR) in leaves of wild pear (*Pyrus pyraster*) plants growing under conditions of water stress. The correlation is quite close, with a correlation coefficient (r) = 0.973681 and statistically significant p value = 0.0002.

service tree leaves also increased; however, under water stress conditions, the chlorophyll content was lower than in the leaves of the control plants.

We confirmed a statistically significant relationship between SLA values and chlorophyll content in the leaves of the service tree under conditions of water stress, and this relationship was described by a polynomial curve of the second order (Figure 11). These data showed that the service tree maintained a balanced content of chlorophyll in leaves with a lower specific leaf area. In the stress variant, the chlorophyll concentration in service tree leaves varied between 340-470 mg.mm^{-2} within a 95% confidence level.

The relationship between SLA and chlorophyll content in the leaves of the wild pear under water stress conditions was also described as a polynomial function of the second order (Figure 12). However, this relationship was not significant. The leaf chlorophyll concentration ranged between 490-610 mg.mm^{-2} in the wild pear plants under conditions of lower substrate saturation (water stress).

Plot of Fitted Model for Sorbus domestica with 40% saturation of the substrate

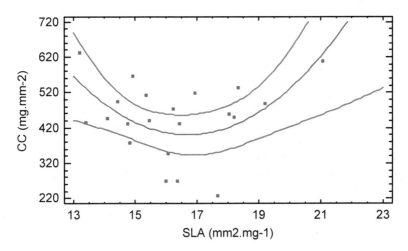

Fig. 11. Polynomial regression of the second order between specific leaf area (SLA) and chlorophyll content (CC) in the leaves of service tree (*Sorbus domestica*) plants grown under conditions of water stress. $R^2 = 30.92\%$; $p = 0.0358$.

Plot of Fitted Model for Pyrus pyraster with 40% saturation of the substrate

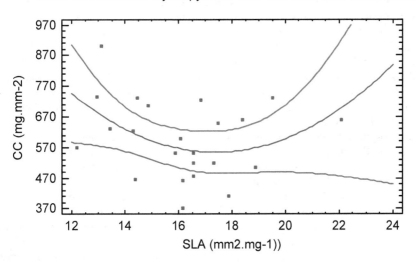

Fig. 12. Polynomial regression of second order between specific leaf area (SLA) and chlorophyll content (CC) in the leaves of wild pear (*Pyrus pyraster*) plants growing under conditions of water stress. $R^2 = 18.3086\%$; $p = 0.1324$

According to the results obtained from experiments with the differentiated water regime, we found a non-significant influence of low substrate saturation on the metabolic processes related to chlorophyll production in both of the analysed woody plant species.

6. Conclusion

With regard to progressive aridization, the research of resistant autochthonous woody plants that survive in extreme drought conditions is considerable. We have studied two taxa that naturally grow in the cultural landscape of Slovakia – the wild pear and service tree. Both species are light-demanding woody plants and occur in similar stands. Compared to the wild pear, the service tree prefers stands at lower altitudes, and is prevalent in warm and arid climates. The wild pear has wider ecological amplitude, and also grows at higher altitudes in stands with a different water regime and climate extremes.

Two-year old plants of the studied taxa were used in experiments with a regulated water regime. The plant material was grown from seeds collected directly from original stands in Slovakia, and the plants were maintained under semi-controlled conditions with 60% and 40% substrate saturation. Under these conditions, we analysed the following parameters: leaf dry mass, size of leaf area, leaf water content, specific leaf area, and the complex of assimilatory pigments.

Assessment of the analysed parameters confirmed interspecific differences in the physiological reactions of the woody plants under regulated conditions of a water regime. Each of the studied taxa utilized unique drought tolerance strategies. Under a differentiated water regime, the wild pear produced and increased leaf dry mass regardless of the level of substrate saturation (water regime). Based on these findings, the wild pear uses this mechanism to resist drought conditions.

Interspecific differences between the wild pear and service tree were confirmed by measuring the specific leaf area (SLA) and leaf water content (LWC). Compared to the service tree, the wild pear had higher SLA values when provided with a sufficient water supply. The SLA values of both taxa had a positive linear correlation with the leaf water content (LWC). Under water stress conditions, the wild pear reduced SLA, which was influenced not only by water deficit, but also by different morphogensis of the assimilation apparatus. During the experiment with the regulated water regime, the service tree had lower values of SLA than the wild pear and maintained them without significant changes, even under conditions of water stress.

A statistically significant relationship was confirmed between SLA values and chlorophyll concentration in the leaves of the service tree under conditions of water stress. This relationship was described as a polynomial curve of the second order. The relationship between SLA and chlorophyll concentration in the leaves of the wild pear under water stress conditions was also described as a polynomial function of the second order; however, this relationship was not significant. The low level of substrate saturation did not significantly influence metabolic processes related to chlorophyll production in both of the analysed taxa. The water regime of the analysed woody plants is the decisive factor that affects their distribution and survival in conditions of progressive aridization. Considering the natural distribution of these woody plants and their tolerance to a wide range of water supply, the wild pear exhibits good adaptability to decreasing humidity. The service tree has similar qualities and the potential to adapt to arid conditions; however, it is generally found on slopes of forest-steppe stands.

In the future, studies will focus on strategies of water utilization used by xerotermic woody plants under conditions of aridization. The photosynthetic activity and transpiration of woody plants will also be analysed under conditions of limited water supply. The research will focus on the photosynthesis, transpiration, stomatal resistance, structural leaf elements, and root system of woody plants.

7. Acknowledgment

The research was supported by research grant project VEGA 1/0426/09 „Plant adaptability and vitality as criteria of their utilization in urban environment and in the landscape" from the Slovak Grant Agency for Science.

8. References

Blattný, T &, Šťastný, T. (1959). The natural distribution of forest woody plants in Slovakia. Bratislava (in Slovak), Slovenské vydavateľstvo pôdohospodárskej literatúry, 402 pp.

Benčať, F. (1995). Distribution and originality of Sorbus domestica L. in Slovakia (in Slovak). Zb. ref." Výsledky botanických záhrad a arborét pri záchrane domácej dendroflóry a II. Dendrologické dni".Vyd. TU vo Zvolene, Zvolen, p. 136–149.

Bignami, C. (2000). Der Speierling in Süditalien: Erforschung der Hänge des Vesuvs (Kampanien). Bovenden. Corminaria, 14: 7–10.

Bouček, B. (1954). Pear (in Czech). In: Lesnická práce, roč. 33, č.2, p. 57-62.

Brütsch, U. (1993). Der Speierling, ein Vielfältiger Baum mit Zukunft. Wald und Holz, 13:8 - 11.

Brütsch,U. & Rotach, P. (1993). Der Speierling (Sorbus domestica L.) in der Schweiz: Verbreitung, Ökologie, Standortsansprüche, Konkurrenzkraft und waldbauliche Eignung. Schweiz. Z. Forswes., 144, 12: 967–991.

Dijkstra, P. (1989). Cause and effect of differences in SLA. In: Lambers, H., Cambridge, M.,L. Konings, I.L. Pons, eds, Causes and Consequencess of variation in Growth rate and productivity of Higher Plants, SPB Academic Publishing, TheHague, This volume, pp.125-140.

Ellenberg, H. (1978). Vegetation Mitteleuropas mit den Alpen in okologischer Sicht. Phytologie IV/2. - Stuttgart: Ulmer

Essau, K. (1977). Anatomy of seed plants, 2nd ed. John Wiley, new York, New York, USA., 576 pp. ISBN 9780471245209

Fahn, A. (1990). Plant anatomy, Pergamon Press, Oxford,UK, , 600 pp., ISBN 978-0080374918

Geiger, R. (1961). Das Klima der bodennahen Luftschicht. Ed. 4. Vieweg and Sohn, Braunschweig, 646 pp.

Gielen, B., Brognolas, F., Jian, G., Johnson, J.D., Karnosky, D.F., Polle, A., Scarascia-Mugnozza, G., Schroeder, W.R., Ceulemans, R. (2008). Poplars and willows in the world. International Poplar Commission Thematic Papers, Workong Paper IPC/9-7, FAO, Rome, Italy

Haeupeler, H. & Schönfelder, P. (1988). Atlas der Farn-und Blütenpflanzen der Bundesrepublik Deutschland. Eugen Ulmer Verlag, Stuttgart, 95 pp.

Hofmann, H. (1993). Zur Verbreitung und Ökologie der Wildbirne (Pyrus communis L.) in Süd-Niedersachsen und Nordhessen sowie ihrer Abgrenzung von verwilderten

Kulturbirnen (*Pyrus domestica* Med.). In Mitteilungen der Deutschen Dendrologischen Gesellschaft. *81*, p. 27-69.

Itai, Ch., Nejidat, A., Bar-Zvi, D. (2002). Stress Physiology in the New Millenium. In: Plant Physiology in the New Millenium, Belgrade, p.31-39 ISBN 86-7384-011-2,

Jones, H.G. (1993). Drought tolerance and water-use efficiency. In: J.A.C. Smith & Griffiths, eds. Water deficit: Plant Responses From Cell To Community, pp.193-201. Oxford, UK, Bios Scientific Publishers. 345 pp.

Jones, H.G. (2004). What is water use efficiency ? In: M.A. Bacon& J.A. Roberts eds. Water use Efficiency in Plant Biology. Blackwell Publishing, CRC Press, Oxford, UK, pp.27-41

Kárpáti, Z. (1960). Die Sorbus-Arten Ungarns und der angrenzenden Gebeite. – Feddes Repertorium 62:71-334

Kausch, W. (1992). Der Speierling. Goltze-Druck & Co. GmbH., Göttingen, 224 pp.

Kausch, W. (2000) Der Speierling. Eigenverlag , Bovenden, 184 pp.

Kirisits, T. (1992). Der Speierling – Die seltensteeinheimische Baumart. Österreichische Forstzeitung, *10*: 55–56.

Kozlowski, T.T. & Pallardy, S.G. (1997). Physiology of Woody Plants. Ed.2. Academic Press, San Diego, CA.

Kozlowski, T.T. & Pallardy, S.G. (2002). Acclimation and Adaptive responses of Woody plants to Environmental Stresses. In: The Botanical Review 68(2):207-334.

Kleinschmit, J. & Svolba, J. (1998). Auslese von Wildbirne (*Pyrus pyraster*) und Rückführung in den Wald. In: Kleinschmit, J., Soppa, B. & Fellenberg,U. (eds) 1998: Die Wildbirne, *Pyrus pyraster* (L.) Burgsd. Tagung zum Baum des Jahres am 17. und 18.3. 1998 in Göttingen. Frankfurt am Main, J.D.Sauerländers, p.83-96

Landolt, E. (1977) Okologisce Zeigerwerte zur Scheier Flora. Verrff. Geobot. Inst. ETH, Stifung Riibel, Zürich 64:1-208

Larcher, W. (2003). Physiological Plant Ecology. Berlin : Springer Verlag, Berlin Heidelberg, 2003. 488 p. ISBN 3-540-43516-6

Liakoura, V., Fotelli, M. N., Rennenberg, H., Karabourniotis, G. 2009: Should structure-function relations be considered separately for homobaric vs. Heterobaric leaves?. American Journal of Botany 96(3) : 612-619,2009

Lin, L.,Z. & Harnly, J.N. (2008). Phenolic compounds and chromatographic profiles of pear (Pyrus). J.Agric. Food Chem.2008,56,9094-9101.

Májovský, J. (1992). *Sorbus* L. emend. Crantz. In Bertová L. (ed.), Flora of Slovakia, IV/3. (in Slovak) Veda, Bratislava, p. 405–408.

Mc Dowell, N., Pockman, W., Allen,D., Breshears, D., Cobb,N., Kolb, T., Plaut, J., Sperry, J.,West, A., Williams, D. & Yepez, E.A. (2008). Mechanisms of plant survival and mortality during drought: why do some plants survive while others succumb to drought? New Phytol.178: 719-739.

Michalko, J. (1961). Originality of the service tree (*Sorbus domestica* /L./) in oak communities of the Carpathian Mountains (in Slovak). Biológia, Bratislava, *16*, 4: 241–248.

Mooney, H.A., Gulmon, S.L. (1982). Constrains on leaf structure and function in reference to herbivory. Bioscience,32,198-206.

Namvar, K. & Spethmann,V. (1986). Die Wild- oder Holzbirne (*Pyrus pyraster*). In Allgemeine Forstzeitschrift, 21, p. 520-522.

Niclas,K. & Cobb, E.D. (2008). Evidence for "diminishing returns" from the scaling of stem diameter and specific leaf area. American Journal of Botany 95: 549-557.

Pagan, J., (1996). Forestry Dendrology (in Slovak).Vysokoškolské skriptá. Zvolen, Vyd. TU vo Zvolene, 378 pp. ISBN 80-228-0534-3.

Paganová, V. (2003). Wild pear *Pyrus pyraster* (L.) Burgsd. requirements on environmental conditions. In: Ecology (Bratislava) Vol. 22, č. 3 (2003), s. 225-241.

Paganová, V. (2008). Ecology and distribution of service tree *Sorbus domestica* (L.) in Slovakia. In: Ekológia, Bratislava, 2008, 27(2):152-168.

Passioura, J., B. (2002b). Soil conditions and plant growth. Plant, Cell and Environment 25:311-318

Passioura, J., B. (2002a). Environmental biology and crop improvement. Funct.Plant Biol.,29:537-546

Peniašteková, M, (1992). *Pyrus* L. Hruška. In Bertová L. (ed.): Flóra Slovenska, IV/3, Bratislava, Veda, p. 381-388.

Reich, P. B., Wright, I. J., Cavender-Bares, J., Craine, J., Oleksyn, M., Westoby, M. & Walters, M. B.(2003). The evolution of plant functional variation: Traits, spectra and strategies. International Journal of Plant Sciences 164:S143-S164.

Rittershoffer, B., (1998). Förderung seltener Baumarten im Wald. Auf den Spuren der Wildbirne. InAllgemeine Fosrtzeitschrift /Der Wald, 16, p. 860-862.

Roloff, A., (1998). Der Baum des Jahres 1998: die Wildbirne (*Pyrus communis* L. sp. *pyraster* Gams.) In Kleinschmit,J., Soppa, B., Fellenberg,U. (eds): Die Wildbirne, *Pyrus pyraster* (L.) Burgsd. Tagung zum Baum des Jahres am 17.und 18.3. 1998 in Göttingen. Frankfurt am Main, J.D.Sauerländers, p. 9-15.

Schmitt, H. P. (1998). Wildbirnen-Vorkommen in Westfalen-Lippe. In Kleinschmit, J., Soppa, B., Fellenberg, U. (eds), 1998: Die Wildbirne, *Pyrus pyraster* (L.) Burgsd. Tagung zum Baum des Jahres am 17. und 18.3. 1998 in Göttingen. Frankfurt am Main, J.D.Sauerländers, p.57-59.

Steffens, R., (2000). Der Speierling in Sachsen-Anhalt – Verbreitung, Ökologie und genetische Variation. Bovenden. Corminaria, *14*: 14–17.

Steiner, M., (1995). Speierlingskartierung im südöstlichen Wienerwald. Österreichische Forstzeitung, *6*: 50–51.

Šály, R., (1988). Pedology and microbiology (in Slovak). Vyd. VŠLD Zvolen, 378 pp.

Šesták, J. & Čatský J., (1966). Metody studia fotosyntetické produkce rostlin, Akademia, ČSAV Praha, 393 s.

Šiška, B. & Takáč, J. (2009). Drought naalyses of agricultural regions as influenced by climatic conditions in the slovak Republic. In: Időjárás. Quarterly Journal of the Hungarian Meteorological Service. Vol. 113, No. 1-2, January-June 2009, pp. 135-143.

Škvarenina, J., Križová, E. & Tomlain, J. (2004). Impact of the climate change on the water balance of altitudinal vegetation stages in Slovakia. Ekológia 23 (Suppl. 2/2004):13-19

Škvarenina, J., Tomlain, J., Hrvoľ, J. & Škvareninová, J. (2009a). Ocurrence of dry and wet periods in Altitudinal vegetation stages of West Carpatians in Slovakia: Time-series analysis 1951-2005. In: Strelcova, K; Matyas, C; Kleidon, A; Lapin, M; Matejka, F; Blazenec, M; Skvarenina, J; Holecy, (eds), Bioclimatology and natural hazards, Springer Science+Business Media B.V. 2009, p. 97-106. ISBN: 978-1-4020-8875-9

Škvarenina, J., Tomlain, J., Hrvoľ, J., Škvareninová, J. & Nejedlík, P. (2009b). Progress in dryness and wetness parameters in altitudinal vegetation stages of West Carpatians in Slovakia: Time-series analysis 1951-2007. In: Időjárás. Quarterly Journal of the Hungarian Meteorologicl Service. Vol. 113, No. 1-2, January-June 2009, pp. 47-54.

Špánik, F., Antal, J., Tomlain,J., Škvarenina,J., Repa,Š., Šiška, B. & Mališ, J., (1999). Applied agrometheorology.(in Slovak). Vyd. SPU, Nitra, 194 pp.

Tarábek, K., (1980). Climate-geographic types.(in Slovak). In Atlas SSR. Slovenský úrad geodézie a kartografie, Bratislava, 64 pp.

Terashima, I. (1992). Anatomy of non-uniform leaf photosynthesis. Photosynthesis research 31: 195-212.

Wagner, I. (1995). Identifikation von Wildapfel (*Malus sylvestris* (L.) M i l l.) und Wildbirne (*Pyrus pyraster* (L.) B u r g s d.). In Forstarchiv, 66, p. 39-47.

Wilhelm, G., J., (1998). Im Vergleich mit Elsbeere und Speierling Beobachtungen zur Wildbirne. AFZ. Der Wald, *16*: 856–859.

Wilhite, D.A. & M.H. Glantz. (1985). Understanding the Drought Phenomenon: The Role of Definitions. Water International 10:111-120.

Wright , I. J., Reich. P.B. & Villar, R. (2004). The worldwide leaf economics spectrum. Nature 428: 821-827.

Zlatník, A. (1976). Forest phytocenology. State Agricultural. Publishing-House Praha. 495. (in Czech).

12

Possibilities of Deriving Crop Evapotranspiration from Satellite Data with the Integration with Other Sources of Information

Gheorghe Stancalie and Argentina Nertan
National Meteorological Administration 97,
Soseaua Bucuresti-Ploiesti, Bucharest
Romania

1. Introduction

After precipitation, evapotranspiration is one of the most significant components in terrestrial water budgets.

Evapotranspiration (ET) describes the transport of water into the atmosphere from surfaces (including soil - soil evaporation) and from vegetation (transpiration). Those are often the most important contributors to evapotranspiration. Other contributors to evapotranspiration are the e from wet canopy surface (wet-canopy evaporation) and evaporation from vegetation-covered water surface in wetlands the process of evapotranspiration is one of the main consumers of solar energy at the Earth's surface. The energy used for evapotranspiration is generally referred to as latent heat flux. The term latent heat flux includes other related processes unrelated to transpiration including condensation (e.g., fog, dew), and snow and ice sublimation.

There are several factors that affect the evapotranspiration processes: energy availability; the humidity gradient away from the surface(the rate and quantity of water vapor entering into the atmosphere are higher in drier air); the wind speed at the soil level (wind affects evapotranspiration by bringing heat energy into an area); Water availability (it is well known that the evapotranspiration cannot occur if water is not available); Vegetation biophysical parameters (many physical parameters of the vegetation, like cover plant height, leaf area index and leaf shape and the reflectivity of plant surfaces can affect evapotranspiration); Stomatal resistance (the transpiration rate is dependent on the diffusion resistance provided by the stomatal pores, and also on the humidity gradient between the leaf's internal air spaces and the outside air); soil characteristics which includes its heat capacity, and soil chemistry and albedo. For a given climatic region the evapotranspiration follows the seasonal declination of solar radiation and the resulting air temperatures: minimum evapotranspiration rates generally occur during the coldest months of the year and maximum rates generally coincide with the summer season (Burba, 2010). Even so evapotranspiration depends on solar energy; the availability of soil moisture and plant maturity, the seasonal maximum evapotranspiration actually may precede or follow the seasonal maximum solar radiation and air temperature by several weeks (Burba, 2010). If the moisture is available, evapotranspiration is dependent mainly on the availability of solar energy to vaporize water: evapotranspiration varies with latitude, season, time of day, and cloud cover. Most of the evapotranspiration of water at the Earth's surface level occurs in the subtropical regions (Fig.1). In these areas, high quantities of solar radiation provide the energy necessary to convert liquid water into a gas. Usually, evapotranspiration exceeds precipitation on middle and high latitude large areas during the summer season. As a result of climate change it is expected to induce a further intensification of the global water cycle, including ET (Huntington, 2006). Therefore accurate estimates of evapotranspiration are needed for weather forecasting and projecting the long-term effects of land use change and global climate change, irrigation scheduling and watershed management.

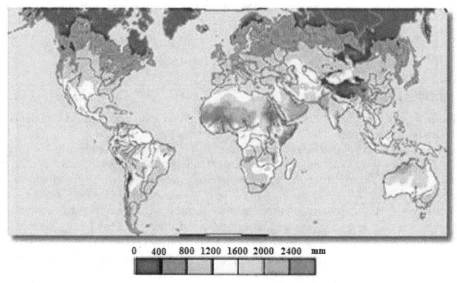

0   400   800 1200 1600 2000 2400   mm

Fig. 1. Mean Annual Potential Evapotranspiration (UNEP World Atlas of Desertification)

In this regard, remote sensing data with the increasing imagery resolution is a useful tool to provide ET information over different temporal and spatial scales. During the last decades important progresses were made in the determination of ET using remote sensing techniques. Some studies have classified the methods of ET estimation in two categories: semi- empirical methods - use empirical relationship and a minimum set of meteorological data; analytical methods – consist in the establishment of the physical process at the scale of interest. A study done by Courault (2007) proposed a few methods which can be classified as follows: empirical direct methods, residual methods of the energy budget, deterministic methods, and vegetation index methods.

In agriculture, an accurate quantification of ET is important for effective and efficient irrigation management. When evaporative demand exceeds precipitation, plant growth and quality may be unfavorably affected by soil water deficit. A large part of the irrigation water applied to agricultural lands *(Fig. 2)* is consumed by evaporation and transpiration. In a given crop, evapotranspiration process is influenced by several factors: plant species, canopy characteristics, plant population, degree of surface cover, plant growth stage, irrigation regime (over irrigation can increase ET due to larger evaporation), soil water availability, planting date, tillage practice, etc. As it can be observed from Fig. 2 the movement of the water vapor from the soil and plant surface, a t a field level is influenced mainly by wind speed and direction although other climatic factors also can play a role. Evapotranspiration increases with increasing air temperature and solar radiation. Wind speed can cause ET increasing. For high wind speed values the plant leaf stomata (the small pores on the top and bottom leaf surfaces that regulate transpiration) close and evapotranspiration is reduced. There are situations when wind can cause mechanical damage to plants which can decrease ET due to reduced leaf area. Hail can reduce also leaf area and evapotranspiration. Higher relative humidity decreases ET as the demand for water vapor by the atmosphere surrounding the leaf surface decreases. If relative humidity (dry air) has lower values, the ET increases due to the low humidity which increases the vapor pressure deficit between the vegetation surface and air. On rainy days, incoming solar radiation decreases, relative humidity increases, and air temperature usually decreases, generation ET decreasing. But, depending on climatic conditions, actual crop water use usually increases in the days after a rain event due to increased availability of water in the soil surface and crop root zone.

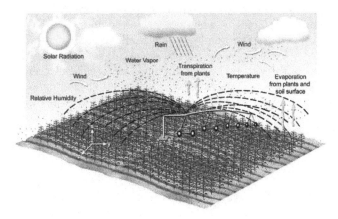

Fig. 2. Evaporation and transpiration and the factors that impact these processes in an irrigated crop.

2. Evapotranspiration and energy budget

The estimation of ET parameter, corresponding to the latent heat flux (λE) from remote sensing is based on the energy balance evaluation through several surface properties such as albedo, surface temperature (T_s), vegetation cover, and leaf area index (LAI). Surface energy balance (SEB) models are based on the surface energy budget equation. To estimate regional crop ET, three basic types of remote sensing approaches have been successfully applied (Su, 2002).

The first approach computes a surface energy balance (SEB) using the radiometric surface temperature for estimating the sensible heat flux (H), and obtaining ET as a residual of the energy balance. The single-layer SEB models implicitly treat the energy exchanges between soil, vegetation and the atmosphere and compute latent heat flux (λE) by evaluating net (all-wave) radiant energy (R_n), soil heat flux (G) and H. For instantaneous conditions, the energy balance equation is the following:

$$\lambda E = R_n - H - G \qquad (1)$$

where: R_n = net radiant energy (all-wave); G = soil heat flux; H = sensible heat flux (Wm^{-2}); λE = latent energy exchanges (E = the rate of evaporation of water (kg m^{-2} s^{-1}) and λ = the latent heat of vaporization of water (J kg^{-1})). λE is obtained as the residual of the energy balance contain biases from both H and (R_n - G). There are several factors which affect the performance of single-source approaches, like the uncertainties about atmospheric and emissivity effects. LST impacts on all terms of the energy balance in particular on long wave radiation. The radiative surface temperatures provided by an infrared radiometer from a space borne platform are measured by satellite sensors such as LANDSAT, AVHRR, MODIS and ASTER. Converting radiometric temperatures to kinetic temperature requires considerations about surface emissivity (λE), preferably from ground measurements. Remotely LST is subject to atmospheric effects which are primarily associated with the absorption of infrared radiation by atmospheric water vapor and which lead to errors of 3–5 K. A wide range of techniques have been developed to correct for atmospheric effects, including: single-channel methods; split-window techniques; multi-angle methods and combinations of split-window and multi-channel methods. Radiant and convective fluxes can be described: by considering the observed surface as a single component (single layer approaches); by separating soil and vegetation components with different degrees of canopy description in concordance with the number of vegetation layers (multilayer approaches). Net radiant energy depends on the incident solar radiation (R_g), incident atmospheric radiation over the thermal spectral domain (R_a), surface albedo (α_s), surface emissivity (ε_s) and surface temperature (T_s), according to the following equation:

$$R_n = (1 - \alpha_s)R_g + \varepsilon_s R_a - \varepsilon_s \sigma T_s^4 \qquad (2)$$

For single layer models, R_n is related to the whole surface and in the case of multiple layer models, R_n is linked with both soil and vegetation layers. For single approaches, sensible heat flux H is estimated using the aerodynamic resistance between the surface and the reference height in the lower atmosphere (usually 2 m) above the surface. Aerodynamic resistance (r_a) is a function of wind speed, atmospheric stability and roughness lengths for momentum and heat. For multiple layer models, H is characterized taking into account the soil and canopy resistance, with the corresponding temperature:

$$H = \rho c_p \frac{(T_s - T_a)}{r_a} \qquad (3)$$

Eq. (3) shows that the estimation of λE parameter can be made using the residual method, which induces that λE is linearly related to the difference between the surface temperature (T_s) and air temperature (T_a) at the time of T_s measurement if the second order dependence of r_a on this gradient is ignored.

$$\lambda E = R_n - G - \rho cp \frac{(T_s - T_a)}{r_a} \qquad (4)$$

Possibilities of Deriving Crop
Evapotranspiration from Satellite Data with the Integration with Other Sources of Information

263

Equation (4) is usually used to estimate λE. At midday, it provides a good indicator regarding the plant water status for irrigation scheduling. For λE estimation over longer periods (daily, monthly, seasonal estimations), the use of ground-based ET from weather data is necessary to make temporal interpolation. Some studies have used the trend for the evaporative fraction (EF), such as the ratio of latent heat flux to available energy for convective fluxes, to be almost constant during the daytime. This allows estimating the daytime evaporation from one or two estimates only of EF at midday, for example at the satellite acquisition time (Courault et al., 2005).

$$EF = \frac{\lambda E}{R_n - G}, \quad ET_{24} = EF * R_{n24} \tag{5}$$

ET can be estimated from air vapor pressure (p_a) and a water vapor exchange coefficient (h_s) according to the following equation:

$$\lambda E = \rho c_p h_s (p_s^*(T_s) - e_a) \tag{6}$$

Usually this method is used in models simulating Soil–Vegetation–Atmosphere Transfers (SVAT). $p_s^*(T_s)$ represent the saturated vapor pressure at the surface temperature T_s and h_s is the exchange coefficient which depends on the aerodynamic exchange coefficient ($1/r_a$), soil surface and stomatal resistances of the different leaves in the canopy. Katerji & Perrier (1985) estimated a global canopy resistance (r_g) including both soil and canopy resistances (equation 6)

$$r_g = \cfrac{1}{\cfrac{1}{r_{veg} + r_w} + \cfrac{1}{r_0 + r_s}} \tag{7}$$

where: r_{veg} is the resistance due to the vegetation structure, r_w the resistance of the soil layer depending on the soil water content, r_0 the resistance due to the canopy structure and r_s the bulk stomatal resistance. To calculate this parameters it necessary to have information regarding the plant structure like LAI and fraction of vegetation cover (FC), the minimum stomatal resistance (r_{smin}). Many studies proposed various parameterizations of the stomatal resistance taking into account climatic conditions and soil moisture (Jacquemin & Noilhan, 1990). This proves that the ($T_s - T_a$) is related to ET term, and that Ts can be estimated using thermal infrared measurements (at regional or global scale using satellite data, and at local scale using ground measurements).

The second approach uses vegetation indices (VI) derived from canopy reflectance data to estimate basal crop coefficient (K_{cb}) that can be used to convert reference ET to actual crop ET, and requires local meteorological and soil data to maintain a water balance in the root zone of the crop. The VIs is related to land cover, crop density, biomass and other vegetation characteristics. VIs such as the Normalized Difference Vegetation Index (NDVI), the Soil Adjusted Vegetation Index (SAVI), the Enhanced Vegetation Index (EVI) and the Simple Ratio (SR), are measures of canopy greenness which may be related to physiological processes such as transpiration and photosynthesis. Among the relatively new satellite sensors it has to be mentioned the advantages of using MODIS/Aqua that offer improved spectral and radiometric resolution for deriving surface temperatures and vegetation indices, as well as   increased frequency of evaporative fraction and evaporation estimates when compared with other sensors. The observed spatial variability in radiometric surface temperature is used with reflectance and/or vegetation index observations for evaporation estimation. For ET estimation from agricultural crops the most direct application is to substitute the VIs for crop coefficients (defined as the ratio between actual crop water use and reference crop evaporation for the given set of local meteorological conditions). Negative observing correlations between the NDVI and radiometric surface temperature could be linked to evaporative cooling, although for most landscapes variations in fractional vegetation cover, soil moisture availability and meteorological conditions will cause considerable scatter in those relationships. The methods associated with this approach generate spatially distributed values of K_{cb} that capture field-specific crop development and are used to adjust a reference ET (ET_o) estimated daily from local weather station data.

The third approach uses remotely sensed LST with Land Surface Models (LSMs) and Soil-Vegetation–Atmosphere (SVAT) models, developed to estimate heat and mass transfer at the land surface. LSMs contain physical descriptions of the transfer in the soil–vegetation-atmosphere continuum, and with proper initial and boundary conditions provide continuous simulations when driven by weather and radiation data. The energy-based LSMs are of particular interest because these approaches allow for a strong link to remote sensing applications. The use of the spatially distributed nature of remote sensing data as a calibration source has been limited, with the focus placed on data assimilation approaches to update model states, rather than inform the actual model structure. Data assimilation is the incorporation of observations into a numerical model(s) with the purpose of providing the model with the best estimate of the current state of a system. There are two types of data assimilation: (i) sequential assimilation which involves correcting state variables (e.g. temperature, soil moisture) in the model whenever remote sensing data are available; and (ii) variation assimilation when unknown model parameters are changed using data sets obtained over different time windows. Remotely sensed LSTs have been assimilated at point scales into various schemes for estimating land surface fluxes by comparing simulated and observed temperatures and adjusting a state variable (e.g. soil moisture) or model parameters in the land surface process model. Such use of remote sensing data has highlighted problems of using spatial remote sensing data with spatial resolutions of tens or hundreds of kilometers with point-scale SVAT models and has led to the search for "effective" land surface parameters. There exist no effective means of evaluating ET spatially distributed outputs of either remote sensing based approaches or LSMs at scales greater than a few kilometers, particularly over non-homogeneous surfaces. The inability to evaluate remote sensing based estimates in a distributed manner is a serious limitation in broader scale applications of such approaches. It must be noted here that ET evaluation of remote sensing based approaches with ground based data tends to favour those few clear sky days when fluxes are reproduced most agreeably, and on relatively flat locations.

In this case the radiation budget is given by the following equation (Kalma et al., 2008):

$$R_n = K\downarrow -K\uparrow +L\downarrow -L\uparrow \tag{8}$$

where K↓ is the down-welling shortwave radiation and it depends on atmospheric transmissivity, time of the day, day of the year and geographic coordination. K↑ represents the reflected shortwave radiation which depends on K↓ and surface albedo (a), L↓ is the down-welling long wave radiation and L↑ is the up-welling long wave radiation. L↓ depends on the atmospheric emissivity (which in turn is influenced by amounts of atmospheric water vapor, carbon dioxide and oxygen) and by air temperature. L↑ si influenced by land surface temperature and emissivity

3. Direct methods using difference between surface and air temperature

Mapping daily evapotranspiration over large areas considering the surface temperature measurements has been made using a simplified relationship which assumes that it is possible to directly relate the daily (λE_d) to the difference $(T_{rad} - T_a)_i$ between (near) mid-day observations (i) of surface temperature and near-surface air temperature (Ta) measured at midday as follows:

$$\lambda E_d = (R_n)_d - B(T_{rad} - T_a)_i^n \tag{9}$$

B is a statistical regression coefficient which depends on surface roughness. n depends on atmospheric stability. Equation 9 was derived from Heat Capacity Mapping Mission (HCMM) observations over fairly homogeneous irrigated and non-irrigated land surfaces, with areas between 50 and 200 km^2 (Seguin et al. 1982a, b). Some authors as Carlson et al. (1995a) proposed a simplified method based on Eq. 9 which uses the difference $(T_{rad} - T_a)$ at 50 m at the time of the satellite overpass. They showed that B coefficient and n are closely related to fractional cover f_c that can be obtained from the NDVI-T_{rad} plots. B values vary from 0.015 for bare soil to 0.065 for complete vegetation cover and n decreased from 1.0 for bare soil to 0.65 for full cover.

4. Surface energy balance models

Surface energy balance models (SEBAL) assume that the rate of exchange of a quantity (heat or mass) between two points is driven by a difference in potential (temperature or concentration) and controlled by a set of resistances which depend on the local atmospheric environment and the land surface and vegetation properties. In the review made by Overgaard et al. (2006) regarding the evolution of land surface energy balance models are described the following approaches: the combination approach by Penman (1948) which developed an equation to predict the rate of ET from open water, wet soil and well-watered grass based on easily measured meteorological variables such as radiation, air temperature, humidity, and wind speed; the Penman–Monteith "one-layer", "one-source" or "big leaf" models (Monteith 1965) which recognize the role of surface controls but do not distinguish between soil evaporation and transpiration; this approach estimates ET rate as a function of canopy and boundary layer resistances; "two-layer" or "two-source" model such as described by Shuttleworth and Wallace (1985) which includes a canopy layer in which heat and mass fluxes from the soil and from the vegetation are allowed to interact; multi-layer models which are essentially extensions of the two-layer approach.

4.1 The Penman–Monteith, "one-source" SEB models

The Penman–Monteith (PM) approach combines energy balance and mass transfer concepts (Penman, 1948) with stomatal and surface resistance (Monteith, 1981). Most "one source" SEB models compute λE by evaluating R_n, G and H and solve for λE as the residual term in the energy balance equation (see Eq. 10). The sensible heat flux (H) is given by:

$$H = \rho C_p \left[\frac{(T_{ad} - T_a)}{r_a} \right] \tag{10}$$

Where: ρ = air density (kg*m^{-3}); C_p = specific heat of air at constant pressure (J kg^{-1} K^{-1}); T_{ad} = aerodynamic surface temperature at canopy source height (K); T_a = near surface air temperature (K); r_a = aerodynamic resistance to sensible heat transfer between the canopy source height and the bulk air at a reference height above the canopy (s m^{-1}). The r_a term is usually calculated from local data on wind speed, surface roughness length and atmospheric stability conditions. According to Norman and Becker (1995), the aerodynamic surface temperature (T_{ad}) represent the temperature that along with the air temperature and a resistance calculated from the log-profile theory provides an estimate H. The key issue of PM approach is to estimate an accurately sensible heat flux. T_{ad} is obtained by extrapolating the logarithmic air temperature profile to the roughness length for heat transport (z_{oh}) or, more precisely, to (d + z_{oh}) where d = zero-plane displacement height. Usually, due to the fact that T_{ad} cannot be measured using remote sensing, it is replaced with T_{rad}. As it is demonstrated by Troufleau et al. (1997), for dense canopy T_{rad} and T_{ad} may differ with 1-2 K and much more for sparse canopy. Surface temperature ($T_{rad)}$ is related to the kinetic temperature by the surface emissivity (ε) (Eq, 11) and it depends on view angle (θ) (Norman et. al, 2000). On the other hand T_{ad} and aerodynamic resistance are fairly difficult to obtain for non-homogenous land surfaces.

$$T_{rad} = \varepsilon^{1/4} * T_k \tag{11}$$

The aerodynamic resistance r_a can be calculated with the following equation:

$$r_a = \frac{1}{k^2} u \left[ln \frac{z-d}{z_{oh}} - \Psi_h \frac{z-d}{L} \right] \left[ln \frac{z-d}{z_{om}} - \Psi_m \frac{z-d}{L} \right] \tag{12}$$

where: k = 0.4 (von Karman's constant); u = wind speed at reference height z (m s^{-1}); d = zero-plane displacement height (m); z_{oh} and z_{om} = roughness lengths (m) for sensible heat and momentum flux, respectively; Ψ_h and Ψ_m = stability correction functions for sensible heat and momentum flux, respectively; L = Monin-Obukhov length L (m). The Ψ_h = 0 and Ψ_m = 0 if near surface atmospheric conditions are neutrally stable. Usually, the aerodynamic resistance is estimated from local data, even that area averaging of roughness lengths is highly non-linear (Boegh et al. 2002). Several studies, such as Cleugh at al. (2007) used these equations for evapotranspiration landscape monitoring. Their approach estimates E at 16-day intervals using 8-day composites of 1 km MODIS T_{rad} observations and was tested with 3 years of flux tower measurements and was obtained significant discrepancies between observed and simulated land surface fluxes, generated by the following factors: the estimation of H with Eqs. 9 and 10 is not constrained by the requirement for energy conservation; errors in z_{oh} determination; use of unrepresentative emissivities; using time-averages of instantaneous T_{rad}, T_a and R_n, the non-linearity of Eq. 9 may cause significant errors; standard MODIS data processing eliminates all cloud-contaminated pixels in the composite period. Bastiaanssen et al. (1998a) developed a calibration procedure using image data to account for the differences between T_{aero} and T_{rad}, which are important, mainly for incomplete vegetation covers. Other authors, such as Stewart et al. (1994) and Kustas et al. (2003a), made empirical adjustments to aerodynamic resistance, related to z_{oh} (eq. 13).

$$H = \rho C_p \left[\frac{T_{rad}(\Theta) - T_a}{r_a - r_{ex}} \right] \tag{13}$$

where: T_{rad} (θ) = radiometric surface temperature (K) at view angle θ derived from the satellite brightness temperature; r_{ex} = excess resistance (s m^{-1}) (reflects differences between momentum and sensible heat transfer. According to Stewart et al. (1994) r_{ex} is function of the ratio of roughness lengths for momentum z_{om} and for sensible heat z_{oh} and the friction velocity u^* (m s^{-1}) (eq. 14):

$$r_{ex} = \frac{kB^{-1}}{ku^*} = ln\frac{z_{om}/z_{oh}}{ku^*} \tag{14}$$

where kB^{-1} = dimensionless ratio determined by local calibration. Eq. 14 assumes that the ratio z_{om}/z_{oh} may be treated as constant for uniform surfaces, although kB^{-1} has been found to be highly variable (Brutsaert 1999).

In the case of the one source Surface Energy Balance System (SEBS) (Su, 2002) the surface heat fluxes are estimated from satellite data and available meteorological data. There are three sets of input data in SEBS: the first set includes the following parameters: α, ε, T_{rad}, LAI, fractional vegetation coverage and the vegetation height (if the vegetation information is not explicitly available, SEBS can use as input data the Normalized Difference Vegetation Index (NDVI)); the second set includes T_a, u, actual vapour pressure (e_a) at a reference height as well as total air pressure; the third set of data consists of measured (or estimated) $K\downarrow$ and $L\downarrow$. For R_n, G, and the partitioning of (R_n - G) into H and λE, SEBS use different modules (Fig. 3): H is estimated using Monin–Obukhov similarity theory; in the case of u and vegetation parameters (height and LAI) is used the Massman (1997) model to to estimate the displacement height (d) and the roughness height for momentum (z_{om}); the equations proposed by Brutsaert (1982, 1999) are used when only the height of the vegetation is available. The SEBS was successfully tested for agricultural areas, grassland and forests, across various spatial scales. Several studies used flux tower method and data from Landsat, ASTER ad Modis sensors (Su et al. 2005, 2007, McCabe and Wood 2006).

The Fig. 4 shows the time series, determined during the Soil Moisture Atmosphere Coupling Experiment 2002 (SMACEX-02) (Kustas et al. 2005). These time series illustrates latent heat fluxes and sensible heat fluxes measured with in situ eddy-covariance equipment (closed) together with SEBS model (open) over a field site (corn) from Iowa. The gaps in the time series are caused either the missing ancillary data or absence of flux measurements. Many factors influence the single-source approach: there are uncertainties due to atmospheric and emissivity effects; because of the vegetation properties and of the angle view, the relationship between T_{ad} and T_a is not unique; this approach requires representative near-surface T_a and other meteorological data measured (or estimated) at the time of the satellite overpass at a location closely with the T_{rad} observation. This can generate errors in defining meteorological parameter for each satellite pixel from a sparse network of weather stations (at the time of satellite overpass), mainly for areas with high relative relief and slopes. Another important factor is that the accuracy of any of the estimates depends on the performance of the algorithm used for temperature retrieval.

The major advantages of SEBS are: uncertainty due to the surface temperature or meteorological variables can be limited taking into account the energy balance at the limiting cases; through the SEBS was formulated a new equation for the roughness height for heat transfer, using fixed values; a priori knowledge of the actual turbulent heat fluxes is not required. Another single-source energy balance models, developed based on the conception of SEBAL, are S-SEBI (Simplified-SEBI), METRIC (Mapping EvapoTranspiration at high Resolution with Internalized Calibration), etc. The main difference between such kinds of models is the difference in how they calculate the sensible heat, i.e. the way to define the dry (maximum sensible heat and minimum latent heat) and wet (maximum latent heat and minimum sensible heat) limits and how to interpolate between the defined upper and lower limits to calculate the sensible heat flux for a given set of boundary layer parameters of remotely sensed data (T_s, albedo, NDVI, LAI) and ground-based air temperature, wind speed, humidity. The assumptions in all these models are that there are few or no changes in atmospheric conditions (especially the surface available energy) in space and sufficient surface horizontal variations are required to ensure dry and wet limits existed in the study area.

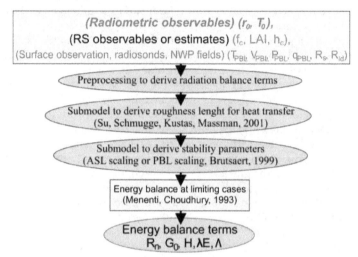

Fig. 3. Schematic representation of SEBS (after Su, 2008)

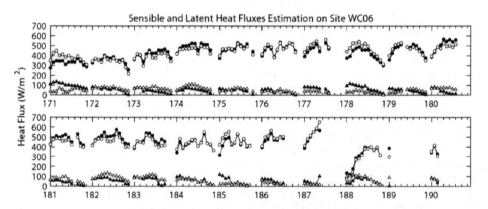

Fig. 4. Reproduction of surface flux development with a one-source model (SEBS) (after Kalma, 2008)

4.2 Two-source SEB models
The equations 10 and 13 make no difference between evaporation soil surface and transpiration from the vegetation and from this reason the resistances are not well defined.

To solve this problem two-source models have been developed for use with incomplete canopies (e.g. Lhomme et al. 1994; Norman et al. 1995; Jupp et al. 1998; Kustas and Norman 1999). These models consider the evaporation as the sum of evaporation from the soil surface and transpiration from vegetation. For example, Norman et. Al. (1995) developed a two-source model (TSM) based on single-time observations which eliminate the need for r_{ex} as used in equations 13 and 14. They reformulated the equation 10 as:

$$ H = \rho C_p \frac{T_{rad}(\theta) - T_a}{r_r} \tag{15} $$

where: T_{rad} = directional radiometric surface temperature obtained at zenith view angle θ; r_r = radiometric-convective resistance (s m^{-1}). The radiometric convective resistance is calculated according to the following formula:

$$ r_r = \frac{T_{rad}(\theta) - T_a}{\left[\frac{(T_c - T_a)}{r_a} + \left(\frac{(T_s - T_a)}{r_a + r_s} \right) \right]} \tag{16} $$

where: T_c = canopy temperature; T_s = soil surface temperature; R_s = soil resistance to heat transfer (s m-1). To estimate the T_c and T_s variables, Norman et al. used fractional vegetation cover (fc) which depends on sensor view angle (Eq. 17):

$$ T_{rad}(\theta) \approx [f_c(\theta)T_c^4 + \{1 - f_c(\theta)\}T_s^4]^{\frac{1}{4}} \tag{17} $$

H variable is divided in vegetated canopy (H_c) and soil (H_s) influencing the temperature in the canopy air-space. Other revisions of TSM compared flux estimates from two TSM versions proved that thermal imagery was used to constrain T_{rad} and H and microwave remote sensing was employed to constrain near surface soil moisture. The estimations resulting from those two models were compared with flux tower observations. The results showed opposing biases for the two versions that it proves a combination between microwave and thermal remote sensing constraints on H and λE fluxes from soil and canopy. Compared to other types of remote sensing ET formulations, dual-source energy balance models have been shown to be robust for a wide range of landscape and hydro-meteorological conditions.

5. Spatial variability methods using vegetation indices

Visible, near-infrared and thermal satellite data has been used to develop a range of vegetation indices which have been related to land cover, crop density, biomass or other vegetation characteristics (McVicar and Jupp 1998). Several vegetation indices as the Normalized Difference Vegetation Index (NDVI), the Soil Adjusted Vegetation Index (SAVI), the Enhanced Vegetation Index (EVI) and the Simple Ratio (SR), are indicators of canopy greenness which can be related to physiological processes such as transpiration and photosynthesis (Glenn et al., 2007).

5.1 Vegetation indices, reflectance and surface temperature
The SEBAL approach used remotely sensed surface temperature, surface reflectivity and NDVI data. It has been developed for the regional scale and it requires few ground level observations from within the scene. K↓ and L↓ are computed using a constant atmospheric

transmissivity, an appropriate atmospheric emissivity value and an empirical function of T_a, respectively. G is calculated as a fraction of R_n depending on T_{rad}, NDVI and α (Bastiaanssen 2000). The instantaneous values of sensible heat flux are calculated in three main steps. First step makes the difference between T_{ad} and T_{rad} and assumes that the relationship between T_{rad} and the near-surface temperature gradient ($\Delta T = T_{ad} - T_a$) is quasi-linear. Therefore wet and dry extremes can be identified from the image. These extremes fix the quasi-linear relationship relating ΔT to T_{rad}, allowing ΔT to be estimated for any T_{rad} across the image. In the second step, a scatter plot is obtained for all pixels in the entire image of broadband α values versus T_{rad}. Low temperature and low reflectance values correspond to pixels with large evaporation rates, while high surface temperatures and high reflectance values correspond to the areas with little or no evaporation rates. Scatter plots for large heterogeneous regions frequently show an ascending branch controlled by moisture availability and evaporation rate, and a radiation-controlled descending branch where evaporation rate is negligible. The ascending branch indicates that the temperatures increase with increasing α values as water availability is reduced and evaporation rate becomes more limited. For the descending branch the increasing of α induce a decreasing of surface temperature. If the radiation-controlled descending branch is well defined, r_a may be obtained from the (negative) slope of the reflectance–surface temperature relationship. The last step use the local surface roughness (z_{om}) based on the NDVI; is assumed that the z_{om}/z_{oh} ratio has a fix value and H can be calculated for every pixel with λE as the residual term in Eq. 1. The SEBAL models have been used widely with satellite data in the case of relatively flat landscapes with and without irrigation.

The Mapping EvapoTranspiration with high Resolution and Internalized Calibration (METRIC) models, derived from SEBAL are used for irrigated crops (Allen et al. 2007a, b). METRIC model derive ET from remotely sensed data (LANDSAT TM) in the visible, near-infrared and thermal infrared spectral regions along with ground-based wind speed and near surface dew point temperature. In this case extreme pixels are identified with the cool/wet extreme comparable to a reference crop, the evaporation rates being computed wit Penman-Monteith method. The ET from warm/dry pixel is calculated using soil water budget having local meteorological data as input parameters. METRIC model can be used to produce high quality and accurate maps of ET for areas smaller than a few hundred kilometers in scale and at high resolution (Fig. 5). In their study, Boegh et al. (1999) presented an energy balance method for estimating transpiration rates from sparse canopies based on net radiation absorbed by the vegetation and the sensible heat flux between the leaves and the air within the canopy. The net radiation absorbed by the vegetation is estimated using remote sensing and regular meteorological data by merging conventional method for estimation of the land surface net radiation with a ground-calibrated function of NDVI.

SEBAL and METRIC methods assume that the temperature difference between the land surface and the air (near-surface temperature difference) varies linearly with land surface temperature. Bastiaanssen et al. (1998) and Allen and al. (2007) derive this relationship based on two anchor pixels known as the hot and cold pixels, representing dry and bare agricultural fields and wet and well-vegetated fields, respectively. Both methods use the linear relationship between the near-surface temperature difference and the land surface temperature to estimate the sensible heat flux which varies as a function of the near-surface. temperature difference, by assuming that the hot pixel experiences no latent heat, i.e., ET = 0.0, whereas the cold pixel achieves maximum ET.

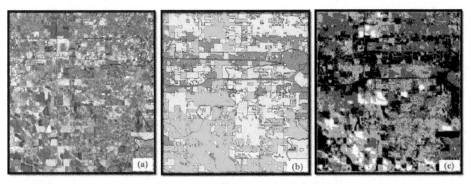

Fig. 5. (a) Landsat color infrared image of T3NR1E of the Boise Valley; (b) Land use/land cover polygons in T3NR1E of the Boise Valley; (c) ET image of T3NR1E the Boise Valley (after R.G. Allen et al., 2007)

The sensible heat flux is assessed like a linear function of the temperature difference between vegetation and mean canopy air stream. The surface temperature recorded by satellite comprises information from soil and from vegetation; therefore the vegetation temperature is estimated taking into account the linear relationship between NDVI and surface temperature. The difference between the surface temperature and the mean canopy air stream temperature is linearly related to the difference between surface temperature and the air temperature above the canopy with the slope coefficient which depend on the canopy structure. This relationship was used to evaluate the mean canopy air stream temperature. The method was used in the Sahel region for agricultural crops, natural vegetation, forest vegetation, with ground based, airborne and satellite remote sensing data and validated with sapflow and latent heat flux measurements. Agreement between remote sensing based estimates and ground based measurements of λE rates is estimated to be better than 30–40 W m-2.

5.2 Reflectance and surface temperature

The Simplified Surface Energy Balance Index (S-SEBI) proposed by Roerink et al. (2000) estimate the instantaneous latent heat flux (λE_i) with (Kalma, 2008):

$$\lambda E_i = \Lambda_i (R_{ni} - G_i) \tag{18}$$

where: ($R_{ni} - G_i$) = available energy at the time of the satellite overpass; Λ_i = the evaporative fraction. The S-SEBI algorithm has two limitations: the atmospheric conditions have to be almost constant across the image and the image has to contain borh dry and wet areas. Λ_i was obtained from a scatter plot of observed surface temperature (T_{rad}) and Landsat TM derived broadband a values across the single scene. Λ_i is with:

$$\Lambda_i = \frac{T_H - T_{rad}}{T_H - T_{rad}} \tag{19}$$

where: T_{rad} = observed surface temperature for a given pixel; T_H = temperature for the upper boundary (dry radiation controlled conditions - all radiation is used for surface heating and α decreases with increasing surface temperature (T_H - where $\lambda E = 0$ (W m-2)); $T_{\lambda E}$ = temperature at the lower boundary (evaporation controlled wet conditions - all energy is used for λE and α increases with an increase of surface temperature ($T_{\lambda E}$ -where H = 0 W m^{-2})). This method does not need any additional meteorological data.

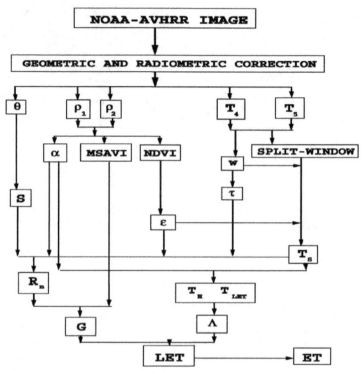

Fig. 6. Flowchart of the proposed methodology to obtain ET from NOAA–AVHRR data (after Sobrino et al., 2007)

Sobrino et. al (2007) use S-SEBI algorithm to estimate the daily evapotranspiration from NOAA-AVHRR images for the Iberian Penisnula. The Figure 6 present the flowchart used by Sobrino et al. (2007) to obtain ET from NOAA-AVHRR. Daily evapotranspiration (ET$_d$) is given by:

$$ET_d = \frac{\Lambda_i C_{di} R_{ni}}{L} \tag{20}$$

where: R_{nd} = daily net radiation; R_{ni} = instantaneous net radiation: L = 2.45 MJ kg^{-1} = latent heat vaporization; $C_{di} = R_{nd} / R_{ni}$. In this case the daily ground heat flux was considered close to 0. There are several studies which proposed methods for C_{di} calculation. For example Seguin and Itier (1983) proposed a constant value for C_{di} = (0.30±0.03). Wassenaar et al. (2002) showed that this ratio have a seasonal variation 0.05 in winter to 0.3 in summer, following a sine law. In the Sobrino et al. (2007) study, C_{di} was calculated using net radiation fluxes measured at the meteorological station of located on the East coast of the Iberian Peninsula (El Saler area). The ET estimation from high spectral and spatial resolution data (~5 m) was adapted to the low resolution data NOAA-AVHRR (1 km spatial resolution) based on the evaporative fraction concept proposed by Roerink et al. (2007). The main advantage of the Sobrino et al. (2007) methodology is that the method requires only satellite data to estimate ET.

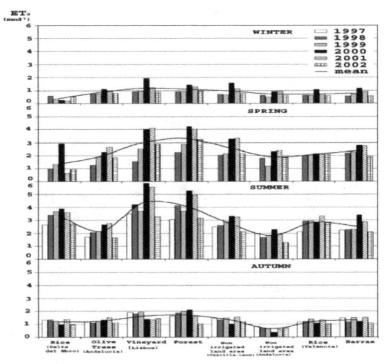

Fig. 7. Monthly evolution (from June 1997 to November 2002) of the daily evapotranspiration (ET_d) in the eight selected zones. There is represented also the temporal mean for the six years of analyzing (after Sobrino et al., 2007).

Its major disadvantage is represented by the requiring that satellite images must have extreme surface temperatures. The method was tested over agricultural area using high resolution values, with errors lower than 1.4 mm d^{-1}. As it can be observed from Fig. 7, regarding the monthly and seasonal evolution of ET the highest values (~6 mm d^{-1}) were obtained in the West of the Iberian Peninsula, which is the most vegetated area. Taking into account the impact of incoming solar energy the higher values of ET was obtained in spring and summer and the lower values in autumn and winter. Seasonal ET was obtained by averaging daily ET over the season. Figure 8 shows as an example the monthly ET maps obtained from the NOAA-AVHRR images acquired in 1999. Fig. 9 also indicates that the highest ET values were obtained in the summer and spring, in the north and west of Iberian Peninsula. To map land surface fluxes and surface cover and surface soil moisture, Gillies and Carlson (1995) combined two model, SVAT and ABL and run it for vegetative cover with the maximum known NDVI and for bare soil conditions with the minimum known NDVI in the scene for a range of soil moisture values until AVHRR observed (T_{rad}) and simulated (T_{ad}) surface temperatures corrected, at which stage the actual fractional vegetation cover (f_c) and surface soil moisture were estimated.

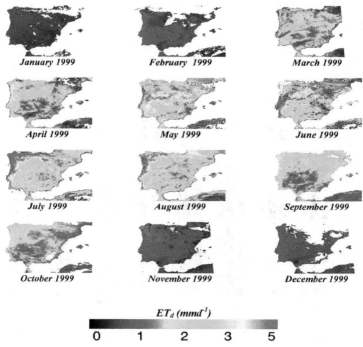

Fig. 8. Monthly mean for the daily evapotranspiration obtained from NOAA–AVHRR data over the Iberian Peninsula in 1999. Pixels in black color correspond to sea and cloud masks and red correspond to higher value of ET (after Sobrino et al., 2007).

5.3 Vegetation indices and surface temperature

Several studies shown the efficiency of "triangle method" (Carlson et al. (1995a, b); Gillies et al. 1997; Carlson 2007) to estimate soil moisture from the NDVI-T_{rad} relationship. The major advantages of the remotely sensed VI-T_s triangle method are that: the method allows an accurate estimation of regional ET with no auxiliary atmospheric or ground data besides the remotely sensed surface temperature and vegetation indices; is relatively insensitive to the correction of atmospheric effects. Its limitations are: determination of the dry and wet edges requires a certain degree of subjectivity; to make certain that the dry and wet limits exist in the VI-T_{rad} triangle space most of pixels over a flat area with a wide range of soil wetness and fractional vegetation cover are required. So, the boundaries of this triangle are limiting conditions for H and λE. Other studies suggest the dependence of T_{rad} variability on the remote sending data resolution, thus higher resolution data means that the variations of T_{rad} and NDVI is more related to the land cover type. Lower resolution data show the dependency of the NDVI and T_{rad} variations to agricultural practices and rainfall. Jiang and Islam (2001) proposed a triangle method based on the interpolation of the Priestley–Taylor method (Priestley and Taylor, 1972) using the triangular (T_{rad}, NDVI) spatial variation. The Priestley–Taylor expression for equilibrium evaporation from a wet surface under conditions of minimal advection (λE_{PT}) is given by:

$$\lambda E_{PT} = \alpha_{PT}(R_n - G)\frac{\Delta}{\Delta + \gamma} \tag{21}$$

where: Δ = slope of the saturated vapour pressure curve at the prevailing Ta ((Pa K^{-1}); γ = psychrometric constant (Pa K^{-1}); α_{PT} = Priestley-Taylor parameter defined as the ratio between actual E and equilibrium E. For wet land surface conditions, α_{PT} = 1.26. Its value is affected by global changes in air temperature, humidity, radiation and wind speed. Jiang and Islam (2001) replaced α_{PT} with parameter ϕ which varies for a wide range of r_a and r_c values. The warm edge of the (T_{rad}, NDVI) scatter plot represents pixels with the highest T_{rad} and minimum evaporation from the bare soil component, while E_a can vary function of the vegetation type. Linear interpolation between the sides of the triangular distribution of T_{rad} - NDVI allows to derive ϕ for each pixel using the spatial context of remotely sensed T_{rad} and NDVI. The ϕ values are related to surface wetness, r_s and T_{rad}. Therefore, the minimum value of ϕ is 0 for the driest bare soil pixel and the maximum value is 1.26 for a densely vegetated, well-watered pixel. Thus the actual ϕ value for each pixel in a specified NDVI interval is obtained from the observed $(T_{rad})_{obs}$ with the following:

$$\phi = \phi_{max} \frac{(T_{rad})_{max} - (T_{rad})_{obs}}{(T_{rad})_{max} - (T_{rad})_{min}} \quad (22)$$

where $(T_{rad})_{min}$ and $(T_{rad})_{max}$ are the lowest and highest surface temperatures for each NDVI class, corresponding to the highest and lowest evaporation rates, respectively. The evaporative fraction can be calculated with:

$$\Lambda = \phi \frac{\Delta}{\Delta + \gamma} \quad (23)$$

Based on the Jiang and Islam (2001) approach, Wang et al. (2006) obtained better results using the spatial variation (ΔT_{rad}, NDVI), where ΔT_{rad} represent the day–night difference in T_{rad}, obtained from MODIS data. However, to convert Λ into E, the method described above still requires estimation/ measurement of net radiation (R_n) and soil heat flux (G). In a later work, Jiang and Islam (2003) consider the fractional vegetative cover (f_c) as a more suitable generalized vegetation index calculated from the normalized NDVI with (Kalma et al. 2008):

$$f_c = \left(\frac{NDVI - NDVI_{min}}{NDVI_{max} - NDVI_{min}} \right)^2 \quad (24)$$

They assumed that the evaporative fraction $\Lambda = \lambda E/(R_n - G)$ is linearly related to $\Delta T = T_{rad} - T_a$, inside a certain class f_c. The reason for this assumption is that the ΔT is more representative for sensible heat flux H. Thus the evaporative fraction can be estimated from f_c and ΔT, for a given set of ΔT_{max}, ΔT_e ($\Delta T_e = \Delta T_{max}$ for f_c = 1) and a stress factor (β). In their study, they used NOAA-AVHRR data and obtained better results using the aerodynamic resistance-energy balance method represented by Eq. 13, this equation including atmospheric stability corrections and using an iterative procedure to reach the most appropriate kB^{-1} value.

Serban et al. (2010) used the Priestly-Taylor equation modified by Jiang and Islam (2001) in their study to estimate the evapotranspiration using remote sensing data and Grid Computing. The most advantage of Priestly-Taylor equation is that the all terms can be calculated using remotely sensed data. Grid computation procedure has two major advantages: strong data processing capacity and the capability to use distributed computing resources to process the spatial data offered by a satellite image. According to Jiang and Islam (2001) the parameter α_{PT} parameter is obtained by two-step linear interpolation: in the first step is obtained upper and lower bounds of α_{PT} for each specific NDVI class (determined from the land use/land cover map); in the second step the parameter α_{PT} is ranged within each NDVI class between the lowest temperature pixel and the highest temperature pixel. According to land use/land cover map, for this paper, was considered four main land uses: vegetation, water, barren land and urban. Each NDVI value corresponds to a certain NDVI class. In this case the relationship between LST and NDVI is used. Thus, the parameter α_{PT} is calculated with:

$$\alpha_{PT} = \left(\frac{\Delta + \gamma}{\Delta}\right)\left(\frac{LST_i^{max} - LST}{LST_i^{max} - LST_i^{min}}\right)\left(\frac{NDVI_i^{max} - NDVI_i^{min}}{NDVI_i^{max}}\right) + \left(\frac{\Delta + \gamma}{\Delta}\right)\left(\frac{NDVI_i^{min}}{NDVI_i^{max}}\right) \quad (25)$$

where: LST = surface temperature for current pixel; LST_i^{max} and LST_i^{min} = maximum and minimum surface temperature within NDVI class which has the current pixel; $NDVI_i^{max}$ and $NDVI_i^{min}$ are the maximum and minimum NDVI within NDVI class which has the current pixel. They calculated the daily value of ET with the following (Fig. 9):

$$\lambda E_{daily} = \alpha_{PT}\frac{2DL(R_i - G_i)}{\pi sin\left(\pi\frac{t}{DL}\right)} \quad (26)$$

where: DL = total day length (hours); t = time beginning at sunrise. To obtain the 24 hours totals, the daily ET values are multiplied by 1.1 for all days. LST was computed using Jimenez- Munoz and Sobrino's algorithm which requires a single ground data (the total atmospheric water vapor content – w) (Fig. 10):

$$LST = \gamma[LSE^{-1}(\psi_1 L_{sensor} + \psi_2) + \psi_3] + \delta \quad (27)$$

$$\gamma = \left[\frac{c^2 L_{sensor}}{T_{sensor}^2}\left(\frac{\lambda^4}{c_1}L_{sensor} + \lambda^{-1}\right)\right] \quad (28)$$

$$\delta = \gamma L_{sensor} + T_{sensor} \quad (29)$$

$$L_{sensor} = gain * DN + bias - spectral\ radiance \quad (30)$$

$$T_{sensor} = \frac{K_2}{ln\left(\frac{K_1}{L_{sensor}} + 1\right)} \quad (31)$$

where: LSE = land surface emissivity = 1.0094+0.047*ln(NDVI); λ = effective wavelength; DN = digital number of a pixel; T_{sesnor} = brightness temperature; $c1$ = 1.19104*10^8 Wμm^4 m^{-2}sr^{-1}; $c2$ = 14387.7μmK; ψ_i (i = 1, 2, 3) = atmospheric parameters, which depend on total atmospheric water vapor content (w). Besides satellite data, this study uses two ground meteorological data: the total atmospheric water vapor content - w, used in LST estimation algorithm, and the air temperature - T_{air}. To estimate evapotranspiration, Serban et al. (2010) used one subset of Landsat ETM+ (7th June 2000) for Dobrogea area corresponding to Constanta weather station, which was atmospherically corrected.

From the bands ETM+ 3 and 4 were analyzed the NDVI values, the band ETM+ 6 was processed to determine LST, and the other bands (ETM+ 1, 2, 5 and 7) were used to estimate the albedo values. The difference between the actual mean soil surface temperature at the

Possibilities of Deriving Crop
Evapotranspiration from Satellite Data with the Integration with Other Sources of Information

277

time when satellite passed and the remote sensed mean land surface temperature (0.73°C) is considered acceptable. The evapotranspiration (Fig. 10) ranges between 0.33 and 5.24mm/day. According to Constanta weather station, the multi-annual average of the evapotranspiration in June is between 4.5 and 5.6 mm/day, so the estimation error is eligible.

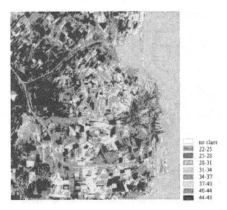

Fig. 9. LST Image - Dobrogea region, 2000 (After Serban et al., 2010)

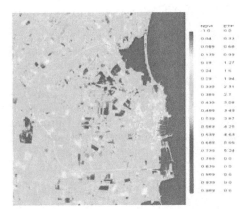

Fig. 10. ETP Image - Dobrogea region, 2000 (After Serban et al., 2010)

6. ET estimation using meteorological data

6.1 Crop evapotranspiration

At a crop level, ET may not occur uniformly because variations in crop germination, soil water availability, and other factors such as non-uniform water and nutrient applications and an uneven distribution of solar radiation within the canopy. Usually, the top leaves are more active in transpiration than the lower leaves because they receive more light. Also, the bottom leaves mature and age earlier and they may have lower transpiration rates than the greener and younger top leaves. Thus, weather parameters, crop characteristics, environmental and management aspects are the factors which influence the evaporation and transpiration processes. The main weather parameters influencing evapotranspiration are radiation, air temperature, humidity and wind speed. Several algorithms have been developed to estimate the evaporation rate from these parameters. The evaporation power of the atmosphere is expressed by the reference crop evapotranspiration (ET_o) which represents the evapotranspiration from a standardized vegetated surface (Allen et al., 1998). The reference surface is a hypothetical grass reference crop with specific characteristics. Because ET_o is affected by only climatic parameters, it is a climatic parameter and may be computed from weather data. Thus ET_o is the evaporating power of the atmosphere at a specific location and time of the year and does not take into account the crop characteristics and soil factors.

Crop water requirement is defined as the amount of water required to compensate the evapotranspiration loss from the cropped field. Even the values for crop evapotranspiration are identical with crop water requirement (CWR), crop evapotranspiration refers to the amount of water that is lost by evapotranspiration, while CWR refers to the amount of water that needs to be supplied. Thus, the irrigation water requirement represents the difference between the crop water requirement and effective precipitation and also includes additional water for leaching of salts and to compensate for non-uniformity of water application (Allen et al., 1998). Several empirical methods have been developed over the last five decades in order to estimate the evapotranspiration from different climatic variables. Testing the accuracy of the methods under a new set of conditions is laborious, time-consuming and costly, and yet evapotranspiration data are frequently needed at short notice for project planning or irrigation scheduling design. To meet this need, guidelines were developed and published in the FAO Irrigation and Drainage Paper No. 24 'Crop water requirements'. From different data availability, four methods are usually used to estimate the reference crop evapotranspiration (ET_o): the Blaney-Criddle, radiation, modified Penman and pan evaporation methods. From these four methods, the modified Penman-Monteith method offer the best results with minimum possible error in relation to a living grass reference crop. The radiation method can be used for areas where available climatic data include measured air temperature and sunshine, cloudiness or radiation, but not measured wind speed and air humidity. The Blaney-Criddle method is better to be applying for areas where available climatic data cover air temperature data only. The pan method gives acceptable estimates, depending on the location of the pan. Based on the original Penman- FAO proposed a standard parameterization of the Penman–Monteith method for estimating the evaporation from a -irrigated, homogenous, 0.12 m grass cover considered as a "reference crop" (Allen et al., 1998) (Fig. 11).

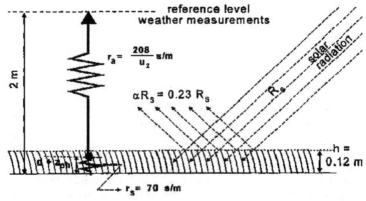

Fig. 11. Characteristics of the hypothetical reference crop (after Allen et al., 1998)

Possibilities of Deriving Crop
Evapotranspiration from Satellite Data with the Integration with Other Sources of Information

279

Monteith equation and the equations of the aerodynamic and surface resistance, the FAO
Penman-Monteith method to estimate ET_0 is the following:

$$ET_0 = \frac{0.408\Delta(R_n - G) + \gamma\left(\frac{900}{T + 273}\right)u_2(e_s - e_a)}{\Delta + \gamma(1 + 0.34u_2)} \qquad (32)$$

where: ET_0 = reference evapotranspiration [mm day^{-1}]; R_n = net radiation at the crop surface
[MJ m^{-2} day^{-1}]; G = soil heat flux density [MJ m^{-2} day^{-1}]; T = mean daily air temperature at 2
m height [°C]; u_2 = wind speed at 2 m height [m s^{-1}]; e_s = saturation vapour pressure [kPa];
e_a = actual vapour pressure [kPa]l; e_s - e_a = saturation vapour pressure deficit [kPa]; Δ =
slope vapour pressure curve [kPa °C^{-1}]; γ = psychrometric constant [kPa °C^{-1}]. The equation
uses standard climatological records of solar radiation (sunshine), air temperature, humidity
and wind speed. To obtain correct estimations of ET_0, the weather measurements should be
made at 2 m (or converted to that height) above an extensive surface of green grass, shading
the ground and not short of water. The psychrometric constant, γ, is calculated with:

$$\gamma = \frac{C_p P}{\varepsilon\lambda} = 0.665 * 10^{-3} \qquad (33)$$

Where: P = atmospheric pressure [kPa]; λ = latent heat of vaporization, 2.45 [MJ kg^{-1}]; c_p =
specific heat at constant pressure, 1.013 10-3 [MJ kg^{-1} °C^{-1}]; ε = ratio molecular weight of
water vapour/dry air = 0.622. For standardization, T_{mean} for 24 hour is defined as the mean
of the daily maximum (T_{max}) and minimum temperatures (T_{min}) rather than as the average of
hourly temperature measurements.

$$T_{mean} = \frac{T_{max} - T_{min}}{2} \qquad (34)$$

The temperature is given in degrees Celsius (°C), Fahrenheit (°F) or in Kelvin (K =C° + 273,16).

$$P = 101.3\left(\frac{293 - 0.0065z}{293}\right)^{526} \qquad (35)$$

where: z = elevation above sea level [m].

6.2 CROPWAT model

CROPWAT is a decision support system developed by the Land and Water Development
Division of FAO for planning and management of irrigation. The main functions of
CROPWAT model are: to calculate the reference evapotranspiration, crop water
requirements and crop irrigation requirements; to develop irrigation schedules under
different management conditions and water supply schemes; to estimate the rainfed
production and drought effects; to evaluate the efficiency of irrigation practices.
The input data of the model are the following climatic, crop and soil data: reference crop
evapotranspiration: (ET_0) values measured or calculated using the FAO Penman–Montieth
equation based on monthly climatic average data of the minimum and maximum air
temperature (°C), relative humidity (%), sunshine duration (h) and wind speed (m/s);
rainfall data: (daily/monthly data); monthly rainfall is divided for each month into a
number of rainstorms; a cropping pattern: crop type, planting date, crop coefficient data
files (including K_c values, stage days, root depth, depletion fraction, K_y values) and the area
planted (0- 100% of the total area); a set of typical crop coefficient data files are provided in
the program; soil type: total available soil moisture, maximum rain infiltration rate,

maximum rooting depth, and initial soil moisture depletion (% of the total available moisture);scheduling criteria: several options can be selected regarding the calculation of the application timing and application depth.

The output parameters for each crop are crop reference crop evapotranspiration Et_0 (mm/period), crop K_c (average values of crop coefficient for each time step, effective rain (mm/period) (the amount of water that enters in the soil; water requirements (CWR) or ET_m (mm/period); irrigation requirements (IWR - mm/period); actual crop evapotranspiration (ET_c - mm); effective rain (mm/period) which represents the amount of water that enters into the soil; daily soil moisture deficit (mm); estimated yields reduction due to crop stress (when ET_c/ET_m falls below 100%).

The CROPWAT model can compute the actual evapotranspiration using the FAO Penman–Monteith equation or using directly the evapotranspiration measurements values. The crop water requirements (CWR) or maximum evapotranspiration (ET_m) (mm/period) are calculated as:

$$CWR = ET_0 * CropK_c \tag{36}$$

This means that the peak CWR in mm/day can be less than the peak Et_0 value when less than 100% of the area is planted in the cropping pattern.

The average values of the crop coefficient (K_c) for each time step are estimated by linear interpolation between the K_c values for each crop development stage. The "Crop K_c" values are calculated as:

$$CropK_c = K_c * CropArea \tag{37}$$

where CropArea is the area covered by the crop. So, if the crop covers only 50% of the area, the "Crop Kc" values will be half of the Kc values in the crop coefficient data file.

The CROPWAT model operates in two modes: computing the actual evapotranspiration using climatic parameters and using directly the evapotranspiration measurements values.

Possibilities to use the satellite-based data as input into the CROPWAT model are limited, because this model was not developed to use satellite-derived information directly. But this information can be useful for the comparison/validation procedures of some model input/output data, as precipitation, sunshine duration and evapotranspiration. Satellite based data can be used by CROPWAT model in different ways: measured evapotranspiration may be replaced with estimations derived from satellite data; for comparison and validation procedures; satellite-derived evapotranspiration values may bring better accuracy for the specialization of the punctual computing values; satellite information may be used for the assessment of the some reference parameters of the actual evapotranspiration (e.g. Land surface temperature, vegetation indexes, etc.).

6.3 Using earth observation data and CROPWAT model to estimate the actual crop evapotranspiration

There is a strong dependence between evapotranspiration and surface temperature on the, thus thermal images meteorological satellites (METEOSAT, NOAA, MODIS, LANDSAT) adequate for mapping of regional evapotranspiration. Several works have been done to determine regional evapotranspiration from satellite data (Batra et al., 2006; Courault et al., 2005; Wood et al., 2003). The application of NOAA AVHRR data seems to be more successful because of the higher spatial and spectral resolution (Stancalie et al., 2010). Multichannel algorithms are routinely used for atmospheric correction of the AVHRR data.

Efforts are directed towards the estimation of surface temperatures by considering the effects of emissivity (Lagouarde and Brunet, 1991; Li and Becker, 1993). The method used for the estimation of the daily crop actual evapotranspiration, ET_{cj}, is based on the energy balance of the surface. The method uses the connection between evapotranspiration, net radiation and the difference between surface and air temperatures measured around 14:00 h (the time of the satellite passage), local time. The first version of the method used a simplified linear relationship as:

$$ET_{cj} - R_{nj} = A - B * (T_s - T_{amax})$$ (38)

where R_{nj} is the daily net radiation; T_s and T_{amax} is the surface and air maximum temperature; A, B are coefficients which depend on the surface type and the daily mean wind speed. Coefficients A and B may be determined either analytically, on the basis of the relationships given by Lagouarde and Brunet (1991), or statistically. The coefficients A and B are stable in the case of mature crop vegetation cover and in clear sky conditions. The coefficient B vary considerably, function of the land vegetation cover percent. In case of soil with great thermal inertia, the heat flux changed by conduction at the soil-atmosphere interface can be neglected and the computing relationship for daily actual crop evapotranspiration can be expressed in a version 2 of the proposed method:

$$ET_{cj} = R_{nj} - B' * (T_s - T_{amax})$$ (39)

$$B' = 0.0253 + \left[\frac{1.0016}{log2(2/zh)}\right] v$$ (40)

$$zh = [1 - exp(-LAI)]\left[exp\left(-\frac{LAI}{2}\right)\right]$$ (41)

where: v = daily average wind speed; zh = vegetation roughness and LAI the foliar index.
One possible use of satellite information is to replace the measured evapotranspiration by estimations made from satellite information. Because the estimations made from satellite information are available only for clear sky conditions, it was not possible to estimate the monthly average evapotranspiration, as input data in the CROPWAT model. For this reason, the satellite-derived data have been used for comparison/validation procedures of the CROPWAT model output data, like evapotranspiration. Fig. 12 presents the comparison between daily crop evapotranspiration values computed by the CROPWAT model and those computed through the energy balance method (Version 1), using remotely sensed data at the Alexandria and Craiova test-areas (situated in the south-western part of Romania), in the conditions of the year 2000 (Stancalie et al., 2010, 2010).
Analysis of model results concerning comparison of daily actual crop evapotranspiration calculated by using climatic data vs. satellite estimations based on the surface energetic balance (Version 1) showed that ET_c values from satellite information are in general higher than those simulated by the model, the differences being from +0.45 - 1.9 mm/day. Preliminary results highlighted a good correlation between the simulated values (CROPWAT) and those derived from the satellite data; with relative errors from +20% - 18% at Craiova site and from +13% -17% at Alexandria site (Stancalie et al., 2010).
Fig. 13 shows a comparison between ET_c simulated daily by the CROPWAT model over the whole maize-growing season and by the energy balance method (Version 2) respectively, using satellite data, at Alexandria and Craiova test-areas. The ET_c calculated by the model is very similar to the estimated one. The results obtained can constitute the premise of an ET_c data validation process, determined by the CROPWAT model (Stancalie et al., 2010).

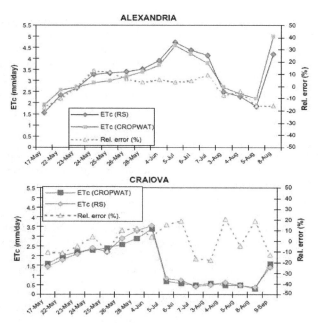

Fig. 12. Comparison between daily crop evapotranspiration values computed by the CROPWAT model and by the energy balance method (Version 1) using satellite data at the Alexandria and Craiova test-areas (after Stancalie et al., 2010)

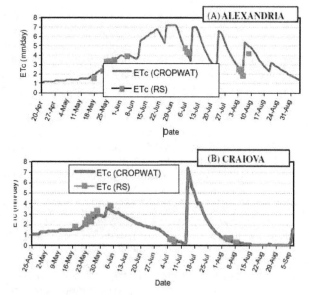

Fig. 13. Comparison between daily crop et values computed by the CROPWAT model and by the energy balance method (Version 2) using satellite data, at Alexandria (A) and Craiova (B) test-areas, for the maize vegetative development period in 2000 (Stancalie et al., 2010).

7. Conclusions

The use of the multispectral satellite data can improve the classical methods applied in determining the agrometeorological parameters, including evapotranspiration.

Estimating evapotranspiration using remote sensing methodologies have a significant role in irrigation management and crop water demand assessment, for plant growth, carbon and nutrient cycling and for production modeling in dry land agriculture and forestry. Also it can have an important role in catchment hydrology, and larger scale meteorology and climatology applications. In the last years, due to the exceptional developments of satellite technology, a wide range of remote sensing-based evapotranspiration (ET) methods/models have been developed and evaluated. The use of remote sensing data for ET estimation is mainly based on land surface temperature (LST) and reflectivity (using different spectral regions) due to satellite ability to spatially integrate over heterogeneous surfaces at a range of resolutions and to routinely generating areal products once long time-series data availability issues are overcome. The chapter reviews some main methods for estimating crop evapotranspiration based on remotely sensed data, and highlights uncertainties and limitations associated with those estimation methods. This paper is focused on Surface Energy Balance models (SEB), spatial variability methods using vegetation indices and ET estimation using meteorological data through CROPWAT model. The analysis and critical issues are supported by the dedicated literature and specific case-studies. This review provides information of temporal and spatial scaling issues associated with the use of optical and thermal remote sensing for estimating evapotranspiration. Improved temporal scaling procedures are required to extrapolate estimates to daily and longer time periods and gap-filling procedures are needed when temporal scaling is affected by intermittent satellite coverage. It is also noted that analysis of multi-resolution data from different satellite/sensor systems is able to assist the development of spatial scaling and aggregation approaches. Approaches differ in: (i) type and spatial extent of application (e.g. irrigation, dry-land agriculture); (ii) type of remote sensing data; and (iii) use of ancillary (micro-) meteorological and land cover data. The integration of remotely sensed data into methods/models of ET facilitates the estimation of water consumption across agricultural regions. There are important limitations for using remote sensing data in estimating evapotranspiration.

Usually evapotranspiration is computed using land surface temperature and air temperatures. All this methods are affected by errors induced by estimation or measurements of those temperatures. The accuracy of T_{rad} observations is influenced by atmospheric factors, surface emissivity or view angle. Emissivity information is useful in estimating of the radiative temperature of the land surface. Several direct methods (which atmospheric variables are coupled with radiative transfer models) or indirect algorithms (use only remote sensing data) to make atmospheric corrections in order to obtain the brightness temperature that represents the temperature of a black body that would have the same radiance as that observed by the radiometer. The uncertainties of surface temperature have a strong influence in determination of sensible heat flux H. The difference between surface and air temperatures depends on many factors, including vegetation type, fractional cover f_c and view angle. Another important limitation of various spatial variability methods is considered the fact according to the highest and lowest surface temperatures observed in the one scene are assumed to represent very dry and very wet pixels. Usually the available energy (R_n - G) is obtained from ground based point observations of R_n: R_n is estimated based on observations of K↓, α, LAI, emissivity of land surface and atmosphere, and T_{rad}. Such kind of estimation generates errors in the calculation of long and short wave components. G can be estimated for example as function of NDVI. An alternative method would be to assume that soil heat flux is a constant fraction of net radiation flux, but this estimation doesn't take into account the diurnal variation. Many models for ET estimation need ground based meteorological data, mainly air temperature and wind speed. For that models which based on computing the difference between Tad and Ta, the time and location of air temperature (Ta) observations and their spatial representativeness are very important).

Incomplete vegetation cover generates also errors in evapotranspiration estimation. The two source models require parameterizations for the segmentation of the computed surface temperature between vegetation and soil, for the turbulent exchange of heat and mass between soil and atmosphere and between vegetation and atmosphere. Also, these models require some assumptions regarding solar transmittance, extinction coefficients and canopy emissivity in order to compute the variation of net radiation flux inside the canopy.

Another important limitation, regarding the spatial variability methods is that a large number of pixels are required over the area of interest with a wide range of soil wetness and fractional vegetation cover. The identification of vegetation limits for bare soil or full vegetation cover can be easily done using high resolution images which display a wide range in surface wetness conditions and land cover conditions

Remote sensing data is a useful tool that provides input data in land surface model (NDVI, LAI, f_c – fraction cover) and can be used to correct the state variables of the models.

The frequency of spatial resolution imagery is also very significant: satellites which provide high resolution data usually have lower temporal frequency while low spatial resolution images have higher temporal frequency. Some applications require different spatial and temporal coverage rates and need different "turn-around" times. If acquiring the satellite data and ET estimation method are more time consuming, the method are not very convenient for operational applications like determining water requirements for irrigated agriculture.

Another significant limitation for using remote sensing is the presence of clouds that generates intermittent coverage. Cloudy days are characterized by a diffuse light, whereas while direct light is dominant on clear days when most TIR data are acquired for use in modeling applications. Most SEB models have been developed for use in cloud-free conditions and do not makes difference between direct and diffuse radiation; they use only daytime data obtained for clear-sky conditions. For a continuously monitoring of water balance, the effects of an increased diffuse fraction should be taking into account, because the diffuse radiation is used by vegetation more efficiently than direct radiation. For water use efficiency, to ignore difference between direct and diffuse radiation can induce significant differences in ET estimations.

8. References

Allen RG, Pereira LS, Raes D, Smith M (1998), Crop evapotranspiration - guidelines for computing crop water requirements. FAO irrigation and drainage paper 56, Rome, Italy http://www.fao.org/docrep/X0490E/X0490E00.htm

Allen RG, Tasumi M, Trezza R (2007a) Satellite-based energy balance for mapping evapotranspiration with internalized calibration (METRIC): model. J Irrig Drain Eng 133(4):380–394. doi:10.1061/(ASCE) 0733-9437(2007)133(4):(380)

Allen RG, Tasumi M, Trezza R (2007b) Satellite-based energy balance for mapping evapotranspiration with internalized calibration (METRIC): applications. ASCE J Irrig Drain Eng 133(4):395–406

Bastiaanssen WGM, Menenti M, Feddes RA, Holtslag AAM (1998a) A remote sensing surface energy balance algorithm for land. I. Formulation. J Hydrol (Amst) 212/213:198–212. doi:10.1016/S0022-1694(98)00253-4

Bastiaanssen WGM (2000) SEBAL-based sensible and latent heat fluxes in the irrigated Gediz Basin, Turkey. J Hydrol (Amst) 229:87–100. doi:10.1016/S0022-1694(99)00202-4

Batra N, Islam S, Venturini V, Bisht G, Jiang L (2006) Estimation and comparison of evapotranspiration from MODIS and AVHRR sensors for clear sky days over the Southern Great Plains. Remote Sens Environ 103:1–15. doi:10.1016/j.rse.2006.02.019

Boegh E, Soegaard H, Hanan N, Kabat P, Lesch L (1999) A remote sensing study of the NDVI-Ts relationship and the transpiration from sparse vegetation in the Sahel based on high-resolution satellite data. Remote Sens Environ 69:224–240. Doi: 10.1016/S0034-4257(99)00025-5

Boegh E, Soegaard H, Thomsen A (2002) Evaluating evapotranspiration rates and surface conditions using Landsat TM to estimate atmospheric resistance and surface resistance. Remote Sens Environ 79:329–343. doi:10.1016/S0034-4257(01)00283-8

Brutsaert W (1999) Aspects of bulk atmospheric boundary layer similarity under free-convective conditions. Rev Geophys 37:439–451. Doi: 10.1029/1999RG900013.

Burba G., Hubart J.A., Pidwirny M. (2010), Evapotranspiration, Encyclopedia of Earth, Eds. Cutler J. Cleveland (Washington, D.C.: Environmental Information Coalition, National Council for Science and the Environment, August 3, 2010, http://www.eoearth.org/article/Evapotranspiration

Carlson TN, Capehart WJ, Gillies RR (1995a) A new look at the simplified method for remote sensing of daily evapotranspiration. Remote Sens Environ 54:161–167. Doi: 10.1016/0034-4257(95)00139-R

Carlson TN (2007) An overview of the "triangle method" for estimating surface evapotranspiration and soil moisture from satellite imagery. Sensors 7:1612–1629

Cleugh HA, Leuning R, Mu Q, Running SW (2007) Regional evaporation estimates from flux tower and MODIS satellite data. Remote Sens Environ 106:285–304. doi:10.1016/j.rse.2006.07.007

Courault D, Seguin B, Olioso A (2005) Review on estimation of evapotranspiration from remote sensing data: from empirical to numerical modelling approaches. Irrig Drain Syst 19:223–249. Doi: 10.1007/s10795-005-5186-0

Gillies RR, Carlson TN (1995) Thermal remote sensing of surface soil water content with partial vegetation cover for incorporation into climate models. J Appl Meteorol 34:745–756. doi :10.1175/1520-0450(1995)034\0745:TRSOSS[2.0.CO;2

Gillies RT, Carlson TN, Cui J, Kustas WP, Humes KS (1997) A verification of the "triangle" method for obtaining surface soil water content and energy fluxes from remote measurements of the Normalized Difference Vegetation Index (NDVI) and surface radiant temperatures. Int J Remote Sens 18(15):3145–3166. doi:10.1080/ 014311697217026

Glenn EP, Huete AR, Nagler PL, Hirschboeck KK, Brown P (2007) Integrating remote sensing and ground methods to estimate evapotranspiration. Crit Rev Plant Sci 26(3):139–168. doi:10.1080/07352680701402503

Hope AS, Petzold DE, Goward SN, Ragan RM (1986) Simulated relationships between spectral reflectance, thermal emissions, and evapotranspiration of a soybean canopy. Water Resour Bull 22:1011–1019

Huntington, T. 2006. Evidence for intensification of the global water cycle: review and synthesis. J Hydr. 319: 83–95.

Hutley, L., O'Grady, A., and Easmus, D. 2001. Monsoonal influences on evapotranspiration of savanna vegetation of northern Austalia. Oecologia 126: 434–443.

Huxman, T., Smith, M., Fay, P., Knapp, A., Shaw, M., Loik, M., Smith, S., Tissue, D., Zak, J., Weltzin, J., Pockman, W., Sala, O., Haddad, B., Harte, J., Koch, G., Schwinning, S., Small, E.,Williams, D. 2004. Convergence across biomes to a common rain-use efficiency. Nature 429: 651–654.

Huxman, T.,Wilcox, B., Breshears, D., Scott, R., Snyder, K., Small, E., Hultine, K., Pockman, W., and Jackson, R. 2005. Ecohydrological implications of woody plant encroachment. Ecology 86: 308–319.

Jacquemin, B.&Noilhan, J. 1990. Sensitivity study and validation of land surface parametrization using the Hapex-Mobilhy data set. Bound-Layer Meteorology 52: 93–134.

Jiang L, Islam S (2001) Estimation of surface evaporation map over southern Great Plains using remote sensing data. Water Resour Res 37:329–340. doi:10.1029/ 2000WR900255

Jupp DLB, Tian G, McVicar TR, Qin Y, Fuqin L (1998) Soil moisture and drought monitoring using remote sensing I: theoretical background and methods. CSIRO Earth Observation Centre, Canberra http://www.eoc.csiro.au/pubrep/scirpt/jstc1.pdf

Kalma, J. D., McVicar, T. R. and McCabe, M. F. (2008). "Estimating land surface evaporation: A review of methods using remotely sensed surface temperature data." Surveys in Geophysics 29(4-5): 421-469.

Katerji N., Perrier A.1985 - Détermination de la résistance globale d'un couvert végétal à la diffusion de vapeur d'eau et de ses différentes composantes. Approche théorique et vérification expérimentale sur une culture de luzerne. Agric. For Meteorol., 34, 2-3, 105-120.

Kustas WP, Norman JM (1996) Use of remote sensing for evapotranspiration monitoring over land surfaces. Hydrol Sci J 41(4):495–516

Kustas WP, Norman JM (1999) Evaluation of soil and vegetation heat flux predictions using a simple two source model with radiometric temperatures for partial canopy cover. Agric For Meteorol 94:13–29. doi:10.1016/S0168-1923(99)00005-2

Kustas WP, French AN, Hatfield JL, Jackson TJ, Moran MS, Rango A (2003a) Remote sensing research in hydrometeorology. Photogramm Eng Remote Sensing 69(6):613–646

Kustas WP, Hatfield JL, Prueger JH (2005) The Soil Moisture–Atmosphere Coupling Experiment (SMACEX): background, hydrometeorological conditions, and preliminary findings. J Hydrometeorol 6:825–839. doi:10.1175/JHM460.1

Lagouarde, J.P., Brunet, Y., 1991. A simple model for estimating the daily upward longwave surface radiations from NOAA–AVHRR data. International Journal of Remote Sensing 12, 1853–1864.

Lhomme JP, Monteny B, Amadou M (1994) Estimating sensible heat flux from radiometric temperature over sparse millet. Agric For Meteorol 44:197–216

Li, Z.L., Becker, F., 1993. Feasibility of land surface temperature and emissivity determination from AVHRR data. Remote Sensing of Environment 43, 67–85.

Li F, Kustas WP, Prueger JH, Neale CMU, Jackson TJ (2005) Utility of remote sensing based two-source energy balance model under low and high vegetation cover conditions. J Hydrometeorol 6(6):878–891. doi:10.1175/JHM464.1

Li Z.L., Tang R., Wan Z., Bi Y., Zhou C., Tang B., Yan G. and Zang X. (2009), A Review of Current Methodologies for Regional Evapotranspiration Estimation from Remotely Sensed Data, Sensors 2009, 9, 3801-3853; doi:10.3390/s90503801

McCabe MF, Wood EF (2006) Scale influences on the remote estimation of evapotranspiration using multiple satellite sensors. Remote Sens Environ 105(4):271–285. doi:10.1016/j.rse.2006.07.006

McVicar TR, Jupp DLB (1998) The current and potential operational uses of remote sensing to aid decisions on drought exceptional circumstances in Australia: a review. Agric Syst 57:399–468. doi:10.1016/S0308-521X(98)00026-2

Monteith JL (1965) Evaporation and the environment. In: Fogg GE (ed) The state and movement of water in living organisms, 19th symposium of the society for experimental biology. University Press, Cambridge, pp 205–234

Monteith, J.L., 1981. Evaporation and surface temperature. Quart. J. Roy. Meteorolog. Soc., 107, 1–27.

Norman JM, Kustas WP, Humes KS (1995) A two-source approach for estimating soil and vegetation energy fluxes from observations of directional radiometric surface temperature. Agric For Meteorol 77:263–293. doi:10.1016/0168-1923(95)02265-Y

Norman JM, Kustas WP, Prueger JH, Diak GR (2000) Surface flux estimation using radiometric temperature: a dual-temperature-difference method to minimize measurement errors. Water Resour Res 36:2263–2274. doi:10.1029/2000WR900033

Overgaard J, Rosbjerg D, Butts MB (2006) Land-surface modelling in hydrological perspective – a review. Biogeosciences 3:229–241

Penman HL (1948) Natural evaporation from open water, bare soil and grass. Proc R Soc Lond A Math Phys Sci 193:120–146. doi:10.1098/rspa.1948.0037

Roerink GJ, Su Z, Menenti M (2000) S-SEBI: a simple remote sensing algorithm to estimate the surface energy balance. Phys Chem Earth, Part B Hydrol Oceans Atmos 25(2):147–157. doi:10.1016/S1464-1909(99)00128-8

Seguin B, Baelz S, Monget JM, Petit V (1982a) Utilisation de la thermographie IR pour l'estimation de l'e´vaporation re´gionale I Mise au point me´thodologique sur le site de la Crau. Agronomie 2(1):7–16. doi:10.1051/agro:19820102

Serban C., Maftei C., Barbulescu A. (2010), Assessment of Evapotranspiration Using Remote Sensing Data and Grid Computing and Apglication, WSEAS Transactions on computers, ISSN: 1109-2750, Issue 11, Volume 9, November 2010, pg.1245-1254

Shuttleworth WJ, Wallace JS (1985) Evaporation from sparse crops-an energy combination theory. Q J R Meteorol Soc 111:839–855. doi:10.1256/smsqj.46909

Smith, R. C. G. & Choudhury, B. J. (1991) Analysis of normalized difference and surface temperature observations over southeastern Australia. Int. J. Remote Sens. 12, 2021-2044.

Sobrino JA, Gomez M, Jimenez-Munoz JC, Olioso A (2007) Application of a simple algorithm to estimate daily evapotranspiration from NOAA-AVHRR images for the Iberian Peninsula. Remote Sens Environ 110:139–148. doi:10.1016/j.rse.2007.02.017

Stancalie GH, Marica A, Toulios L, (2010) Using earth observation data and CROPWAT model to estimate the actual crop evapotranspiration. Physics and Chemistry of the Earth, 35:25-30, doi:10.1016/j.pce.2010.03.013.

Stewart JB, Kustas WP, Humes KS, Nichols WD, Moran MS, De Bruin HAR (1994) Sensible heat flux–radiometric surface temperature relationships for eight semi-arid areas. J Appl Meteorol 33:1110–1117. doi :10.1175/1520-0450(1994)033\1110:SHFRST [2.0.CO;2

Su Z (2002) The Surface Energy Balance System (SEBS) for estimation of turbulent heat fluxes. Hydrol Earth Syst Sci 6(1):85–99 (HESS)

Su H, McCabe MF, Wood EF, Su Z, Prueger JH (2005) Modeling evapotranspiration during SMACEX: comparing two approaches for local- and regional-scale prediction. J Hydrometeorol 6(6):910–922. doi:10.1175/JHM466.1

Su Z (2008) The Surface Energy Balance System (SEBS) for estimation of turbulent heat fluxes and evapotranspiration, Dragon 2, Advanced Trainig Course in Land Remote Sensing, http://dragon2.esa.int/landtraining2008/pdf/D3L2b_SU_SEBS.pdf

Wang K, Li Z, Cribb M (2006) Estimation of evaporative fraction from a combination of day and night land surface temperature and NDVI: a new method to determine the Priestley–Taylor parameter. Remote Sens Environ 102:293–305. doi:10.1016/j.rse.2006.02.007

Wood, E.F., Hongbo, Su, McCabe, M., Su, B., 2003. Estimating evaporation from satellite remote sensing. In: Geoscience and Remote Sensing Symposium 2003. IGARSS Proceedings of the IEEE International, vol. 2, pp. 163–1165.

Permissions

The contributors of this book come from diverse backgrounds, making this book a truly international effort. This book will bring forth new frontiers with its revolutionizing research information and detailed analysis of the nascent developments around the world.

We would like to thank Dr. Ayse Irmak, for lending their expertise to make the book truly unique. They have played a crucial role in the development of this book. Without their invaluable contribution this book wouldn't have been possible. They have made vital efforts to compile up to date information on the varied aspects of this subject to make this book a valuable addition to the collection of many professionals and students.

This book was conceptualized with the vision of imparting up-to-date information and advanced data in this field. To ensure the same, a matchless editorial board was set up. Every individual on the board went through rigorous rounds of assessment to prove their worth. After which they invested a large part of their time researching and compiling the most relevant data for our readers. Conferences and sessions were held from time to time between the editorial board and the contributing authors to present the data in the most comprehensible form. The editorial team has worked tirelessly to provide valuable and valid information to help people across the globe.

Every chapter published in this book has been scrutinized by our experts. Their significance has been extensively debated. The topics covered herein carry significant findings which will fuel the growth of the discipline. They may even be implemented as practical applications or may be referred to as a beginning point for another development. Chapters in this book were first published by InTech; hereby published with permission under the Creative Commons Attribution License or equivalent.

The editorial board has been involved in producing this book since its inception. They have spent rigorous hours researching and exploring the diverse topics which have resulted in the successful publishing of this book. They have passed on their knowledge of decades through this book. To expedite this challenging task, the publisher supported the team at every step. A small team of assistant editors was also appointed to further simplify the editing procedure and attain best results for the readers.

Our editorial team has been hand-picked from every corner of the world. Their multi-ethnicity adds dynamic inputs to the discussions which result in innovative outcomes. These outcomes are then further discussed with the researchers and contributors who give their valuable feedback and opinion regarding the same. The feedback is then collaborated with the researches and they are edited in a comprehensive manner to aid the understanding of the subject.

Apart from the editorial board, the designing team has also invested a significant amount of their time in understanding the subject and creating the most relevant covers. They scrutinized every image to scout for the most suitable representation of the subject and create an appropriate cover for the book.

The publishing team has been involved in this book since its early stages. They were actively engaged in every process, be it collecting the data, connecting with the contributors or procuring relevant information. The team has been an ardent support to the editorial, designing and production team. Their endless efforts to recruit the best for this project, has resulted in the accomplishment of this book. They are a veteran in the field of academics and their pool of knowledge is as vast as their experience in printing. Their expertise and guidance has proved useful at every step. Their uncompromising quality standards have made this book an exceptional effort. Their encouragement from time to time has been an inspiration for everyone.

The publisher and the editorial board hope that this book will prove to be a valuable piece of knowledge for researchers, students, practitioners and scholars across the globe.

List of Contributors

Christiaan van der Tol and Gabriel Norberto Parodi
University of Twente, Faculty of ITC, The Netherlands

Georgeta Bandoc
University of Bucharest, Department of Meteorology and Hydrology, Center for Coastal Research and Environmental Protection, Romania

L.O. Lagos
Universidad de Concepción Chile, Chile
University of Nebraska-Lincoln, USA

D. Martin, S. Verma and A. Suyker
University of Nebraska-Lincoln, USA

G. Merino
Universidad de Concepción Chile, Chile

Martina Eiseltová
Crop Research Institute, Czech Republic
Environment and Wetland Centre, Czech Republic

Petra Hesslerová
Enki, o.p.s., Czech Republic
Faculty of Environmental Sciences, Czech University of Life Sciences, Prague, Czech Republic

Wilhelm Ripl
Aquaterra System Institute, Germany

Jan Pokorný
Enki, o.p.s., Czech Republic

Sungwon Kim
Dongyang University, Republic of Korea

Giacomo Bertoldi
Institute for Alpine Environment, EURAC Research, Bolzano, Italy

Riccardo Rigon
University of Trento, Italy

Ulrike Tappeiner
Institute for Alpine Environment, EURAC Research, Bolzano, Italy
Institute of Ecology, University of Innsbruck, Austria

Nebo Jovanovic and Sumaya Israel
Council for Scientific and Industrial Research, Stellenbosch, South Africa

Giacomo Gerosa
Dipartimento Matematica e Fisica, Università Cattolica del Sacro Cuore, via Musei 41,
Brescia, Italy

Simone Mereu
Dipartimento di Economia e Sistemi Arborei, Università di Sassari, Italy

Angelo Finco
Ecometrics s.r.l., Environmental Monitoring and Assessment, via Musei 41, Brescia, Italy

Riccardo Marzuoli
Dipartimento Matematica e Fisica, Università Cattolica del Sacro Cuore, via Musei 41,
Brescia, Italy
Fondazione Lombardia per l'Ambiente, Piazza Diaz 7, Milano, Italy

Yann Chemin
International Water Management Institute, Sri Lanka

Ayse Irmak and Ian Ratcliffe
School of Natural Resources University of Nebraska–Lincoln, HARH, Lincoln NE, USA

Richard G. Allen, Jeppe Kjaersgaard and Ricardo Trezza
University of Idaho, Kimberly, ID, USA

Justin Huntington
Desert Research Institute, Raggio Parkway, Reno, NV, USA

Baburao Kamble
University of Nebraska-Lincoln, Lincoln, NE, USA

Viera Paganová and Zuzana Jureková
Slovak University of Agriculture in Nitra, Slovak Republic

Gheorghe Stancalie and Argentina Nertan
National Meteorological Administration 97, Soseaua Bucuresti-Ploiesti, Bucharest, Romania

Printed in the USA
CPSIA information can be obtained
at www.ICGtesting.com
JSHW011502221024
72173JS00005B/1170

9 781632 393326